Samuel L. Oppenheimer

FUNDAMENTALS OF ELECTRIC CIRCUITS

PRENTICE-HALL, INC., Englewood Cliffs, New Jersey 07632

Library of Congress Cataloging in Publication Data

Oppenheimer, Samuel L.
   Fundamentals of electric circuits.

   Includes index.
   1. Electric circuits. 2. Electronic circuits.
I. Title.
TK454.O67 1984   621.391'2   83-19141
ISBN 0-13-336974-9

*Editorial/production supervision and
   interior design:* Barbara Palumbo
Cover design: Frederick Charles
Manufacturing buyer: Gordon Osbourne

© 1984 by Prentice-Hall, Inc., Englewood Cliffs, New Jersey 07632

All rights reserved. No part of this book may be
reproduced, in any form or by any means,
without permission in writing from the publisher.

Printed in the United States of America

10 9 8 7 6 5 4 3 2 1

ISBN 0-13-336974-9

PRENTICE-HALL INTERNATIONAL, INC., *London*
PRENTICE-HALL OF AUSTRALIA PTY. LIMITED, *Sydney*
EDITORA PRENTICE-HALL DO BRASIL, LTDA., *Rio de Janeiro*
PRENTICE-HALL CANADA INC., *Toronto*
PRENTICE-HALL OF INDIA PRIVATE LIMITED, *New Delhi*
PRENTICE-HALL OF JAPAN, INC., *Tokyo*
PRENTICE-HALL OF SOUTHEAST ASIA PTE. LTD., *Singapore*
WHITEHALL BOOKS LIMITED, *Wellington, New Zealand*

Contents

PREFACE ... xiii

1 COMPUTATIONAL SKILLS ... 1

1-1. Introduction ... 1
1-2. Decimals ... 2
1-3. Decimals and Numbers ... 2
1-4. Addition and Subtraction with Decimals ... 3
1-5. Multiplication of Decimals ... 4
1-6. Division with Decimals ... 5
1-7. Percentage ... 6
1-8. Working with Percentages ... 7
1-9. Ratio and Proportion ... 10
1-10. Powers of Ten ... 11
1-11. Using Powers of 10 ... 12
1-12. Scientific Notation ... 16
1-13. Engineering Notation ... 16
1-14. Physical and Electrical Units ... 18
1-15. Working with Units ... 21
1-16. Using the Scientific Calculator ... 23
Summary ... 27
Problems ... 27

2 MATTER AND ELECTRICITY 30

2-1. Elements 30
2-2. Structure of the Atom 31
2-3. Electricity 33
2-4. Work, Energy, and Power 34
2-5. Conduction of Charge 35
2-6. Conductors 36
2-7. Insulators 36
2-8. Semiconductors 37
Summary 37
Problems 38

3 VOLTAGE AND VOLTAGE SOURCES 39

3-1. Voltage 39
3-2. Voltage Units 40
3-3. Pure dc Voltage 41
3-4. Pulsating dc Voltage 41
3-5. Alternating Voltage 42
3-6. Frequency 44
3-7. Sources of Voltage 46
Summary 48
Problems 49

4 CURRENT AND POWER 50

4-1. Current Units 50
4-2. Current Direction 52
4-3. Types of Currents 53
4-4. Electrical Power 54
4-5. Efficiency 56
4-6. A Practical Unit of Electrical Energy 58
Summary 59
Problems 60

5 RESISTANCE 61

5-1. Introduction 61
5-2. Conductance 61
5-3. Physical Considerations 63
5-4. Physical Standards 64
5-5. Resistance of a Wire 65
5-6. Temperature Effects 67

5-7.	Current Capacity of a Wire	69
5-8.	Resistors	70
5-9.	Types of Fixed Resistors	70
5-10.	Resistance Color Code	72
5-11.	Variable Resistors	74
5-12.	Power Dissipation Ratings	75
	Summary	76
	Problems	76

6 THE BASIC ELECTRIC CIRCUIT 79

6-1.	The Closed Circuit	79
6-2.	Source Voltage and Load Current	80
6-3.	Effect of Resistance on Current	81
6-4.	Open and Short Circuits	81
6-5.	Current and Power	84
6-6.	Current Measurement	85
6-7.	Voltage Measurement	86
6-8.	Voltage Drop	86
	Summary	88
	Problems	88

7 OHM'S LAW 93

7-1.	A Basic Physical Law	93
7-2.	Current Formula	93
7-3.	Resistance Formula	95
7-4.	Voltage Formula	96
7-5.	Learning the Formulas	97
7-6.	Ohm's Law and Entire Circuits	98
7-7.	Ohm's Law Applied to Part of a Circuit	100
7-8.	Computations	102
	Summary	104
	Problems	104

8 SERIES CIRCUITS 108

8-1.	Introduction	108
8-2.	Current in a Series Circuit	108
8-3.	Resistors in Series	109
8-4.	Total Resistance	110
8-5.	Voltage Drops	111
8-6.	Sum of the Voltage Drops	113
8-7.	Voltage Sources in Series	114

8–8.	Circuit Analysis	116
8–9.	Measurements and Troubleshooting	119
	Summary	121
	Problems	122

9 PARALLEL CIRCUITS 126

9–1.	Introduction	126
9–2.	Parallel Resistance Circuit	126
9–3.	Equivalent Resistance	128
9–4.	Equivalent Resistance: Many-Branched Circuit	129
9–5.	Equivalent Resistance: Two-Branched Circuit	132
9–6.	Equivalent Resistance: Equal-Branch Resistance	134
9–7.	More Useful Formulas	135
9–8.	Circuit Analysis	139
9–9.	Troubleshooting	143
	Summary	144
	Problems	144

10 SERIES-PARALLEL CIRCUITS 151

10–1.	Introduction	151
10–2.	Voltage and Current Relationships	155
10–3.	Series-Equivalent Circuits	157
10–4.	Circuit Analysis	160
10–5.	Voltage and Current Measurements	166
10–6.	Voltage Dividers	168
	Summary	173
	Problems	174

11 NETWORK ANALYSIS 180

11–1.	Kirchhoff's Current Law	180
11–2.	Kirchhoff's Voltage Law	180
11–3.	Voltage Division Rule (VDR)	183
11–4.	Current Division Rule (CDR)	184
11–5.	Applications of Kirchhoff's Laws	184
11–6.	Superposition Theorem	193
11–7.	Thevenin's Theorem	199
11–8.	Norton's Theorem	212
	Summary	216
	Problems	216

12 POWER IN DC CIRCUITS   221

    12-1. Basic Power Formulas   221
    12-2. Efficiency   223
    12-3. Maximum Power Transfer Theorem   230
    12-4. Voltage and Current Sources   231
    12-5. Voltage to Current Source Conversion   235
    12-6. Current to Voltage Source Conversion   236
           Summary   237
           Problems   237

13 CAPACITANCE   242

    13-1. Introduction   242
    13-2. Electrostatic Field   242
    13-3. Capacitance   244
    13-4. Determining Capacitance   245
    13-5. Dielectric Strength and Leakage Current   247
    13-6. Types of Capacitors   248
    13-7. Capacitors in Series   251
    13-8. Capacitors in Parallel   254
    13-9. Energy Storage   255
           Summary   256
           Problems   256

14 ELECTROMAGNETISM   259

    14-1. Introduction   259
    14-2. Magnetic Field   260
    14-3. Magnetic Materials   261
    14-4. Molecular Theory of Magnetism   262
    14-5. Flux Density   263
    14-6. Current and Magnetic Flux   263
    14-7. Permeability   266
    14-8. Reluctance   267
    14-9. Magnetic Circuit   268
           Summary   271
           Problems   271

15 APPLICATIONS OF ELECTROMAGNETISM   272

    15-1. Solenoids   272
    15-2. Relays   273
    15-3. Electromagnetic Induction   274

15-4.	Generators and Alternators	279
15-5.	Motors	281
15-6.	Transformers	283
15-7.	DC Meters	285
	Summary	289
	Problems	289

16 INDUCTANCE 290

16-1.	Introduction	290
16-2.	Induced Voltages	291
16-3.	Unit of Inductance	292
16-4.	Mutual Inductance	294
16-5.	Inductors	296
16-6.	Series and Parallel Inductors	297
16-7.	Energy Storage	302
	Summary	302
	Problems	303

17 TRANSIENT RESPONSE AND TIME CONSTANTS 305

17-1.	Transient Voltages and Currents	305
17-2.	Resistance-Capacitance (RC) Circuits with dc Sources	306
17-3.	Voltage Rise in an RC Circuit	307
17-4.	Meaning of RC Time Constant	310
17-5.	Discharge in an RC Circuit	311
17-6.	Pulse Response of RC Circuits	313
17-7.	Current Rise in LR Circuits	317
17-8.	L/R Time Constant	317
17-9.	Discharge in an RL Circuit	319
17-10.	Pulse Response of an LR Circuit	319
17-11.	Using the Scientific Calculator	321
	Summary	328
	Problems	328

18 ALTERNATING VOLTAGES AND CURRENTS 331

18-1.	Frequency and Period	331
18-2.	Types of ac Waves	334
18-3.	Sine Wave	336
18-4.	Equation of a Sine Wave	337
18-5.	Phasor Representation of a Sine Wave	338
18-6.	Phase Angles and Phase Differences	339
18-7.	Effective Value of a Sine Wave	342

18-8.	Average Values	344
18-9.	Periodic Functions	346
18-10.	Using a Scientific Calculator	348
	Summary	351
	Problems	361

19 RESISTANCE, INDUCTANCE, AND CAPACITANCE, IN AC CIRCUITS 354

19-1.	Rate of Change of a Sine Wave	354
19-2.	Resistance in ac Circuits	356
19-3.	Inductance in ac Circuits	360
19-4.	Capacitance in ac Circuits	363
19-5.	Basic Right Triangle Trigonometry	365
19-6.	Polar and Rectangular Forms	368
19-7.	Rectangular to Polar Conversion	369
19-8.	Polar to Rectangular Conversion	371
19-9.	Phasor Arithmetic	373
	Summary	375
	Problems	376

20 SERIES CIRCUITS 379

20-1.	Impedance of a Series RL Circuit	379
20-2.	Impedance of a Series RC Circuit	383
20-3.	Voltage and Current Relationships	384
20-4.	Analyzing Series ac Circuits	389
20-5.	The Meaning of Q	392
20-6.	The Q of an Inductor	393
20-7.	The Q of a Capacitor	393
20-8.	Effective Q	393
	Summary	395
	Problems	396

21 ANALYSIS OF AC CIRCUITS 398

21-1.	Rules of Parallel Circuits	398
21-2.	Two-Branch Parallel Circuits	399
21-3.	Multibranch Parallel Circuits	404
21-4.	Susceptance and Admittance	406
21-5.	Parallel RC Circuits	407
21-6.	Parallel RL Circuits	412
21-7.	Parallel RLC Circuits	415
21-8.	Circuit Analysis	415

	21-9.	*The Q of a Parallel Circuit*	422
	21-10.	*Series-Parallel Circuits*	423
		Summary	423
		Problems	424

22 POWER IN AC CIRCUITS — 427

22-1.	*Introduction*	427
22-2.	*Apparent Power*	428
22-3.	*True Power*	429
22-4.	*Power Factor*	429
22-5.	*Power Factor Correction*	432
	Summary	434
	Problems	434

23 RESONANT CIRCUITS — 436

23-1.	*Introduction*	436
23-2.	*Series Resonance*	436
23-3.	*Circuit Q*	442
23-4.	*Ideal Parallel Resonant Circuit*	447
23-5.	*Practical Parallel Resonant Circuits*	449
23-6.	*Effective Q and Bandwidth*	454
23-7.	*Applications of Resonant Circuits*	457
	Summary	462
	Problems	463

24 TRANSFORMERS — 466

24-1.	*Types of Transformers*	466
24-2.	*Transformer Construction*	467
24-3.	*Voltage Ratios*	468
24-4.	*Current Ratios*	470
24-5.	*Impedance Ratios*	472
24-6.	*Autotransformers*	475
24-7.	*Loosely Coupled Transformers*	478
	Summary	481
	Problems	481

25 NETWORK THEOREMS AND FILTERS — 484

25-1.	*Delta-to-Wye Conversion*	484
25-2.	*Bridge Circuits*	490

25–3.	Thevenin's Theorem	493
25–4.	Norton's Theorem	496
25–5.	Maximum Power Transfer	498
25–6.	Source Conversions	500
25–7.	Mesh Current Analysis	503
25–8.	Superposition	505
25–9.	Node Voltage Analysis	508
25–10.	Millman's Theorem	511
25–11.	Decibels	512
25–12.	Low-Pass Filters	516
25–13.	High-Pass Filters	519
	Summary	521
	Problems	521
	Answers to Even-Numbered Problems	526

INDEX 537

Preface

This text is for students of electricity and circuit analysis. It is designed for use in community colleges, technical schools, and vocational programs. Because of its many examples, it is also suitable for self-study.

It is organized to respond to the needs of teachers and students, without making the material so elementary as to have little value to the learner. The level accommodates the beginning learner but increases as the reader progresses through the textbook.

Electron flow is used rather than conventional current flow. However, it is made clear that either form of current flow is acceptable, and that both are used in the industry.

The first chapter is devoted nearly entirely to computations and engineering units. Many years of experience and actual research have demonstrated the significant correlation between computational skills and *learning* of circuit analysis. The use of a scientific calculator is discussed along with examples illustrating proper operation. Circuit analysis examples using a scientific calculator are found throughout the book. In the matter of computations, trigonometry is used in ac circuit analysis. A scientific calculator and the explanations given in the textbook are sufficient for any student, regardless of background, to solve ac circuits using polar and rectangular quantities.

The beginning chapters are short and designed to bring the reader rapidly to the point that makes the study of Ohm's law and electric circuits meaningful. Concepts of alternating voltages and currents are also covered early on in the text. The text treats circuit analysis in detail, in ac as well as in dc circuits. The frequency response of untuned filters, the use of decibels in problems, the pulse response of *RC* and *LR* circuits, as well as network theorems with examples, are some of the efforts made to improve the learner's ability to understand complex networks. In all examples, the values used have been chosen as practical ones, representative of actual practice.

Early on in the material on circuits, troubleshooting is covered, again with examples. Circuit measurements are included with the troubleshooting material. Concepts of relative opens and shorts are clarified before the study of Ohm's law.

Thorough grounding in the fundamentals of circuit analysis is important for electrical and electronic technicians of all levels. The state of the art is such that the only quantity that is constant is change itself! New devices and applications appear with great regularity and frequency. These require that the technician continue to study and learn, long after completion of formal studies. The circuit analysis techniques covered in this text, in great detail and with many examples, are basic to the understanding of any new circuit or electronic device.

Samuel L. Oppenheimer

Computational Skills

1-1. INTRODUCTION

The author conducted research on those factors that have the greatest effect on, and therefore the greatest relationship to, grades in courses in electrical circuit analysis. Of all the factors tested, computational skills proved to be the most significant. The basis of computational skills is arithmetic, and it is with good reason that it is often referred to as "the queen of the maths."

We recognize that electronic calculators, which are so readily available, make it seem needless to learn computational skills. Nothing could be further from the truth, even though we will make use of calculators. You need only think of what can happen when you keystroke a complicated problem into a calculator. If you have no idea of at least a "ballpark" approximation of the correct answer, how can you know when you have made an error in the way that you keyed in the problem? Besides, electronics technology need not always work with eight-digit readouts. In fact, it never does! Also, in many cases, a reasonably accurate approximation of the exact answer is sufficient for the job at hand.

So we shall begin our studies in basic electricity with computational skills and then develop practical skill, which is as important to the electronics technician as skill with the oscilloscope and other electronic test instruments.

1-2. DECIMALS

A decimal is simply another way of stating a fractional quantity. It is by far the easiest way to work most computations that involve fractions. The solution of problems requires no need for common denominators, nor need we concern ourselves with the reduction of fractions. In fact, the use of electronic calculators makes computations with decimals merely a matter of pushing the right buttons.

Decimals (properly called decimal fractions) really derive from common fractions. *A common fraction is a way of representing a division problem. The dividend is the numerator, and the divisor is the denominator.*

> **EXAMPLE 1-1:**
> Rewrite the following common fractions as division problems:
> (a) 351/483   (b) 75/35
> (c) 2581/10,700   (d) a/b where a and b are any numbers
>
> *Solution:*
> (a) $351/483 = 351 \div 483 = 0.7267$
> (b) $75/35 = 75 \div 35 = 2.1428$
> (c) $2581/10,700 = 2581 \div 10,700 = 0.2412$
> (d) $a/b = a \div b$

Note that when the fraction is a proper fraction (one whose numerator is less than the denominator), the division results in a numerical quantity that is less than 1, as in parts (a) and (c) of Example 1-1. This quantity is referred to as a *decimal fraction* or, simply, a *decimal.*

The reader will find that working with decimal fractions is very much easier than working with common fractions, particularly when a calculator is used to solve problems.

1-3. DECIMALS AND NUMBERS

In writing a number with a decimal point, the first column to the left of the decimal point is the 1's column. The next to the left is the 10's column; the next to the left is the 100's column; and so on. *Notice that each shift to the left multiplies the value of an integer by 10.* (An *integer* is any whole number from 0 through 9. It may be positive or negative.)

In a similar manner, *each shift to the right divides the value of an integer by 10.* In Fig. 1-1, if we move from the sixth column to the left of the decimal point (100,000's) to the third column, it is the same as dividing 100,000 by 10 three times, resulting in the 100's column. If we continue to shift to the right twice more, we arrive at the 1's column. The next shift to the right places us to the right of the decimal point. The column has a value of $1 \div 10 = 0.1$ (one-tenth). The first column to the right of the decimal point is called the tenth's column. Moving right one more column divides 0.1 by

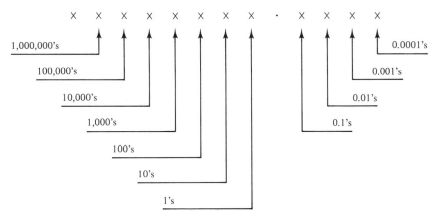

Figure 1-1

10, resulting in 0.01 (one-hundredth), the hundredth's column. The next shift to the right is the thousandth's column (0.01 ÷ 10). The process may be continued for as many places (columns) as are needed.

1-4. ADDITION AND SUBTRACTION WITH DECIMALS

In the addition and subtraction of numbers that include decimals, simply write the problem as you would for the addition or subtraction of whole numbers, but *be sure to line up the decimal points*. The rules for carrying or borrowing apply as they do for whole numbers.

EXAMPLE 1-2:

Addition:
(a) $0.3965 + 0.0035 =$

$$\begin{array}{r} 0.3965 \\ + 0.0035 \\ \hline 0.4000 = 0.4 \end{array}$$

(b) $2.0089 + 1.902 =$

$$\begin{array}{r} 2.0089 \\ + 1.902 \\ \hline 3.9109 \end{array}$$

(c) $2.986 + 1.321 + 5.1 + 0.0833 =$

$$\begin{array}{r} 2.986 \\ 1.321 \\ 5.1 \\ 0.0833 \\ \hline 9.4903 \end{array}$$

(d) $5.4 + 0.082 =$

You should be able to solve this at a glance. None of the integers are added to one another because of their positions with respect to the decimal point. The answer is 5.482.

Subtraction:

(e) $3.26 - 1.804 =$

$$\begin{array}{r} 3.260 \\ -1.804 \\ \hline 1.456 \end{array}$$

(f) $5.091 - 0.83 =$

$$\begin{array}{r} 5.091 \\ -0.830 \\ \hline 4.261 \end{array}$$

(g) $0.0865 - 0.00032 =$

$$\begin{array}{r} 0.08650 \\ -0.00032 \\ \hline 0.08618 \end{array}$$

(h) $100.316 - 85.2 =$

$$\begin{array}{r} 100.316 \\ -\ 85.200 \\ \hline 15.116 \end{array}$$

(i) $0.000528 - 0.000019 =$

$$\begin{array}{r} 0.000528 \\ -0.000019 \\ \hline 0.000509 \end{array}$$

1-5. MULTIPLICATION OF DECIMALS

Multiply the numbers as you would whole numbers. The decimal point in the resulting product is located in two steps: (1) take the sum of the total number of decimal places in the numbers; and (2) place the decimal point of the product the number of places to the left of the last integer by an amount that is equal to the sum of the places found in step 1.

EXAMPLE 1-3:

(a) $300 \times 0.09 = 27$

The product of the integers is 2700 and the total number of decimal places is two.

(b) $3.51 \times 0.03 = 0.1053$

$$\begin{array}{r} 351 \\ \times\ \ \ 3 \\ \hline 1053 \end{array}$$

Total decimal places are four; therefore, the answer is 0.1053.
(c) $6.26 \times 0.021 = 0.13146$

$$\begin{array}{r} 626 \\ \times\ 21 \\ \hline 626 \\ 1252 \\ \hline 13146 \end{array}$$

Total decimal places are five; therefore, the answer is 0.13146
(d) $0.0006 \times 0.3 = 0.00018$
    3×6 is 18 and decimal places are five.
(e) $15.1 \times 0.00082 = 0.012382$
    Total decimal places are six, and 151×82 is 12382.

1-6. DIVISION WITH DECIMALS

Either long or short division may be used in the same way as with whole numbers. *The placement of the decimal point is determined by the number of decimal places in the divisor.* To solve, write the problem as given and then move the decimal point of the dividend to the right by an amount that is equal to the number of places that you must move the decimal point of the *divisor* to the right in order to make the divisor a whole number. This may require that zeros be added to the dividend.

Problem: $\quad 0.00a \overline{)XXX.\dot{X}}$

Step 1: $\quad 0.00a. \overline{)XXX.\dot{X}00.}$

Step 2: $\quad a \overline{)XXXX\dot{0}0.}$

EXAMPLE 1-4:
(a) $200 \div 0.5 = 400$

$$0.5 \overline{)200.} = 5 \overline{)2000.}\ \ (400.)$$

(b) $20 \div 0.00004 = 500{,}000$

$$0.00004 \overline{)20.} = 4 \overline{)2{,}000{,}000.}\ \ (500{,}000.)$$

(c) $0.0086 \div 0.215 = 0.04$

$$0.215 \overline{)0.0086} = 215 \overline{)8.16}\ \ (0.04)$$

(d) $0.0961 \div 0.000025 = 3844$

$$0.000025 \overline{)0.0961} = 25 \overline{)96{,}100.}\ \ (3{,}844.)$$

1-7. PERCENTAGE

Percentage is another way of stating a fraction, one whose numerator may be any number, but whose denominator is *always* 100. The meaning of the word "percent" is from the Latin, meaning per hundredth. The familar percent sign (%) stands for 1/100.

An important difference between percentages and other forms of fractions is that percentages deal with concrete quantities, not abstract ones. Therefore, if we write that 0.21 (21/100) is 21%, we are making a mathematically accurate statement, but one with very little meaning. On the other hand, if we say that car A costs 21% more than car B, and that car B sells for $10,000.00, we are indeed saying something that has meaning. Car A will cost $2100.00 more than car B. We will get much practice in working on percentage in this chapter.

Decimal fractions are changed to percentages by simply *moving the decimal point two places to the right* and then writing the percent sign after the number.

> **EXAMPLE 1-5:**
> Write the following as percentages:
> 0.00356   0.766
> 0.1578   1.35
>
> *Solution:*
> 0.356%   76.6%
> 15.78%   135%

Common fractions are changed to percentages by first finding the decimal equivalent of the common fraction and then restating this decimal equivalent as a percentage, as in Example 1-5. Remember that the decimal equivalent of a common fraction is found by dividing the numerator by the denominator.

> **EXAMPLE 1-6:**
> Write the following common fractions as percentages:
> 3/4   5/8
> 1/3   7/10
> 5/6   14/10
>
> *Solution:*
>
> $$3 \div 4 = 0.75 = 75\%$$
>
> $$1 \div 3 = 0.3333 = 33.33\%$$
>
> $$5 \div 6 = 0.8333 = 83.33\%$$
>
> $$5 \div 8 = 0.625 = 62.5\%$$
>
> $$7 \div 10 = 0.7 = 70\%$$
>
> $$14 \div 10 = 1.4 = 140\%$$

1-8. WORKING WITH PERCENTAGES

There is more genuine confusion in this area of arithmetic than in any other. One finds that this confusion extends to the newspapers and to television reporters, particularly around election time. This section should help you to work successfully with percentages.

If, for example, we use 20% of some quantity, we simply mean that we multiply the original quantity by 0.2. However, if we refer to 20% without relating it to anything, we are simply referring to a fraction whose denominator is five times as large as its numerator.

Suppose that we were to say 150%; what is meant in this case is that the numerator is 1.5 times as large as the denominator. The fraction $3/2$ is 1.5, and expressed in hundredths this is 150%; therefore, 3 is 150% of 2, just as 1 is 20% of 5. Note that we did not say that 3 is 150% greater than 2. "Greater than" means subtraction must take place somewhere in the calculation. The difference between 3 and 2 is 1; 1 is 50% of 2 ($1/2 = 50\%$), and if we add 50% of 2 to 2, we have 3. If this sounds confusing, it is! This is the area where most people have difficulty in working with percentage. When we take a "percent of" something, we do refer to multiplication. In this discussion, 1.5 times 2 equals 3, just as 0.2 times 5 equals 1.

In all problems relating to percentage, you will either know one part of the problem and the percentage and must find the other part of the fraction, or you will be given all numerical data and must determine percentage.

The following examples will demonstrate the many ways we may work with percentage in problems.

EXAMPLE 1-7:

If 15% of a shop's time is devoted to paper work, then, in an 8-hour day, we find that 1.2 hours are used to do paper work. How did we arrive at this answer?

Solution:
(a) We changed 15% to a decimal by dividing by 100.

$$15\% = 0.15$$

(b) We know that the decimal represents a fraction whose denominator is 8, but whose numerator is unknown.

$$\frac{x}{8} = 0.15$$

From the theory of division, we know that the dividend (x) must equal the product of the divisor and the quotient.

$$8 \text{ times } 0.15 = 1.2$$

Note: The solution could have been arrived at directly from the statement of the problem. It was stated that "15% of the time . . ." (8-hour day); therefore, $0.15 \times 8 = 1.2$ hours.

EXAMPLE 1-8:

The selling price for an item is $17.50. As a dealer, you receive a 45% discount. What is your cost for the item?

Solution: Once again, there are two ways to solve the problem.
(a) Find the discount in dollars and cents. Selling price times the percent of discount (as a decimal) equals the discount.

$$\$17.50 \times 0.45 = \$7.875 = \$7.88$$

Your cost is the difference between the selling price and the discount.

$$\$17.50 - \$7.88 = \$9.62$$

(b) A direct solution for the cost may be made, without the need to find the value of the discount. To find the percent of *cost* as it relates to the selling price, simply subtract the percent discount from 100%.

$$\begin{aligned}
\% \text{ cost} &= 100\% - \% \text{ discount} \\
&= 100\% - 45\% \\
&= 55\% \\
\text{cost} &= \text{selling price} \times \% \text{ cost (as a decimal)} \\
&= \$17.50 \times 0.55 \\
&= \$9.62
\end{aligned}$$

EXAMPLE 1-9:

Another way to use a known percentage occurs when both the result and the percentage are known, but we do not know the original number. In this example, suppose that 56% of a certain quantity is 1239. What is the original quantity?

Solution: In this type of problem, the unknown quantity is the denominator of the fraction that resulted in the decimal represented by the percentage. For the moment, assign the letter y to the denominator.

$$\frac{1239}{y} = 0.56$$

Exchange places between the unknown (in the denominator) and the percentage (stated as a decimal).

$$y = \frac{1239}{0.56} = 2212.5$$

The greatest difficulty in percentage problems is to be found in the deciding how to set up the problem. Make sure that you understand the problem, particularly the number that you must use for the denominator.

For example, consider two numbers 55 and 65. Suppose that we wish to know by what percentage 65 is greater than 55. We proceed in the following manner:

1. We select the denominator by examining the question. The question stated ". . . 65 is greater than 55." Therefore, 55 is the reference and becomes the denominator.

2. The numerator is quickly found as the difference of the two numbers, because the problem used the expression "greater than," which requires subtraction between the quantities.
The solution is found:

$$\frac{10}{55} \times 100 = 18.18\%$$

We can check the solution. If 18.18% of 55 is added to 55, the result is 65.

As an additional example, suppose that we state a different problem, using the same two numbers.

By what percentage is 55 less than 65? The numerator is still 10, because "less than" requires that subtraction take place. However, the denominator is now 65, since it is the reference.

$$\frac{10}{65} \times 100 = 15.38\%$$

Let us review the thought processes in both examples. In the first instance, the question really asked, "What percentage of 55 must be added to 55 in order to make the number equal to 65?" Carried out to four decimal places, 18.18% of 55 is 9.9990. If 9.999 is added to 55, the result is 65 for all practical purposes. In the second example, we really asked, "What percentage of 65 must be subtracted from 65 in order to get 55?" Carried out to 4 decimal places, 15.38% of 65 is 9.9970. If this amount is subtracted from 65, the result is close enough to 55 to satisfy most purposes.

EXAMPLE 1-10:
An automobile presently sells for $6581.00. It is predicted that the price will rise by 7.7% next year. What will be (a) the amount of the price increase? (b) the new selling price?

Solution:
(a) $6581.00 × 0.077 = $506.74
(b) The new selling price is the sum of the present price and the price increase.

$$\$6581.00 + \$506.74 = \$7087.74$$

Note that the new selling price could have been found directly. If we consider the present price to represent 100% and then add the increase as a percentage, the sum of the percentages multiplied by the present price results in the new price.

$$100\% + 7.7\% = 107.7\%$$

$$\$6581.00 \times 1.077 = \$7087.74$$

EXAMPLE 1-11:
An electronics plant produces 409,000,000 components each year. It is found that, on average, 5,600,000 parts fail to meet specifications. What percentage of the total production is substandard?

$$\frac{5,600,000}{409,000,000} \times 100 = 1.369\%$$

1-9. RATIO AND PROPORTION

We sometimes refer to a fraction as a ratio. The sequence in which the quantities are stated defines the numerator and the denominator. The ratio of 3 to 4 is the fraction ¾. Ratios may be written in the form $a : b$ (a to b), 3 : 4 (3 to 4), 5 : 8 (5 to 8), 10 : 1, and so on.

Some mathematicians claim that a fraction may be written as a ratio only when both numerator and denominator are in the same units of measure (as in the calculations of percentages). Such a ratio results in an abstract fraction, a "dimensionless quantity." While generally true, there are many exceptions in technology. For example, the fundamental law of electric circuits, Ohm's law is stated as:

Current is the ratio of voltage to resistance.

$$\text{Current} = \frac{\text{voltage}}{\text{resistance}}$$

Knowing the relationship between two quantities means that we know the *proportional relationship* between them. This is of value in technology, finance, construction, and so on. For example, assume that two pulleys are connected by a belt, as in Fig. 1-2. The ratio of pulley A to pulley B is 2.5 : 1. The proportionality is such that A has a diameter 2.5 times that of B. For each revolution of A, there are 2.5 revolutions of B. On the other hand, for B to make one complete revolution, A rotates just four-tenths of a revolution. Suppose that A is rotating at 500 rpm. We then know that B is rotating at 2.5 times the rotation of A or 1250 rpm.

When ratios are equal to each other, their cross products are also equal. By cross product, we mean the product of the numerator of each fraction times the denominator of the other fraction.

Given that 3 : 4 = 9 : 12 = 15 : 20:

3/4 = 9/12 and 9 × 4 = 3 × 12 = 36

3/4 = 15/20 and 3 × 20 = 4 × 15 = 60

9/12 = 15/20 and 9 × 20 = 12 × 15 = 180

This knowledge of the equality of the cross products leads us into another concept relating to proportions. Suppose that we have the statement

$$1.5 = \frac{X}{Y}$$

where X and Y represent any numbers whose ratio is 1.5 : 1.

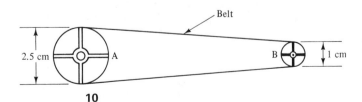

Figure 1-2 An example of proportionality. For each rotation of pulley A, there are 2.5 rotations of pulley B. If, for example, A turns at 1000 rpm, B turns at 2500 rpm.

Suppose that we hold Y constant and cause X to change. Immediately, the value of the ratio must change in order to keep the cross products equal. Assume that X is doubled. The value of the ratio becomes doubled.

Originally, we had

$$\frac{1.5}{1} = \frac{X}{Y} \quad \text{therefore} \quad 1.5Y = X$$

When the value of X is multiplied by 2, while Y is held constant, it becomes clear that 1.5 must also double in order to keep the cross products equal.

$$2(1.5Y) = 2X$$

If the example is repeated, but X is held constant while Y is changed, the ratio changes, but *inversely* (in an opposite direction) to the change in Y.

Suppose that Y is doubled while X remains constant; then the ratio changes from 1.5 to 0.75, half of the original value of the ratio. On the other hand, if the new value of Y is half of that in the original ratio, then the value of the ratio changes from 1.5 to 3, or double its original value.

When the quantity that is changed is in the numerator of the fraction, the change in the value of the ratio is directly proportional to the change in the numerator.

That is, if the numerator is *multiplied* or *divided* by some amount, the value of the ratio is multiplied or divided by the same amount.

When the quantity varied is in the denominator of the fraction, the value of the ratio is inversely proportional to the change in the denominator.

That is, if the denominator is *multiplied* by some amount, the value of the ratio is *divided* by the same amount. When the denominator is *divided* by some quantity, the value of the ratio is *multiplied* by the same quantity.

1-10. POWERS OF 10

The reader will find that proficiency in the solution of problems by the use of 10 is of great help, with or without electronic calculators. First, good approximations may be made mentally; and, second, calculator operations are made more accurately and more rapidly if the powers of 10 are not always keystroked into the calculator.

The power of 10 indicates the number of times that some other quantity is multiplied or divided by 10. The power is a number located to the right and slightly above the number 10. If the power is positive, then repeated multiplication by 10 is called for. The number of the power states the number of times that this multiplication shall take place. On the other hand, if there is a minus sign ahead of the power, meaning a negative power of 10, then repeated division by 10 takes place. Once again, the number of the power indicates the number of times that division by 10 is to be done.

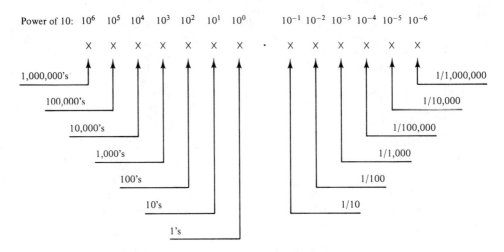

Figure 1-3

EXAMPLE 1-12:

$2 \times 10^4 = 2 \times 10 \times 10 \times 10 \times 10 = 20,000$
$5 \times 10^{-2} = 5 \div 10 \div 10 = 0.05$

If we make use of positional notation, using powers of 10 rather than tens, hundreds, and the like, tenths, Fig. 1-3 is obtained. The reader will note that the 1's column is represented by 10^0. *Any number to the zero power equals 1.*

Note that as we move from one column to the next the power of 10 changes by 1. Movement to the right is negative, whereas movement to the left is positive.

1-11. USING POWERS OF 10

Powers of 10 cannot be used in such a way as to change the value of the original number. When we use powers of 10 with a number, we are writing a product with 10 to some power as one of the factors of the product. The result of the multiplication must be a number that is equal to the original number.

To write a large number as a smaller one, move the decimal point to the left and assign a positive power of 10 to the smaller number as a multiplier. *The power of 10 is equal to the number of places that the decimal point has been moved.*

EXAMPLE 1-13:

$9,950,000 = 99.5 \times 10^5 = 9.95 \times 10^6 = 0.995 \times 10^7$
$125.78 = 12.578 \times 10^1 = 1.2578 \times 10^2 = 0.12578 \times 10^3$

To write a small number as a larger one, move the decimal point to the right and assign a negative power of 10 to the new number as a multiplier. *The negative power of 10 is equal to the number of places that the decimal point has been moved.*

EXAMPLE 1-14:

$$0.0000062 = 6.2 \times 10^{-6} = 0.62 \times 10^{-5} = 62 \times 10^{-7}$$
$$1.534 = 15.34 \times 10^{-1} = 1534 \times 10^{-3} = 1{,}534{,}000 \times 10^{-6}$$

To multiply numbers with powers of 10, multiply the numbers and add the power of 10. If the powers of 10 have different signs, take the difference of the powers and use the sign of the larger power of 10. When the powers have the same signs, simply add and use the sign.

EXAMPLE 1-15:

$$0.003 \times 8500 = 3 \times 10^{-3} \times 8.5 \times 10^{3}$$

Solution:
(a) $3 \times 8.5 = 25.5$
(b) $10^{-3} \times 10^{3} = 10^{-3+3} = 10^{0} = 1$
Because the resulting power of 10 is zero, it is not written in the answer. The resulting answer is simply 25.5.

EXAMPLE 1-16:

$$0.00005 \times 900 = 5 \times 10^{-5} \times 9 \times 10^{2}$$

Solution:
(a) $5 \times 9 = 45$
(b) $10^{-5} \times 10^{2} = 10^{-5+2} = 10^{-3}$
The result is 45×10^{-3}.

EXAMPLE 1-17:

$$0.0044 \times 0.00061 = 4.4 \times 10^{-3} \times 6.1 \times 10^{-4}$$

Solution:
(a) $4.4 \times 6.1 = 26.84$
(b) $10^{-3} \times 10^{-4} = 10^{-3+-4} = 10^{-7}$
The result is 26.84×10^{-7}.

EXAMPLE 1-18:

$$52{,}000 \times 4000 = 5.2 \times 10^{4} \times 4 \times 10^{3}$$

Solution:
(a) $5.2 \times 4 = 20.8$
(b) $10^{4} \times 10^{3} = 10^{4+3} = 10^{7}$
The result is 20.8×10^{7}.

To divide numbers with powers of 10, divide the numbers; then take the sign of the power of 10 in the denominator, change it, and add it to the power of 10 in the numerator. The rules for the addition of powers are the same as for multiplication.

EXAMPLE 1-19:

$$45{,}000 \div 0.009 = \frac{45 \times 10^3}{9 \times 10^{-3}}$$

Solution:
(a) $45/9 = 5$
(b) The power of 10 in the denominator is -3. Change this power to $+3$ and add it to the power of 10 in the numerator.

$$10^3/10^{-3} = 10^{3+3} = 10^6$$

The resulting answer is 5×10^6.

EXAMPLE 1-20:

$$52{,}000/2600 = \frac{5.2 \times 10^4}{2.6 \times 10^3}$$

Solution:
(a) $5.2/2.6 = 2$
(b) The power of 10 in the denominator is $+3$. Change this power to -3 and add it to the power of 10 of the numerator.

$$10^4/10^3 = 10^{4+(-3)} = 10^1$$

The result is 2×10^1.

EXAMPLE 1-21:

$$250 \div 50{,}000 = \frac{2.5 \times 10^2}{5 \times 10^4}$$

Solution:
(a) $2.5/5 = 0.5$
(b) The power of 10 in the denominator is $+4$. Change this power to -4 and add it to the power of 10 in the numerator.

$$10^2/10^4 = 10^{2+(-4)} = 10^{-2}$$

The resulting answer is 0.5×10^{-2}. This answer may be changed to 5×10^{-3} if it is more convenient in this form.

The reader will note that, in these examples of division with numbers and powers of 10, the numbers were selected so as to permit easy division, allowing full attention to be paid to handling the powers of 10. The following example will demonstrate the value of using powers of 10 in the solution of difficult problems.

EXAMPLE 1-22:
Perform the indicated division using powers of 10 with the numbers.

$12560 \div 125786$	$0.00321 \div 986000$
$568000 \div 0.765$	$0.00043 \div 0.0000055$
$3750 \div 0.000455$	$8908000 \div 0.005$

Solution:

$$\frac{1.256 \times 10^4}{1.25786 \times 10^5} = 0.9985 \times 10^{-1} = 9.985 \times 10^{-2}$$

$$\frac{5.68 \times 10^5}{7.65 \times 10^{-1}} = 0.7425 \times 10^6 = 7.425 \times 10^5$$

$$\frac{3.75 \times 10^3}{4.55 \times 10^{-4}} = 0.8242 \times 10^7 = 8.242 \times 10^6$$

$$\frac{3.21 \times 10^{-3}}{9.86 \times 10^5} = 0.3256 \times 10^{-8} = 3.256 \times 10^{-9}$$

$$\frac{4.3 \times 10^{-4}}{5.5 \times 10^{-6}} = 0.7818 \times 10^2 = 7.818 \times 10^1$$

$$\frac{8.908 \times 10^6}{5 \times 10^{-3}} = 1.7816 \times 10^9$$

Note that all solutions were changed, as needed, so that the answers resulted in some number between 1 and 10 with the proper power of 10. This is called *standard scientific notation*.

The addition or subtraction of numbers containing powers of 10 requires that all numbers have the same power of 10. This usually requires that some of the numbers and their powers of 10 be changed. Once all numbers have the same power of 10, *simply add (or subtract) the numbers and use the same power of 10 for the sum (or difference).*

EXAMPLE 1-23:
Add $2.5 \times 10^3 + 8.2 \times 10^2 + 7.6 \times 10^4$.

Solution: Using 10^3 as the common power of 10 we have:

$$\begin{array}{r} 2.5 \times 10^3 \\ 0.82 \times 10^3 \\ \underline{76.0 \times 10^3} \\ 79.32 \times 10^3 \end{array}$$

Using 10^2 as the common power of 10:

$$\begin{array}{r} 25.0 \times 10^2 \\ 8.2 \times 10^2 \\ \underline{760.0 \times 10^2} \\ 793.2 \times 10^2 \end{array}$$

Using 10^4 as the common power of 10:

$$\begin{array}{r} 0.25 \times 10^4 \\ 0.082 \times 10^4 \\ \underline{7.6 \times 10^4} \\ 7.932 \times 10^4 \end{array}$$

Note that all answers are equal to each other and that, if the first two were put into standard scientific form, they would result in the last answer.

EXAMPLE 1-24:

$$3.45 \times 10^{-3} - 10.5 \times 10^{-5}$$

Solution: Using 10^{-5} as the common power of 10:

$$\begin{array}{r} 345.0 \times 10^{-5} \\ - 10.5 \times 10^{-5} \\ \hline 334.5 \times 10^{-5} \end{array}$$

Using 10^{-3} as the common power of 10:

$$\begin{array}{r} 3.45 \times 10^{-3} \\ -0.105 \times 10^{-3} \\ \hline 3.345 \times 10^{-3} \end{array}$$

1-12. SCIENTIFIC NOTATION

We have seen how powers of 10 are a useful tool in arriving at arithmetic solutions involving very large and very small numbers. We have shown that numbers may be modified into any convenient form by using the correct power of 10 with the number. *The standard scientific form of a number with a power of 10 requires that the number be written as some number between 1 and 10 with the correct power of 10.* The following example illustrates this concept.

EXAMPLE 1-25:
Write the following in standard scientific notation.

6,500,000	0.009876	15,350
24.5×10^3	15×10^{-6}	0.05×10^{-2}

Solution:

6.5×10^6	9.876×10^{-3}	1.535×10^4
2.45×10^4	1.5×10^{-5}	5×10^{-4}

The reader is warned that, although using a number between 1 and 10 with the proper power of 10 is indeed correct "standard scientific notation," it is not always the most useful form of the number. We consider this in the following section.

1-13. ENGINEERING NOTATION

The basic physical units used in science and technology are often too large or too small to be of practical value. Therefore, we need to modify the basic units. This is done by the use of powers of 10. Certain powers of 10 are used so frequently in connection with the basic physical units that these powers of 10 are given special names in the form of prefixes to the units. Use of the prefixes eliminates the need to write the related powers of 10.

An example of an overly large unit is the second, the basic unit of time. We have computer circuits capable of performing arithmetic operations in the range of 10^{-8} to 10^{-9} second(s). The prefix corresponding to 10^{-9} is *nano*, and we speak of a typical adder circuit operation taking 24 nanoseconds (24 ns).

Another example, involving an extremely small unit is the hertz (Hz), which is one cycle per second. Suppose that your favorite FM radio station is listed at a frequency of 93.1 megahertz (93.1 MHz). The prefix mega (M) stands for 10^6. Therefore, a frequency of 93.1 MHz is 93.1×10^6 or 93,100,000 cycles per second.

The basic unit of length is the meter. For long distances we use the kilometer (10^3 meters). For very short distances we make use of the centimeter (10^{-2} meter), the millimeter (10^{-3} meter), and others that we shall discuss when we cover the basic physical units.

Table 1-1 illustrates powers of 10 ranging from -12 to $+12$; the table indicates the value of the power of 10 as a number, and lists the prefixes and symbols for those that are used in technology. Usually, the engineering nota-

TABLE 1-1

Power of 10	Number	Prefix	Symbol
10^{-12}	0.000000000001	pico	p
10^{-11}	0.00000000001		
10^{-10}	0.0000000001		
10^{-9}	0.000000001	nano	n
10^{-8}	0.00000001		
10^{-7}	0.0000001		
10^{-6}	0.000001	micro	μ*
10^{-5}	0.00001		
10^{-4}	0.0001		
10^{-3}	0.001	milli	m
10^{-2}	0.01	centi	c
10^{-1}	0.1	deci	d
10^0	1		
10^1	10	deca	dk
10^2	100	hecto	h
10^3	1000	kilo	k
10^4	10000		
10^5	100000		
10^6	1000000	mega	M
10^7	10000000		
10^8	100000000		
10^9	1000000000	giga	G
10^{10}	10000000000		
10^{11}	100000000000		
10^{12}	1000000000000	tera	T

*This is the Greek letter μ, pronounced "mu."

Sec. 1-13. Engineering Notation

tions vary by 10^3 or 10^{-3}. The exceptions are deci (10^{-1}), deca (10^1), centi (10^{-2}), and hecto (10^2).

> **EXAMPLE 1-26:**
>
> A current is measured as 750 microamperes (750 μA). What is the current in milliamperes (mA) and in amperes (A)?
>
> *Solution:*
> (a) 750 μA = 750 × 10^{-6} A
> (b) The power of 10 for milliamperes is 10^{-3}. Therefore, we simply rewrite the number with a 10^{-3}.
>
> $$750 \times 10^{-6} = 0.75 \times 10^{-3}$$
> $$750 \ \mu A = 0.75 \ mA$$
>
> (c) The answer in amperes may be read directly from part (a) of the solution.
>
> $$750 \ \mu A = 0.00075 \ A$$

> **EXAMPLE 1-27:**
>
> A certain resistor is listed as 2.2 MΩ. (The unit of electrical resistance is the ohm and the symbol is the Greek letter, Ω). State the resistance in kilohms (kΩ).
>
> *Solution:*
> (a) 2.2 MΩ = 2.2 × 10^6 Ω
> (b) The power of 10 for kilo (k) is 10^3. Therefore, we rewrite the number with a power of 10^3.
>
> $$2.2 \times 10^6 = 2200 \times 10^3 = 2200 \ k\Omega$$

1-14. PHYSICAL AND ELECTRICAL UNITS

The basic system of units in electrical and electronic technology is the International System of Units (SI*). Let us begin with four of the basic units, those relating to distance, time, mass, and electric charge.

The unit of distance is the meter (m) (39.3701 inches). Originally, the meter was designated as one ten-millionth (10^{-7}) of the distance from the equator to the pole of the earth at sea level. The meter has been modified by the use of engineering prefixes. For large distances, we use the kilometer (km), which is equal to 1000 meters. For smaller units of distance, we work with centimeters (cm), millimeters (mm), micrometers (μm), nanometers (nm), and picometers (pm). We should note that it is common practice to use *micron* (μ) in place of micrometer. It is also a practice among some to use double prefixes, such as millimicron to mean 10^{-9} meter in place of nanometer. We hope that this practice will eventually disappear as more and more use is made of the standard prefixes. In making references to various units of distance, it is possible to make use of any of the proper engineering prefixes. When working with distances that pertain to the universe and outer

*From the French name, Le Système International d'Unités.

space, we use the *light-year*. One light-year is the distance that light will travel in one year.

> **EXAMPLE 1-28:**
> Given that the speed of light is approximately 300,000,000 meters per second (186,000 miles per second), calculate the number of (a) meters and (b) miles equal to a light-year.
>
> *Solution:*
> 1. Calculate the number of seconds in a year.
>
> $$60 \text{ s} \times 60 \text{ min} \times 24 \text{ h} \times 365 \text{ days}$$
>
> $$\text{seconds in a year} = 31{,}536{,}000 = 31.536 \times 10^6$$
>
> 2. Distance = speed (velocity) × time
>
> $$\text{light-year} = 3.0 \times 10^8 \text{ m/s} \times 3.1536 \times 10^7 \text{ s}$$
> $$= 9.4608 \times 10^{15} \text{ m}$$
> $$\text{light-year} = 0.186 \times 10^6 \text{ mi/s} \times 31.536 \times 10^6 \text{ s}$$
> $$= 5.8657 \times 10^{12} \text{ miles}$$

> **EXAMPLE 1-29:**
> A distance is given as 12.6 km. What is the distance in meters, centimeters, and millimeters?
>
> *Solution:*
>
> $$12.6 \text{ km} = 12.6 \times 10^3 \text{ m}$$
> $$= 12.6 \times 10^5 \text{ cm}$$
> $$= 12.6 \times 10^6 \text{ mm}$$

Mass is the quantity of inertia of any substance that occupies space. All matter has mass. In fact, mass is truly defined in terms of the force developed as a result of the mass being accelerated. The standard unit for mass is the kilogram (kg). One kilogram is approximately 2.205 pounds (lb). The gram (g) is one-thousandth of a kilogram and is approximately 0.03527 ounce (oz).

Electric charge is a property of the electron. However, the electric charge of a single electron is very small. Therefore, the practical unit of electric charge is the coulomb (C), corresponding to the charge of 6.24×10^{18} electrons.

> **EXAMPLE 1-30:**
> A charge is given as 1.63 coulomb. What is the equivalent electron charge?
>
> *Solution:* One coulomb equals 6.24×10^{18} electron charges. Given that we have 1.63 C, the result is found by multiplication.
>
> $$1.63 \times 6.24 \times 10^{18} = 10.1712 \times 10^{18} \text{ electron charges.}$$

The basic unit of time is the second. In electronics technology we shall not be concerned with minutes, hours, days, and so on. We are concerned with the second and its fractions. So we shall work with milliseconds (ms), microseconds (μs), and nanoseconds (ns). We have not yet reached the point where the picosecond (ps) is a commonly used unit.

Additional units in the SI system may be defined. The unit of force is the newton (N) and is equal to the product of mass in kilograms and acceleration in meters per second per second.

The unit of work is the joule (J) and is the work done in displacing a body one meter by a constant force of one newton.

Power is the rate of doing work and is defined as the ratio of work to time in *joules per second.* The unit of power is the *watt* (W) and *one watt equals one joule per second.*

The fundamental electrical units are of particular interest to us. These are the units for current, voltage, and opposition to current.

Current is the rate of directed drift of electric charge. We define current in this manner because at normal temperatures electric charges are in constant, but random motion. If there is some net drift of these charges, it is this drift that is considered to be an electric current (Fig. 1-4).

The unit of current is the ampere (A), which is a drift rate of 1 coulomb per second. The symbol for the unit, ampere, is A, while the symbol for current is I (from the French word *intensité*. Current values range from nanoamperes to kiloamperes (10^{-9} to 10^3 A).

The force that produces an electrical current in a circuit is voltage. The symbol for voltage is V, although E is sometimes used. The unit is the volt (V). Voltage values range from microvolts to megavolts (10^{-6} to 10^6 V).

The unit of opposition to current is the ohm (Ω). Frequently, this opposition is referred to as resistance (R). The material on alternating-current circuits develops the concept of reactance, which is also an opposition to current. For the present, the most important difference between these oppositions to current is that resistance dissipates electrical energy (power), whereas reactance is the opposition to current offered by an energy storage device. Both forms of opposition use the unit ohm. Such opposition ranges from a fraction of an ohm up to hundreds and thousands of megohms (10^{-3} to 10^9 Ω).

There are other physical and electrical units. For the present we have developed a sufficient number so that their use may be demonstrated.

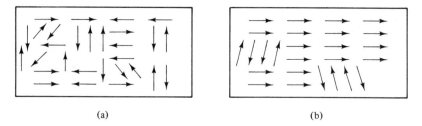

(a) (b)

Figure 1-4 Charge motion in conductor. (a) Random motion. Charges are always in motion at ordinary temperatures. However, the net drift of charge is zero. Although there is charge movement, there is no current flow. (b) Directed drift of charge. There is a net drift of charge. There is current flow.

1-15. WORKING WITH UNITS

The solution of problems in technology requires that we work with physical quantities. These quantities always have units associated with them. It is necessary that a few fundamental rules be followed to obtain correct solutions.

1. Differing systems of units are in use. We have the SI, the c.g.s. (centimeter-gram-second), and the English (foot-pound-second) systems of units. *When solving a problem, all units must be in the same system.* It will be our practice to use SI; when needed, we will convert units that are stated in other systems into SI units.
2. *When adding or subtracting quantities, the units of each term must be of the same type and prefix.* If the units are not the same we must change terms as needed.

The following examples illustrate these two rules.

EXAMPLE 1-31:
Newton's second law states that force is the product of mass and acceleration. It is usually written

$$F = m \times a$$

where m = mass in kilograms (kg)
a = acceleration in meters per second^2 (m/s^2)
F = force in newtons (N)

Suppose that we wished to know the force (weight) exerted by a mass of 12 kg when under the influence of the gravitational acceleration of the Earth. We are given the acceleration due to gravity in the English system: 32.174 ft/s^2. If we were to use mass in kilograms and acceleration in feet per second per second, the answer would be in kg-ft/s^2, a nonexistent unit! Let us change the English system value for gravitational acceleration into the SI unit.

$$g = \frac{32.174 \text{ ft}}{s^2} \times \frac{1 \text{ m}}{3.2808 \text{ ft}} = 9.8067 \text{ m/s}^2$$

Note: In the conversion from English to SI, we multiplied the original unit feet by the conversion factor for feet into meters. (There are 3.2808 ft in 1 m.) Also, note that feet cancel, leaving the SI unit of distance, the meter.

Solution:

$$\begin{aligned} F &= m \times a \\ &= 12 \text{ kg} \times 9.8067 \text{ m/s}^2 \\ &= 117.68 \text{ kg-m/s}^2 \\ &= 117.68 \text{ N} \end{aligned}$$

EXAMPLE 1-32:
The distance from the Sun to the planet Saturn is 886 × 10^6 miles. Given that the Earth is 149.67 × 10^6 km from the Sun, what is the distance between Saturn and the Earth in (a) kilometers and (b) miles?

Solution: Clearly, it becomes necessary to change the systems of units so that both terms use the same unit.
1. There are 1.60934 km in 1 mile.
2. There is 0.62137 mile in 1 km.

(a) $886 \times 10^6 \text{ mi} \times \dfrac{1.60934 \text{ km}}{1 \text{ mi}} = 1425.87 \times 10^6 \text{ km}$

$$1425.87 \times 10^6 \text{ km} - 149.67 \times 10^6 \text{ km} = 1276.2 \times 10^6 \text{ km}$$

(b) Changing the distance between Earth and Saturn into miles, we have

$$1276.2 \times 10^6 \text{ km} \times \dfrac{0.62137 \text{ mi}}{1 \text{ km}}$$
$$= 792.99 \times 10^6 \text{ mi}$$

Frequently, in solving physical problems one arrives at a solution that results in a unit different from those at the start of the problem. For example, if we know the velocity of some object given in terms of distance per unit of time, and we wish to know the time required to travel some distance, we would take the ratio of distance to velocity.

$$t = \dfrac{\text{distance}}{\text{velocity}}$$

EXAMPLE 1-33:
The velocity of light is approximately 300×10^6 meters per second (m/s). Given that the Earth is 149.67×10^6 km from the Sun, what time is required for light from the Sun to reach the Earth?

Solution:

$$t = \dfrac{\text{distance}}{\text{velocity}}$$
$$= \dfrac{149.67 \times 10^9 \text{ m}}{300 \times 10^6 \text{ m/s}}$$
$$= 0.4989 \times 10^3 \text{ s}$$
$$= 498.9 \text{ s} \quad \text{(answer)}$$
(in minutes: $498.9/60 = 8.315$ min)

In this solution we were left with 1/s in the denominator, after meters in the numerator and denominator canceled. Recall that division by a fraction requires that we invert the fraction and multiply. In this example, we obtained seconds in the numerator.

Fortunately, despite these examples, our work in electricity and electronics will rarely ever require that we convert between systems of units. As we said before, the units used in electricity and electronics are in the SI; hence no conversions will be required.

EXAMPLE 1-34:
What is the charge in coulombs at a point if the number of electron charges is 5.261×10^{19}?

Solution: One coulomb equals 6.24 × 10¹⁸ electron charges; therefore, the given electron charge must be divided by this unit quantity.

$$Q = \frac{5.261 \times 10^{19}}{6.24 \times 10^{18}} = 0.843 \times 10^1$$
$$= 8.43 \text{ C}$$

EXAMPLE 1-35:

Charges equal to 2.15 C drift past a point in 0.38 s. What is the current in amperes?

Solution:

$$I = \frac{Q}{t}$$

where I is in amperes, Q is in coulombs, and t is in seconds.

$$I = \frac{2.15}{0.38} = 5.658 \text{ A}$$

1-16. USING THE SCIENTIFIC CALCULATOR

A scientific calculator is a necessary tool for anyone studying electricity and electronics. This type of calculator ranges in cost from a little over ten dollars to several hundred dollars. The inexpensive scientific calculators contain all the necessary functions that you will use in your studies. The more expensive units have additional features but offer no special advantages to the student beginning the study of electricity and electronics.

What is a scientific calculator? The essential features of a scientific calculator are that it can do computations with powers (exponents), perform problems using the trigonometric functions (in degrees and radians), solve problems requiring the use of the Naperian base, and work with powers of 10. You need not be concerned that some of the functions we have just listed are strange to you. We shall have no trouble learning how to use the calculator to solve problems with logarithms and trigonometric functions even if you have never had math courses on this material. We shall introduce material on advanced topics in those parts of your studies where it will be most useful to you. In this section, we examine some of the special keys and functions relating to computational skills.

Figure 1-5 illustrates the keyboard of a typical scientific calculator. We shall examine some of the special keys and their functions.

Change Sign $\boxed{+/-}$. Frequently, the polarity of some number must be changed in order to make a correct solution of a problem. This is especially true in multiplication and division problems where negative quantities may be found. The calculator key with a minus sign on it is a function key. When this key is pressed, whatever number is entered after this step is subtracted from the value in the calculator. Therefore, we cannot use this key to

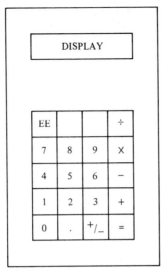

Figure 1.5 A portion of the keyboard of a typical scientific calculator. For the sake of clarity, some functions are not shown.

change a positive value to a negative value. The change sign key $\boxed{+/-}$ is used to allow us to enter negative values into the calculator. Number entries are always positive; when $\boxed{+/-}$ is pressed after the entry, the value of the entry becomes negative. No arithmetic functions take place by use of the change sign key.

EXAMPLE 1-36:
What is the sum of the following voltages?

$$15\,V,\ -35\,V,\ 22\,V,\ -4\,V$$

Solution: The proper sequence of entries is:

$$15\ \boxed{+}\ 35\ \boxed{+/-}\ \boxed{+}\ 22\ \boxed{+}\ 4\ \boxed{+/-}\ \boxed{=}\ -2\,V$$

EXAMPLE 1-37:
A voltage of +28 V is measured between a reference point and some other point in a circuit. At a different point, a voltage of −14 V is measured from the reference. What is the difference of potential (voltage) between these points?

Solution: Difference of potential requires that subtraction take place. We must subtract one voltage measurement from the other.

$$28\ \boxed{-}\ 14\ \boxed{+/-}\ \boxed{=}\ 42\,V$$

If we had begun with −14 V,

$$14\ \boxed{+/-}\ \boxed{-}\ 28\ \boxed{=}\ -42\,V$$

The second solution is the same voltage with the meter leads reversed.

EXAMPLE 1-38:
The product of 60 and −13 is found by the following keystroke sequence:

$$60\ \boxed{\times}\ 13\ \boxed{+/-}\ \boxed{=}\ -780$$

The change sign key is especially useful in working with powers of 10. This will be seen in our discussion on entering numbers with powers of 10 into the scientific calculator.

Powers of Ten $\boxed{EE}$. A number with a power of 10 is keyed into the calculator by use of this key.

1. The number is keyed in as with any number.
2. The $\boxed{EE}$ key is pressed. You will note that the number entered in step 1 skips three places to the left in the display. Two zeros now appear at the right end of the display.
3. The number representing the power of 10 is entered. This entry appears in place of the zeros at the right end of the display.
4. If the power of 10 is negative, enter the power and then press the change sign key.

EXAMPLE 1-39:
(a) Enter 45.67×10^4 into the calculator.
(b) Enter 200×10^{-6} into the calculator.

Solution:

		Display
(a) 1. Enter 45.67	45.67
2. Press $\boxed{EE}$	45.67 00
3. Enter 4	45.67 04
(b) 1. Enter 200	200
2. Press $\boxed{EE}$	200. 00
3. Enter 6	200. 06
4. Press $\boxed{+/-}$	200.−06

EXAMPLE 1-40:
Enter the following quantities into the calculator. After each completed entry, sketch the display.

Number	Display
22.5×10^{-3}	22.5 −03
125×10^5	125. 05
2.228×10^{-12}	2.228 −12
14×10^8	14. 08

The solution of problems with powers of 10 using the calculator is made in the same manner as with any other calculations.

EXAMPLE 1-41:

$$58.6 \times 10^3 \times 76 \times 10^9 =$$

Solution:

58.6 $\boxed{EE}$ 3 $\boxed{\times}$ 76 $\boxed{EE}$ 9 $\boxed{=}$ $4.4536\ 15 = 4.4536 \times 10^{15}$

Sec. 1-16. Using the Scientific Calculator

EXAMPLE 1-42:

$$125 \times 10^{-3} \div 2.866 \times 10^{-9} =$$

Solution:

125 [EE] 3 [+/−] [÷] 2.866 [EE] 9 [+/−] [=] 4.361 × 10⁷

Reciprocal [1/x]. All numbers may be considered to be fractions. In the case of a whole number, the denominator is simply understood to be 1 and is not usually written.

A reciprocal of a number is that quantity which, when multiplied by the number, results in 1. Given that all numbers may be represented by fractions, it follows that the reciprocal of any number (or fraction) is the inverse of that number (or fraction).

Reciprocals of whole numbers are shown as fractions with numerators of 1 and denominators equal to the original number. Reciprocals of fractions are found by simply inverting the fraction.

To take the reciprocal of a number with the scientific calculator, simply enter the number and press the [1/x] key.

EXAMPLE 1-43:

Demonstrate that the product of a number and its reciprocal is 1. Use the following numbers:

$$125, \quad 45 \times 10^{-6}, \quad 15, \quad 25 \times 10^4$$

Solution:

125 [1/x] [×] 125 [=] 1

45 [EE] 6 [+/−] [1/x] [×] 45 [EE] 6 [+/−] [=] 1

15 [1/x] [×] 15 [=] 1

25 [EE] 4 [1/x] [×] 25 [EE] 4 [=] 1

Multiplication and division may be done by the use of reciprocals. Frequently, the use of reciprocals saves time and simplifies a problem.

The multiplication of one number by another by use of reciprocals is done by dividing the multiplicand by the reciprocal of the multiplier.

The division of one number by another by use of reciprocals is accomplished by multiplying the dividend by the reciprocal of the divisor.

EXAMPLE 1-43:

Multiply:

$$54 \times 60 = 54 \div 1/60 = 54 \div 60, \quad [1/x] = 3240$$

$$33 \times 10^{-3} \times 15 \times 10^7 = 33 \times 10^{-3} \div 1/15 \times 10^7$$

$$= 33 \boxed{EE}\ 3\ \boxed{+/-}\ \boxed{\div}\ 15\ \boxed{EE}\ 7\ \boxed{1/x}\ \boxed{=}\ 4.95 \times 10^6$$

Divide:

$$883 \times 10^5 \text{ by } 75 \times 10^{-6} = 883 \times 10^5 \times 1/75 \times 10^{-6}$$
$$= 883\ \boxed{EE}\ 5\ \boxed{\times}\ 75\ \boxed{EE}\ 6\ \boxed{+/-}\ \boxed{1/x}\ \boxed{=}\ 1.17733 \times 10^{12}$$
$$756 \text{ by } 95 \times 10^3 = 756 \times 1/95 \times 10^3$$
$$= 756\ \boxed{\times}\ 95\ \boxed{EE}\ 3\ \boxed{1/x}\ \boxed{=}\ 7.95789 \times 10^{-3}$$

Although the reader cannot really see the value of using reciprocals from these examples, it will be seen in our later efforts with alternating-current circuits.

SUMMARY

1. Powers of 10 are a convenient way of working with very large or very small numbers.
2. Engineering notation enables us to work with certain powers of 10 that relate directly to prefixes used with the basic physical units.
3. The International System (SI) of units is used in electricity and electronics.
4. The unit of electric charge is the coulomb (C): $1\ C = 6.24 \times 10^{18}$ electron charges.
5. Power is the rate of doing work and in electricity is normally measured in watts (W).
6. The unit of electric current (I) is the ampere (A).
7. The unit of electric force or pressure is the volt (V).
8. Opposition to current is measured in ohms (Ω).
9. Resistance is the only form of opposition to current that dissipates power.

PROBLEMS

1-1. (a) Write the following division problems as fractions.
 (b) Use a calculator to solve the problems. Round off answers to four decimal places.

 $476 \div 1321$ $8126 \div 45$
 $50.5 \div 11.86$ $3180 \div 3196$
 $1130 \div 2170$ $17.63 \div 64$
 $96 \div 1451$ $1881 \div 91$

1-2. Add:
 (a) 0.7612   (b) 10.721
   1.0326   9.056
   5.1802   0.831

1-3. Subtract:
 (a) 8.632   (b) 12.08
    −5.971   −3.18

1-4. Multiply:
　　0.0004 × 1.25　　　　　　123.8 × 76.1
　　12.32 × 1.56　　　　　　 12.445 × 1.0076
　　0.0008 × 4　　　　　　　 15.97 × 40.0008
　　0.83 × 0.2　　　　　　　 0.0076 × 5897
　　0.98 × 0.05　　　　　　　10.5 × 25.8

1-5. Divide:
　　0.3618 ÷ 0.4
　　15.6 ÷ 0.00005
　　12.083 ÷ 0.002
　　146.3 ÷ 25.55

1-6. Write the following numbers as percentages:
　　0.0028　　　　　　　　　0.768
　　0.835　　　　　　　　　　2.35
　　0.015　　　　　　　　　　1.05
　　0.25　　　　　　　　　　 0.68

1-7. Change the following fractions to percentages:
　　1/8　　　　　　　　　　　4/5
　　2/3　　　　　　　　　　　4/7
　　1/4　　　　　　　　　　　5/4

1-8. Take 15% of the following quantities:
　　1200　　　　　　　　　　180
　　15　　　　　　　　　　　 200
　　750　　　　　　　　　　　2.8
　　4.5　　　　　　　　　　　3.76

1-9. Find the total cost to the purchaser for each of the following items, assuming a sales tax of 5%.
　　$3.80　　　　　　　　　　$15.75
　　$3500.00　　　　　　　　 $15,750.00

1-10. A microcomputer lists at $5000.00. You are able to purchase it for $3500.00. What is your discount from the list price?

1-11. In each of the following, solve for the unknown number or percent.
　　(a) Eighty-one percent of some number is 66.015.
　　(b) A number that is 12.5% greater than 50.
　　(c) A number that is 150% of 80.
　　(d) A number that is 25% less than 40.
　　(e) Ten dollars is ____% greater than $8.00.
　　(f) Ten dollars ____% less than $15.00.

1-12. Complete the ratios:
$$\frac{3}{4} = \frac{}{12}$$
$$\frac{5}{6} = \frac{15}{}$$
$$\frac{4}{} = \frac{20}{30}$$

1-13. The ratio a/b is equal to 3.2. (a) If a is held constant while b is doubled, what is the new value of the ratio? (b) Suppose that b in the original ratio is halved, while a is held constant. What is the new value of the ratio?

1-14. Change the following to numbers between 1 and 10 with the proper power of 10.
   0.000563   0.00386
   175.62   0.0251
   8761   1610
   9,046,000   0.00000631

1-15. Use powers of 10 to multiply the following. Write the answers as numbers between 1 and 10 with the proper power of 10.
   0.003×150
   $18,000 \times 2000$
   0.004×0.006
   $30,000,000 \times 7000$

1-16. Solve the following, and write the results in scientific notation.
   $45,000 \div 300,000$
   $0.0006 \div 0.02$
   $150 \div 0.0003$
   $27,000 \div 0.009$

1-17. Add the following:
   $36.25 \times 10^2 + 15.1 \times 10^3 = \quad \times 10^3$
   $3.6 \times 10^{-3} + 4.4 \times 10^{-2} = \quad \times 10^{-2}$
   $1.45 \times 10^4 + 3.7 \times 10^3 = \quad \times 10^3$

1-18. Change the following to engineering units:
   15×10^7 Hz = MHz
   0.000055 H = μH
   0.031 A = mA
   0.00004 A = μA
   550,000 W = kW
   10,700,000 W = MW
   0.098 V = mV

1-19. What is the charge in coulombs if the number of electron charges is 1.26×10^{19}?

1-20. If charge drift of 5 C takes place in 0.2 s, what is the current in amperes?

1-21. A current is 3 mA. What is the charge drift in coulombs per second and in electron charges per second?

1-22. Assume that 3.8×10^{20} electron charges drift past a point in 12 s. What is the current in amperes?

1-23. A voltage is given as 680 mV. What is the voltage in volts?

1-24. The current in a certain part of a circuit is directly proportional to the current gain of a transistor. If the transistor is replaced by one with a current gain that is 40% greater, what is the current if the original current was 12 mA?

1-25. A certain transmitter has an output power of 5500 W during daylight hours. With evening, it must reduce its power by 35%. What is the evening power output?

Matter and Electricity

2-1. ELEMENTS

There are some 103 basic substances that either alone or in combination make up all matter. These basic substances are called *elements*. The basic building blocks of all elements are *atoms*. We can define elements in terms of atoms. *An element is a substance consisting of only one type of atom.*

Chemical combinations of elements are known as *compounds*. Compounds have properties entirely different from those of their elements. Table salt is an example of a compound. If we were to reduce salt to the smallest particle that would keep all the properties of salt, the particle would consist of one atom of chlorine and one atom of sodium. Chlorine is a poisonous gas and sodium is a highly explosive element. Yet when combined chemically, the resulting substance is ordinary salt.

Another example is water. If we reduce water to its smallest unit, we find an atom of oxygen and two atoms of hydrogen, the familiar H_2O. Two gases, one highly explosive and the other vital to life, combine chemically to make an entirely different form of matter, water.

The smallest unit of a compound that retains all the properties of the compound is called a *molecule*.

We have defined atoms as the basic building blocks of all matter. Let us now proceed to learn about the properties of atoms, especially those that are important to our knowledge of electrical current and conduction.

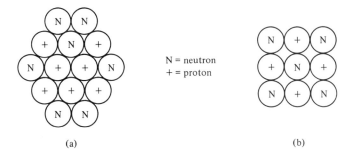

Figure 2-1 (a) A representation of the nucleus of a nitrogen atom. Nitrogen has an atomic number of 7 and an atomic weight of 14. It therefore has 7 protons and 7 neutrons within its nucleus. (b) A representation of the nucleus of a beryllium atom. Beryllium has an atomic number of 4 and an atomic weight of 9, therefore there are 4 protons and 5 neutrons within its nucleus.

2-2. STRUCTURE OF THE ATOM

Modern nuclear theory states that all atoms are made up of a nucleus surrounded by electrons. The nucleus consists of many different types of particles; however, only two are stable, the *proton* and the *neutron* (Fig. 2-1). A proton has a positive electrical charge exactly equal to the negative electrical charge of an electron. The mass of a proton is 1.672×10^{-24} g. Despite this small mass, a proton has 1836 times the mass of an electron. Both are thought to have the same size. A neutron has a mass that is virtually the same as that of a proton and has no electrical charge. Hydrogen is the only known element that has no neutrons in its nucleus. *All atoms normally have equal numbers of protons and electrons and are therefore electrically neutral.*

The atomic number of an atom is determined by the amount of protons in its nucleus. In a neutral atom, this number is also the amount of electrons surrounding the nucleus.

The electrons surrounding the nucleus are considered to be in various orbits, much like the planets orbiting the sun in our solar system. However, the mechanisms governing the behavior of atomic particles cannot be explained by any physical laws relating to the behavior of large masses.

Fortunately, we need not study atomic physics in order to learn those properties of atoms that relate to the study of electricity. The following properties are of value to our learning about conduction of electric charge.

The orbits of the electrons surrounding the nucleus of an atom are confined to fairly specific radial distances from the nucleus. The general region of a permitted radial distance is called a *shell* (see Fig. 2-2).

The shell closest to the nucleus cannot have more than two electrons. This first shell is known as the k shell. The second shell is limited to eight electrons. This l shell consists of two subshells. The first is allowed two

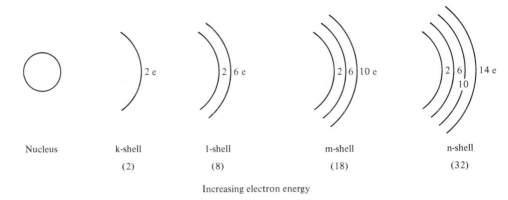

Figure 2-2 Electrons orbit the nucleus of an atom at distances determined by the energies of the different electrons.

electrons and the second subshell is allowed six electrons. The space between the k and l shells is forbidden to electrons, and orbits cannot exist in this region.

The third shell is the m shell. This shell has three subshells in a 2-6-10 electron distribution. Once again, we find a forbidden region between shells.

The n shell, separated from the m shell by a forbidden region, is allowed 32 electrons in a four-subshell arrangement of 2-6-10-14 electrons.

As the complexity of an atom increases with increasing atomic number we find additional forbidden regions, shells, and subshells. The shells and subshells represent energy levels, with those farthest from the nucleus containing the highest-energy electrons.

For an atom to be stable, lower electron energy levels as represented by the shells must fill to their limits before additional shells may be formed. The atomic number of an element gives us the information needed to know the electron distribution of an atom.

Shells always fill according to electron energy levels beginning with the lowest energy level, represented by the k shell.

Thus lithium, atomic number 3, has two k shell electrons and one l shell electron. Carbon, atomic number 6, has two k shell electrons and four l shell electrons. On the other hand, fluorine with atomic number 9, has two k shell electrons and seven l shell electrons. A more complex atom is found in an element such as barium, atomic number 56. A normal atom of barium must have filled k, l, and m shells and a partially filled n shell containing 28 electrons.

Barium	Atomic number 56
k shell	2 electrons
l shell	8 electrons
m shell	18 electrons
n shell	28 electrons

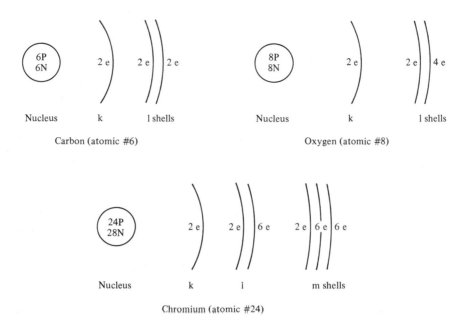

Figure 2-3 The electron distribution of various atoms.

Figure 2-3 illustrates electron distributions for some other common elements.

It is possible by the use of external sources of energy to cause electrons at lower energy states to jump to higher energy states. When this occurs, the atom is considered to be *excited*; however, electrons immediately return to normal energy levels by giving up excess energy in the form of radiation.

Fluorescent lamps are a practical example of this type of atomic behavior. Fluorescent bulbs contain mercury vapor. The atoms of the mercury receive energy from the power line, which causes excitation of some of the electrons. Immediately, these excited electrons return to normal energy states by giving up energy in the form of ultraviolet radiation. The ultraviolet rays cause the fluorescent coating on the inside of the bulb to glow and give off light.

2-3. ELECTRICITY

The earliest historical references to electricity begin at about the sixth century B.C. The Greeks had discovered that amber, when rubbed briskly with fur, had the ability to attract small, light objects such as straw. The Greek word for amber is *elektron*. Therefore, objects touched by the charged amber were "electrified." The words electricity and electronics come from this Greek word for amber.

Static electricity is the electricity of accumulated charge and is not at all useful as a source of electrical power. This is the electricity that varies in amount from that sufficient to give us a slight shock after walking on carpeting and then touching a doorknob, to the enormous amount of charge that is released with a lightning bolt. When early experimenters rubbed glass rods with silk, fur, and the like, they were working with static electricity. The basis of practical electrical power, the *dynamo*, was realized approximately 100 years after Benjamin Franklin and his famous experiments. Franklin was convinced that electricity was a type of "fluid." Franklin concluded that there were positive and negative electric fluids.

Dynamic electricity is the electricity of *moving and controlled charges*. This is the electricity of closed circuits, the form of electricity used to operate machinery, appliances, computers, radio, television, in fact all electrical and electronic equipment.

Electric charge is the same whether we consider static or dynamic electricity. Recall from Chapter 1 that the basic unit of charge is the *coulomb*, which is equal to an electric charge of 6.24×10^{18} electrons.

$$1 \text{ coulomb} = 6.24 \times 10^{18} \quad \text{electron charges} \tag{2-1}$$

2-4. WORK, ENERGY, AND POWER

Energy, work, and power are related physical quantities. *Work* is force acting through a distance (Fig. 2-4).

$$\text{Work} = \text{Force} \times \text{Distance} \tag{2-2}$$

In the SI system the unit of work is the newton-meter (N-m) or joule (J). One newton is the force required to accelerate a one-kilogram mass at the rate of one meter per second per second.

Energy is the capacity to do work and is measured in the same units (joules) as work.

Power is the rate of doing work. The electrical unit of power is the

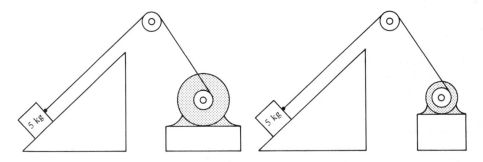

Figure 2-4 Both motors do the same amount of work. However, the smaller motor must turn its pulley more slowly in order to do the same job as the larger, more powerful motor. The larger motor develops more power when given the same load as the smaller motor.

watt (W). Work performed at the rate of one joule per second represents one watt of power.

$$1 \text{ W} = 1 \text{ J/s} \tag{2-3}$$

The horsepower, not an SI unit, is equal to 746 W.

2-5. CONDUCTION OF CHARGE

Current is the directed drift of charge. Clearly, the only atomic particle capable of drift is the electron. Recall that the outermost electrons in complex atoms have high energy states. Additional energy, frequently in the form of light or heat, is sufficient to cause an outer-shell electron to be removed from an atom. When this occurs, the atom is no longer electrically neutral and is an *ion*, in this case a positive ion. The released electron is free to move about and is capable of contributing to current. However, added factors contribute to electron mobility. Consider a solid rather than a gas. In this case, the spacing between atoms is close enough to permit the outermost electrons of atoms to act upon each other. This interaction causes the range of permitted electron energies to spread; so in place of a specific energy level representing an outermost shell or subshell, we now have a band of permitted electron energies. In some solids the outermost band becomes so broad as to overlap the forbidden region that normally separates the outermost shell from the next lower-level shell. In those materials where the overlap between the highest energy shell and the next lower shell occurs, we find that at ordinary room temperature outermost shell electrons acquire sufficient energy to wander in a random manner from atom to atom. This is the case with materials such as copper, silver, and aluminum and is the reason they are useful as *conductors* of electrical current.

Let us examine why the metals listed are good sources of "free" electrons. Silver, atomic number 47, is a complex atom with filled k, l, and m shells and with three filled subshells of its n shell. Thus we find *one* electron in the fourth subshell of each atom of silver. These outermost electrons are at a high energy level, and with a small amount of external energy, usually in the form of heat, the outermost electrons wander freely about in the metal.

Copper, atomic number 29, does not have as complex an atomic structure as silver. The copper atom also has filled k, l, and m shells, with just one electron in the first subshell of the n shell. This electron is at a high energy level, although not as high an energy level as the outermost electron of a silver atom. Once again, a small amount of thermal energy is sufficient to provide large amounts of free electrons in the metal.

Aluminum, atomic number 13, has a less complex atomic structure than either silver or copper. In the aluminum atom we find filled k and l shells, a filled first subshell of the m shell, and one electron in the second subshell. This outermost electron is at a lower energy level than the outermost electrons of copper and silver atoms. We should expect that for a given amount of external energy we would find fewer free electrons than in the case of silver and copper. This is indeed true, and of the three metals aluminum is the poorest contributor of charges, while silver is the best.

2-6. CONDUCTORS

Those materials that are good sources of free electrons are known as *conductors*. A practical conductor must also meet other requirements in terms of physical and economic needs. A conductor must be economical to use in large quantities, a consideration that limits silver to special applications in electronics where the amount of silver used does not add appreciably to the cost of the equipment.

Copper is the most widely used practical conductor. It is relatively inexpensive and has the physical properties that permit it to be drawn and shaped into wire of various sizes. Aluminum, while not as good a conductor of current as copper, is being used in some applications because of weight and cost factors. Some problems have occurred with aluminum wiring, particularly when used in contact with copper connectors. Connections have loosened and caused sparks and fires. Eventually, such problems will be eliminated by the use of new connector designs, and one may expect that many new buildings, particularly homes, will have aluminum wiring.

The property of a material that permits the flow of electric current is defined as *conductivity*. Table 2-1 compares the conductivity of various materials with that of silver.

TABLE 2-1

Material	Relative Conductivity
Silver	1.000
Copper	0.945
Aluminum	0.576
Tungsten	0.297
Nichrome	0.015
Carbon	0.017

2-7. INSULATORS

There is a category of materials that supplies few, if any, free electrons. In fact, the outermost electrons of the atoms of these materials are very tightly bound to the molecules of the material. These compounds are clearly not good sources of free electrons. They do not permit current flow, except for very small leakage currents whose values seldom equal a few nanoamperes (10^{-9} A). This class of materials is known as *insulators*. It is important to understand that, given sufficient external force in the form of voltage, any material will break down and conduct large currents. One measure of an insulator is the ability to withstand voltage without breaking down. This ability is known as *dielectric strength* and is measured in volts per mil of thickness (1 mil = 0.001 in.). Table 2-2 compares the dielectric strength of some common insulators.

TABLE 2-2

Material	Dielectric Strength (V/mil)
Air	21
Porcelain	150
Paper	305
Pyrex	335
Plastic electric tape	1000
Mica	1050

2 8. SEMICONDUCTORS

Some materials have conducting properties that are far greater than those of insulators, but very much less than those of conductors. This broad category of materials is often classified as *semiconductors*.

Silicon and to a much lesser extent germanium are widely used semiconductor materials in electronic devices such as transistors and integrated circuits. Other semiconductor materials are used to make resistors and resistance wire. Resistance and resistors are topics we shall study in Chapter 5.

SUMMARY

1. Elements are the basic chemical substances that make up matter.
2. Atoms are the smallest units of elements.
3. Compounds are chemical combinations of elements.
4. Molecules are the smallest units of compounds.
5. Electric current is the directed drift of electrons.
6. Conductors are materials with many free electrons.
7. The most widely used conductor is copper.
8. Insulators have very few, if any, free electrons. Insulators prevent current flow.
9. The dielectric strength of an insulator describes its ability to withstand voltage without breakdown. The property is stated in volts per mil.
10. Semiconductors are a category of materials with properties between those of insulators and conductors.

PROBLEMS

2-1. The atomic number of lead is 82. What is the number of protons in the nucleus?

2-2. An atom with atomic number 10 has nine electrons in orbits about the nucleus. Does this represent negative or positive ionization?

2-3. Illustrate the electron distribution by shells for an atom of selenium (atomic number 34).

2-4. Using electron distribution as a basis for discussion, explain why carbon is a poor conductor of electric current.

2-5. Suppose that the conductivity of copper is used as a reference. What is the relative conductivity of silver?

2-6. What is the relative conductivity of carbon if copper is the reference conductor?

2-7. What minimum thickness of mica insulator is needed to withstand 8000 V?

2-8. What must be the minimum air space between two terminals if there is to be 16,000 V applied without breakdown of the air gap?

2-9. Suppose that we must insulate a circuit against 5 V. An insulator with a dielectric strength of 40,000 V/cm is used. What is the minimum thickness of the insulator in millimeters?

2-10. A porcelain spacer is 0.5 mm thick (25.4 mm = 1 in.). What is the maximum voltage that may be placed across the material without breakdown?

3

Voltage and Voltage Sources

3-1. VOLTAGE

In our studies so far, we have learned how conductors and conducting materials supply free electrons. The motion of these charges at any instant is random, and is not a current. To have current, we need a force that will cause some of the free charges to move in one direction (at a time), rather than randomly. This force is called *voltage*.

The force represented by voltage is the result of a difference of charge at the terminals of the voltage source. *Energy in some form other than electrical must be used to maintain this difference of charge.* The terminal with negative charges (ions or electrons) is the negative terminal of the source. The terminal with an equal amount of positive charges (ions) is the positive terminal of the source (Fig. 3-1). If a conductor is connected between these terminals, a current will flow as a result of the drift of *some* of the

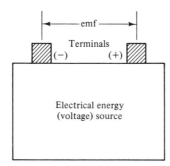

Figure 3-1 Voltage is electrical energy. In order to have voltage, we must convert other forms of energy into electrical energy. In this sketch, which represents a battery, we find that chemical energy is converted into electrical energy.

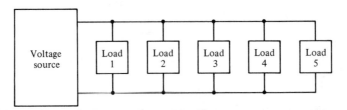

Figure 3-2 A voltage source connected to several loads. The type of load and the quantity of loads depends upon the amount of current and voltage that the source can furnish. Where there is a high capability, as in house wiring, we have many loads connected to a source. With small sources, such as carbon batteries, the number of loads is limited. Usually, just one load at a time is supplied by such a source.

free electrons. This drift of charges in the conductor is away from the negative terminal and to the positive terminal. Within the source of voltage, this drift of charge is from the positive terminal to the negative terminal. The current will have an amount that is determined by the capacities of the voltage source and the conductor.

To have current flow, all circuits no matter how complex, must meet two requirements:

1. There must be a source of voltage.
2. There must be a closed external path for charges connected between the terminals of the source.

Many other expressions are used for voltage: potential, potential difference, pressure, tension, drop, voltage drop, and electromotive force (emf). All are valid synonyms for voltage, although we shall learn that all these names do not have identical meanings.

3-2. VOLTAGE UNITS

The basic unit of voltage is the volt; the abbreviation is V. The symbol for voltage is V or E; both are widely used. In this text we shall use E when referring to voltage sources; for all other meanings we shall use V. Practical amounts of voltage vary depending upon the purposes of the circuits.

In consumer electronics we find that FM receivers are designed to work with as little as 2 μV (2 × 10^{-6} V) of antenna signal. On the other hand, the high voltage needed to operate picture tubes in TV receivers ranges from 16 to 40 kV (16,000 to 40,000 V). Broadcasting equipment requires voltages in the range from 10 to 50 kV.

Electricity is supplied to homes and offices as 120 V and 208-240 V. Some industrial applications require very high voltages, often in the range of 4.8 kV. Electricity is frequently distributed in the 13.5-kV range, although over long distances the voltages used are very much higher. Experimental distribution of electrical power over very long distances is being done with voltages at the million volt (MV) level.

Very high voltages in the megavolt (MV) range are also used in nuclear research for the acceleration of charged particles.

Digital circuits operate at low voltages, with digital integrated circuits requiring as little as 5 V.

This discussion demonstrates that one cannot think of so-called "typical" voltages without being aware of the circuits and system to which the voltage is applied.

3-3. PURE DC VOLTAGE

Direct current (dc) is the voltage we obtain from batteries and from very well filtered rectifier power supplies. It is a voltage that is nonvarying (except) as a battery gradually discharges) in amplitude and has a fixed polarity. Pure dc must be used with electronic amplifier circuits and digital circuits.

3-4. PULSATING DC VOLTAGE

Pulsating dc voltage is obtained from generators and unfiltered or poorly filtered rectifier power supplies. Pulsating dc as in Fig. 3-3, is used in industrial applications; it is used to operate dc motors and battery chargers or

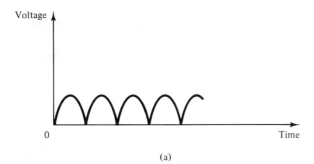

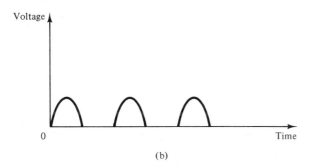

Figure 3-3 Graphs of pulsating dc voltages. We shall learn that a voltage (or current) need only have an *average* value other than zero for it to be considered dc.

in any other applications where fluctuations in amplitude will not affect the performance of the system. Pulsating dc is unidirectional, as is pure dc, because the terminal polarities of the source do not change.

3-5. ALTERNATING VOLTAGE

When a voltage changes polarity on a regular, periodic basis so that the average value for each positive and negative alternation (change in polarity) is the same, the voltage is said to be *alternating*.

A *cycle* of alternating voltage requires that the positive and negative alternations be complete. For example, we may choose to consider a cycle starting at zero amplitude. Then the cycle proceeds to maximum positive amplitude and back to zero. Then it reverses in polarity to maximum negative amplitude and then back to zero. On the other hand, we could have chosen maximum negative amplitude (or any other value) as the start of a cycle. From maximum negative to zero, to maximum positive, to zero, and then to maximum negative completes a cycle. Note that in our discussion a *complete cycle* requires that the wave proceed from any point to the same point on the next wave.

The manner in which a cycle changes with respect to time describes its *wave shape*. Hence, there are square waves, rectangular waves, triangular waves, sawtooth waves, and others (as shown in Figs. 3-4, 3-5, and 3-6). One wave shape of special importance is the *sine wave*. This is the wave shape of voltage supplied by the power companies; it is shown in Fig. 3-7. While it is correct to refer to any alternating wave as *ac* (alternating current), general usage results in the assumption that a wave is a sine wave if simply identified as ac. Reference to any other wave shape of voltage requires that the wave shape be named. The sine wave forms the basis for the mathematical analysis of any alternating wave.

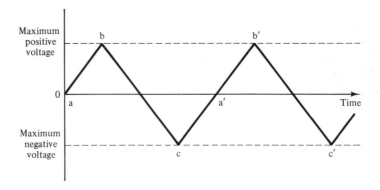

Figure 3-4 A triangular *ac* voltage wave. An *ac* wave can have any waveshape. It is necessary that the average of all the positive values equals the average of all the negative values.

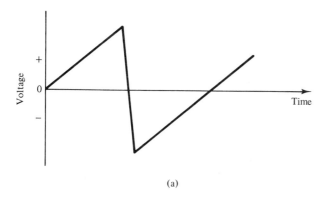

(a)

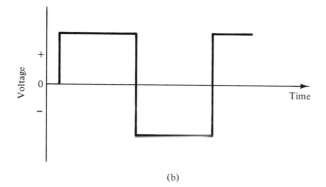

(b)

Figure 3-5 Examples of ac waves frequently found in electronic circuits. The waveform in (a) is called a *sawtooth* wave and the one in (b) is a *square* wave.

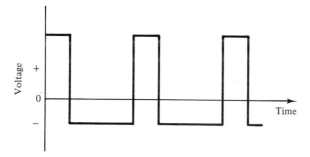

Figure 3-6 A rectangular wave. The wave shown is ac because the average of the positive value is equal to the average of the negative value over a cycle.

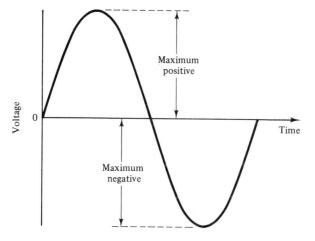

Figure 3-7 A sinewave of ac voltage.

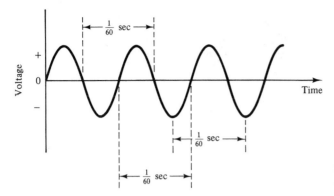

Figure 3-8 Several cycles of 60 Hz voltage. As in any other *ac* waveform, a cycle is measured from one point on a wave to a similar point on the next wave.

3-6. FREQUENCY

The number of cycles per second of an alternating voltage is its *frequency* (Fig. 3-8). The unit for cycles per second is the hertz, abbreviated Hz. Electrical power in the United States is distributed at a frequency of 60 Hz. Audio frequencies cover the range from 20 Hz to 20 kHz. Radio frequencies used in FM broadcasting cover a band from 88 to 108 MHz (88 to 108 × 10^6 cycles per second). Radar operates at very much higher frequencies of greater than 10^9 cycles per second (gigahertz, GHz).

The time of one cycle is defined as the *period*. The period of an alternating voltage (or current) is the reciprocal of the frequency of the wave as shown in Fig. 3-9.

$$T = \frac{1}{f} \tag{3-1}$$

where f = frequency in hertz.

EXAMPLE 3-1:
Given the following frequencies, determine the periods of each: (a) 60 Hz, (b) 10 kHz, (c) 5 MHz.

Solution:

(a) $T = \dfrac{1}{60} = 0.0167$ s $= 16.7$ ms

(b) $T = \dfrac{1}{1 \times 10^4} = 10^{-4}$ s $= 100$ μs

(c) $T = \dfrac{1}{5 \times 10^6} = 0.2 \times 10^{-6}$ s $= 0.2$ μs

If period is known, frequency is readily calculated.

$$f = \frac{1}{T} \tag{3-2}$$

where T is in seconds.

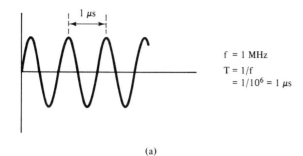

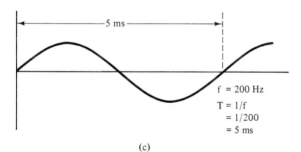

Figure 3-9 Period is inversely related to frequency.

EXAMPLE 3-2:
Determine the frequency of the following ac signals, given the periods: (a) 50 μs, (b) 0.5 ms, (c) 0.02 μs.

Solution:

(a) $f = \dfrac{1}{50 \times 10^{-6}} = 0.02 \times 10^6$ Hz = 20 kHz

(b) $f = \dfrac{1}{0.5 \times 10^{-3}} = 2 \times 10^3$ Hz = 2 kHz

(c) $f = \dfrac{1}{0.02 \times 10^{-6}} = 50 \times 10^6$ Hz − 50 MHz

Sec. 3-6.   Frequency

3-7. SOURCES OF VOLTAGE

For a source of voltage to be useful, it is required that energy in some form other than electrical be used to maintain the difference of charge between its terminals. This clearly means that other forms of energy are used to furnish electrical energy.

Mechanical Energy

Mechanical energy is used for the generation of voltage by the electrical power companies. Machines called *alternators* are used in the generation of electricity. Mechanical energy is used to rotate magnetic fields past stationary coils of wire, or coils of wire are rotated within fixed magnetic fields. In large alternators the magnetic fields are rotated, and in small ones it is usually the coils that are rotated. In either case the reaction as a result of motion between coils and magnetic fields produces voltage at the terminals of the alternator. The name alternator is used because the output is ac. A similar but very much smaller device is used in automobiles as the source of charging voltage for the battery.

In some applications, dc is needed and is produced by machines called *generators*. This is useful for some industrial and control processes. The conversion of mechanical energy into electrical is the same as in alternators, except the magnetic fields are fixed and only coils rotate, and a special coupling between the coils and terminals is used to cause the output to be pulsating dc. The development of voltage as a result of relative motion between magnetic fields and conductors is discussed in detail in Chapter 15.

Chemical Energy

Batteries and fuel cells are examples of the use of chemical energy to produce dc voltage. There are two general categories of batteries, *primary* and *secondary*. Primary cells cannot be recharged; secondary cells are rechargeable. The basic unit of a battery is a cell. Commercial batteries may be single cells or combinations of cells. The cell voltage when no current is drawn (the emf of the cell) is determined by the chemicals and metals used to make up the cell. Thus mercury cells differ from zinc-carbon cells just as nickel oxide cells differ from lead–acid cells.

The terminal voltage of a battery depends upon the number of cells that are connected in series. Given a cell with an emf of 2 V, 6 cells in series supply an emf of 12 V.

The current capacity of a cell is determined by the area of the cell. Batteries may also be made to have increased current capacities by connecting cells in parallel.

An important rating of a battery is its *ampere-hour capacity* (A-h). This is basically a measure of its capability to deliver a *rated* current over a *rated* time interval before replacement or recharging is needed. The ampere-hour capacity is not a simple product of current and time. A battery rated at 70 A-h cannot deliver 70 A for 1 h, nor can it deliver 140 A for ½ h! Its A-h rating is based upon some specified load current. When load current exceeds rated current, the useful life of a battery is greatly reduced. Certainly, load currents less than the rated maximum current used for the A-h rating will extend the useful life of a battery.

Light Energy

Light energy is used with photocells and certain semiconductors to generate voltage. Light is the source of the energy needed for some of the valence band electrons to jump the forbidden region and enter into the conduction band. As a result, a voltage is developed at the terminals of the device. The voltage and current capacities of photoelectric devices are small, which restricts their application. These devices are useful in electronic circuits that are designed to make effective use of their properties. Despite the drawbacks of photoelectric devices in terms of supplying large quantities of electric power, the promise of using the light of the sun to produce commercial levels of electric power has resulted in intense research into the development of larger, more efficient cells, called *solar cells*. Large or small power, the process of using light to produce electricity is a *photovoltaic* process.

Heat Energy

Heat energy may be used directly to produce electricity, as in a *thermocouple*. Two dissimilar metals, connected at one end only, produce a difference of potential at the ends of the metals when the junction of the metals is heated. This potential is the result of the rate of free charge production in each of the metals. Free electrons drift from one metal to the other and tend to concentrate in one more than the other. This difference in charge concentration is the voltage developed at the terminals of the thermocouple. The voltage is small and is used in connection with electronic circuits for process-control applications.

Crystals

An interesting property, piezo electricity, is found in certain ceramics and minerals, both natural and man-made. These crystals produce voltages when stressed mechanically and vibrate when electricity is applied to an axis of the crystal. This is a useful property for microphones, phonograph pickup cartridges, and radio-frequency control circuits. Quartz crystals are shown in Fig. 3-10.

Figure 3-10
Courtesy of Savoy Electronics, Inc.

SUMMARY

1. The unit of voltage is the volt (V). Practical values range from microvolts (μV) to megavolts (MV).
2. Pure dc voltage is a nonvarying quantity with respect to time.
3. Pulsating dc voltage is unidirectional, as is pure dc, but it changes amplitude on a regular, periodic basis. Pulsating dc voltage and current have many industrial applications.
4. Alternating voltage changes amplitude and polarity with each half cycle.
5. The average value of a pure ac wave is zero.
6. The wave shape of any wave describes how it changes over a cycle of operation. Thus we have sine waves, sawtooth waves, and others.
7. The basic ac wave, from which all others may be analyzed, is the sine wave (Chapter 18).
8. Frequency is a measure of the number of cycles per second.
9. Period (T) is the time of one cycle.
10. Alternators and generators change mechanical energy into electrical energy.
11. Batteries convert chemical energy into electrical energy.
12. Photovoltaic devices convert light to electrical energy.
13. Thermocouples convert heat to a voltage. They find extensive use in industrial measurements.
14. When a piezoelectric crystal is mechanically stressed, voltage is developed.
15. Voltage applied to a piezoelectric crystal results in mechanical motion of the crystal.

PROBLEMS

3-1. The input voltage to an amplifier is 0.022 V. What is the voltage in millivolts and microvolts?

3-2. An industrial process requires 1400 V dc. What is the voltage in kilovolts?

3-3. An accelerator, used in nuclear research, has an operating potential of 1200 kV. What is the voltage in megavolts?

3-4. The period of a wave is 0.5 μs. What is the frequency of the wave?

3-5. The period of a radar wave is 0.435 ns. What is the operating frequency of the radar?

3-6. A short-wave diathermy unit operates at 15 MHz. What is the period in nanoseconds of a cycle at this frequency?

3-7. What is the period in milliseconds of a cycle when the frequency is 10 kHz?

3-8. The commercial AM broadcast band is from 550 to 1600 kHz. (a) Which frequency has the longest period? (b) If the period of the frequency of a station is 1 μs, what is the frequency of the station?

3-9. What is the terminal voltage of the battery in Fig. 3-11? The emf of each cell is 1.5 V.

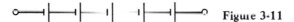

Figure 3-11

3-10. Each battery in Fig. 3-12 can handle loads up to 50 mA for long periods of time. With four batteries connected in parallel, what is the current capacity of the system?

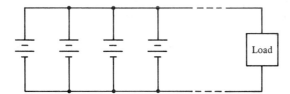

Figure 3-12

Current and Power

4-1. CURRENT UNITS

We have defined current as the directed drift of charge. However, it is the amount of charge drift per second that is of importance in electricity and electronics. Therefore, we work with units of current.

The basic unit of current is the ampere (A). One ampere equals a drift rate of one coulomb per second, where one coulomb equals 6.24×10^{18} electron charges.

Various prefixes are used in association with amperes to identify other useful measures of current. Table 4-1 lists these prefixes and their values related to amperes.

TABLE 4-1

Name	Symbol	Value (A)
Nanoampere	nA	10^{-9}
Microampere	μA	10^{-6}
Milliampere	mA	10^{-3}
Ampere	A	10^{0}
Kiloampere	kA	10^{3}

We find that values of current usually range from nanoamperes to amperes in electronic circuits and from amperes to kiloamperes in electrical circuits.

EXAMPLE 4-1:
(a) A charge of 0.12 C drifts past a point in 1 s. What is the current?

Solution:
$$I = \frac{Q}{t}$$
where Q is in coulombs and t is time in seconds.
$$I = \frac{0.12}{1} = 0.12 \text{ A}$$

(b) The same amount of charge drift is measured in a time of 0.01 s. What is the current?

Solution:
$$I = \frac{0.12}{0.01} = 12 \text{ A}$$

(c) A total charge drift of 0.02 C occurs in 4 s. What is the current?

Solution:
$$I = \frac{0.02}{4} = 0.005 \text{ A} = 5 \text{ mA}$$

(d) A drift of 10^{-4} C takes place in 1 ms. What is the current?

Solution:
$$I = \frac{10^{-4}}{10^{-3}} = 10^{-1} \text{ A} = 0.1 \text{ A} = 100 \text{ mA}$$

Example 4-1 is used to reinforce the concept of the ampere as charge drift per unit time. Current is not measured this way, but is measured directly. The next example is used to illustrate the use of prefixes.

EXAMPLE 4-2:
(a) A current is measured as 15 A. State the current in milliamperes, microamperes, and kiloamperes.

Solution: 1 A = 1000 mA; therefore,
$$15 \text{ A} = 15{,}000 \text{ mA}$$
1 A = 1,000,000 μA; therefore,
$$15 \text{ A} = 15{,}000{,}000 \text{ }\mu\text{A}$$
1 A = 0.001 kA; therefore,
$$15 \text{ A} = 0.015 \text{ kA}$$

(b) A current is 75 mA. What is the current in amperes and microamperes?

Solution: 1 mA = 0.001 A; therefore,
$$75 \text{ mA} = 0.075 \text{ A}$$

1 mA = 1000 µA; therefore,

$$75 \text{ mA} = 75{,}000 \text{ µA}$$

All the answers in the example are correct, but proper usage requires that we choose the prefixes that most clearly state the value with a minimum of zeros. In part (a) the answer is best stated in amperes, and in part (b) the best form is in milliamperes.

This is a good time to clear up some conflicting concepts.

1. We have *defined* current as the directed drift of charge. This drift of charge is *instantaneous* throughout a circuit. That is, the drift in the solid takes place at nearly the speed of light.
2. We *measure* and work with current in amperes. In this case we are concerned with the total amount of charge in coulombs displaced in a specific unit of time, the second. The measurement of current is based upon the ratio of coulombs to time. The reader will recall that amperes equals coulombs per second.

To repeat, charge drift in a solid does not mean that a charge starting at some point in a conductor must drift through the conductor to some other point. There are "clouds" of free electrons within the solid. Drift at any instant is instantaneous throughout the solid, and electrons at some point in a solid do not move more than the distance of a tiny fraction of the conductor. Current in amperes is the result of total drift, represented by the drift at a point.

4-2. CURRENT DIRECTION

We have agreed that those charges capable of drift within a solid are electrons. It would appear to be logical that we would then consider current to consist of the drift of electrons. Indeed, this is the case. However, practice dating back to the eighteenth century established current as the flow of positive charges. While the physics of such a concept are in conflict with modern knowledge of atomic structure, many in electricity and electronics keep to this convention. Some new books still use this convention. There need be no confusion on this point. The eighteenth and nineteenth century researchers made a wrong guess! No harm is done by using this old convention. Solutions to problems, designs, and measurements are made correctly. However, there is a growing trend to adopt electron current and its direction as standard.

Conventional Current. When using conventional current, as in Fig. 4-1, the assumed direction is from the *positive* terminal of the source, through the external circuit, to the *negative* terminal, and then from negative to positive within the source.

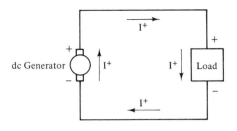

Figure 4-1 Conventional current. The direction of "conventional" current was established in the 18th century.

Voltages developed in the external circuit as a result of current flow (voltage drops) have positive polarity at the point where the conventional current enters a load and negative polarity where current leaves a load. In this text, any use of conventional current will be indicated by I^+.

Electron Current. This is the current direction that is used throughout this text. The direction of this flow is from the *negative* terminal of the source, through the external circuit, to the *positive* terminal, and within the souce from positive to negative, as shown in Fig. 4-2.

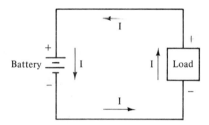

Figure 4-2 Electron current. In this text, we shall use electron flow as current flow; therefore the name "electron current." In this current, simply indicated as *I*, the direction of flow is from the negative terminal of the source, through the circuit and returning at the positive terminal. Within the source, the flow is from positive to negative.

Voltage drops in the external circuit have negative polarity where electron current enters a load and positive polarity where current leaves a load. We shall indicate electron current by *I*, the usual symbol for current.

4-3. TYPES OF CURRENTS

Currents may be *direct* or *alternating*. Names and definitions for currents are similar to those used for voltages.

Pure dc current does not vary in direction and amount with respect to time.

Pulsating current is a dc current that changes in amount periodically, but not in direction. Figure 4-3 shows several examples of pulsating current.

Alternating current, as in our discussion of alternating voltage (Sec. 3-5), varies in amplitude and direction on a regular, periodic basis. As with alter-

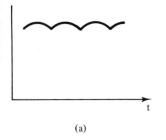
(a)

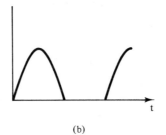
(b)

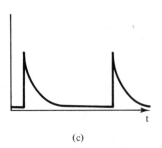
(c)

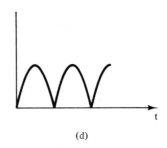
(d)

Figure 4-3 Pulsating *dc* current. Each of the figures is an example of a time-varying current that is still defined as direct current. Remember that we need only have some average value other than zero for the quantity to be considered *dc*.

nating voltages, the average values of each half-cycle are equal. The peak amplitude of the wave, its frequency, and wave shape are all important quantities.

The wave shapes of alternating currents depend upon the type of circuit and the wave shape of the applied alternating voltage. These wave shapes are the same as in alternating voltages. Note that we have made the current wave shape depend upon the circuit as well as upon the voltage. We shall learn that the wave shape of current in a circuit need not always be the same as the wave shape of the applied voltage.

The sine wave is the most important of the wave shapes of alternating current. It is the wave shape of current supplied by the electric power companies. The sine wave is also the basis for the analysis of other, nonsine-wave alternating voltages and currents. The analysis of ac circuits in this text is based upon sinusoidal (shaped like a sine wave) voltages and currents.

4-4. ELECTRICAL POWER

Remember that energy is the *ability* to do work, work is energy *put to use*, and power is the *rate* of doing work (Fig. 4-4).

In electrical circuits, work is done in causing current flow. This requires that electrical energy be put to use. This process causes electrical energy to be transformed into other forms of energy, usually *heat*.

Voltage is electrical energy. That is, a difference of potential has the ability to cause current. Work is performed in moving charges about a circuit. This work has the unit joule (J), as does the basic unit for energy. Our concern is not with energy or work, but with electrical power. The unit of elec-

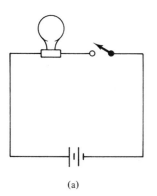
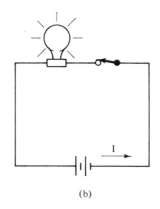

Figure 4-4 (a) The battery has energy but cannot do any work because the switch is open. There is no current flow and the lamp is not lit. (b) When the switch is closed, the energy of the battery is used to do work. Current flows, and the bulb is lit. The rate of doing work is power. The greater the power, the brighter the bulb.

trical power is the watt (W). This is a rate of work equal to one joule per second.

$$1 \text{ W} = 1 \text{ J/s} \qquad (4\text{-}1)$$

$$\text{power} = \frac{\text{work}}{\text{time}} \qquad (4\text{-}2)$$

where power is in watts, work is in joules, and time is in seconds.

Some circuits require very small amounts of power, measured in fractions of a watt. Other circuits require very large amounts of power, as in the case of electric generating systems. Therefore, we find practical values of power range from microwatts (10^{-6} W) to gigawatts (10^9 W).

Another common unit of power is the horsepower. The relationship between watts and horsepower (hp) is given by Eq. (4-3).

$$1 \text{ hp} = 746 \text{ W} \qquad (4\text{-}3)$$

EXAMPLE 4-3:
An appliance performs 1100 joules of work per second. What is the power dissipation of the appliance?

Solution:

$$P = \frac{1100}{1} = 1100 \text{ W} = 1.1 \text{ kW}$$

EXAMPLE 4-4:
What is the power dissipation of a device that does 85 joules of work per hour?

Solution:

$$\text{power (in watts)} = \frac{\text{work in joules}}{\text{time in seconds}}$$

$$P = \frac{85}{3600} = 23.61 \text{ mW}$$

Note that time in hours has been changed to time in seconds.

EXAMPLE 4-5:
An electric motor has a rated output of 1.5 hp. What is the power output in watts?

Solution:
$$P \text{ (watts)} = P \text{ (hp)} \times 746$$
$$= 1.5 \times 746 = 1119 \text{ W} = 1.119 \text{ kW}$$

EXAMPLE 4-6:
The input power to an air conditioner is 2.984 kW. What is the power input in horsepower?

Solution:
$$P \text{ (hp)} = P \text{ (watts)} \div 746$$
$$= 2984/746 = 4 \text{ hp}$$

4-5. EFFICIENCY

All dynamic systems have power losses. We define *power losses* as that part of the input power that does not appear as useful output power, but is instead used to overcome friction and other losses in various parts of a system. Friction is only one of several sources of power loss in a mechanical system. In electrical systems we find losses, but to a much lesser degree than in other systems. The sources of these losses will become clear as we study electrical circuits and electromagnetism.

Regardless of the type of system, efficiency is defined as the ratio of output power to input power. The symbol for efficiency is η and is usually expressed as a percentage, as in Fig. 4-5.

$$\eta = \frac{P_{out}}{P_{in}} \times 100 \qquad (4\text{-}4)$$

$$P_{out} = \frac{\eta P_{in}}{100} \qquad (4\text{-}5)$$

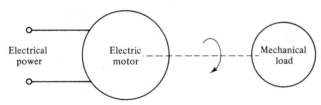

Input power = 500 W

Output power = $\frac{1}{2}$ hp

$\frac{1}{2}$ hp = 373 W

Efficiency $(\eta) = \frac{\text{Power out}}{\text{Power in}}$

$\eta = \frac{373}{500} = 0.746 = 74.6\%$

Figure 4-5 An example of efficiency calculations. The motor losses and the losses in the coupling to the load must be 127 W; the difference between the input power and the output power. All input power must be accounted for in terms of losses plus useful output.

$$P_{in} = \frac{P_{out} \times 100}{\eta} \qquad (4\text{-}6)$$

When η is used as a decimal fraction rather than a percentage, Eqs. (4-5) and (4-6) must be used *without* the number 100.

Efficiency can never be greater than 100%. In fact, 100% efficiency is entirely theoretical because we would be describing a lossless system, a system that cannot exist. We shall learn that electrical devices are highly efficient, but none are 100% efficient.

EXAMPLE 4-7:
An electric motor requires an input power of 1.2 kW to produce an output power of 1.08 kW. What is the efficiency of the motor when operated under these conditions?

Solution:

$$\eta = \frac{P_{out}}{P_{in}} \times 100$$
$$= \frac{1.08 \times 10^3}{1.2 \times 10^3} \times 100 = 90\%$$

EXAMPLE 4-8:
A greater amount of load is placed upon the motor of Example 4-7. It is found that the power input increases to 1.5 kW and the power output increases to 1.2 kW. What is the efficiency under these operating conditions?

Solution:

$$\eta = \frac{P_{out}}{P_{in}} \times 100$$
$$= \frac{1200}{1500} \times 100 = 80\%$$

EXAMPLE 4-9:
A certain electrical system is 42% efficient. What is the output power if the input power is 50 W?

Solution:
(a) Using η as a percentage,

$$P_{out} = \frac{\eta P_{in}}{100}$$
$$= \frac{42 \times 50}{100} = 21 \text{ W}$$

(b) Using η as a decimal,

$$P_{out} = \eta P_{in}$$
$$= 0.42 \times 50 = 21 \text{ W}$$

EXAMPLE 4-10:
Given an electrical system that is 60% efficient, if the output power is 30 W, what is the input power?

Solution:
(a) Using η as a percentage,

$$P_{in} = \frac{P_{out} \times 100}{\eta}$$

$$= \frac{30 \times 100}{60} = 50$$

(b) Using η as a decimal,

$$P_{in} = \frac{P_{out}}{\eta}$$

$$= \frac{30}{0.6} = 50 \text{ W}$$

4-6. A PRACTICAL UNIT OF ELECTRICAL ENERGY

Clearly, we already have a unit of electrical energy, the joule. Recall that one watt is equal to one joule per second and that the product of power in watts and time in seconds will result in the watt-second as a unit of electrical energy. However, the joule is too small a unit of energy for practical usage. A time of one second is also too small a physical unit in this case. We require that work be done for longer periods of time, not seconds.

When power is used over a period of time, we have a practical means of determining the rate of conversion of electrical energy into other forms of energy. We find that the practical unit of power is the kilowatt, and the practical unit of time is the hour. The resultant unit of electrical energy is the *kilowatt-hour* (kWh).

$$\text{kWh} = P \text{ (kW)} \times \text{time (h)} \tag{4-7}$$

EXAMPLE 4-11:
A toaster requires power input of 1200 W. If the toaster is used for a total of 18 min, what is the rate of conversion of electrical energy into heat?

Solution:
(a) Convert minutes to hours.

$$18 \text{ min} \div 60 = 0.3 \text{ h}$$

(b) Convert watts to kilowatts.

$$1200 \text{ W} \div 1000 = 1.2 \text{ kW}$$

(c) kWh = P (kW) $\times$ t (h)
   = 1.2 kW $\times$ 0.3 h
   = 0.36

EXAMPLE 4-12:
A 250-W lamp is used for 8 h each day. What is the daily usage of electrical energy?

Solution:

$$\text{energy usage} = 0.25 \text{ kW} \times 8 \text{ h}$$
$$= 2 \text{ kWh}$$

We purchase electricity from power companies by the kilowatt-hour. Rates vary in different regions of this nation, but all charges are calculated in the same way.

EXAMPLE 4-13:
An electric clothes dryer is rated at 5500 W. The average cost of kilowatt-hour is 7.8 cents. If the dryer is used for 25 h/month, calculate the annual cost of operating the dryer.

Solution:

$$\text{Cost} = \text{kWh charge} \times \text{total kWh}$$
$$= 0.078 \times 5.5 \text{ kW} \times 25 \text{ h} \times 12 \text{ months}$$
$$= \$128.70$$

SUMMARY

1. While the ampere is the unit of current, practical values range from nA (10^{-9} A) to kA (10^3 A).
2. The drift of current in a circuit is nearly at the speed of light. Therefore, we assume current to be instantaneous around a circuit.
3. Current is the directed drift of electrons. Therefore, current direction in a circuit is from the negative terminal of a voltage to the positive terminal. This is called *electron* current.
4. Conventional current assumes a circuit direction from + to −. This is the opposite direction of electron flow. This direction was selected long before science knew of electrons.
5. Voltage drops in a circuit are the result of the force needed across an opposition in order to sustain current.
6. We have as many types of currents as we have types of voltages.
7. The unit of electrical power is the watt (W). Practical values range from 10^{-6} to 10^9 W.
8. Another common unit of power is the horsepower (hp).

$$1 \text{ hp} = 746 \text{ W}$$
$$1 \text{ kW} = 1.34 \text{ hp}$$

9. Efficiency is always the ratio of output power to input power.
10. The most practical unit of electrical energy is the kilowatt-hour.

PROBLEMS

4-1. A current is 15 mA. What is the drift of charge in microcoulombs per second?

4-2. Given a current of 120 µA, what is the current in milliamperes? in amperes?

4-3. In the circuit of Fig. 4-6, indicate current direction for electron flow and for conventional (I^+) flow.

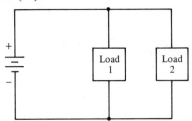

Figure 4-6

4-4. Identify each of the currents in Fig. 4-7 as pure dc, pulsating dc, alternating, and as alternating but with a dc component.

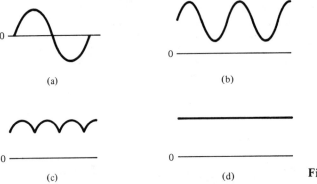

Figure 4-7

4-5. Explain the differences between energy, work, and power.

4-6. A motor is rated at 0.25 hp at 10,000 rpm. What is the rating in watts?

4-7. The output of a microphone is 2.5×10^{-8} W. What is the power in nanowatts?

4-8. Is it possible for a system to have greater output power than its *total* input power? Explain.

4-9. A system has an efficiency of 62%. If the output power is 300 W, what is the input power?

4-10. The input power of the system of Prob. 4.9 is reduced to 250 W. What is the new output power?

4-11. A system develops an output power of 50 W, when the input power is 80 W. What is the efficiency of the system?

4-12. A motor develops a rated power of 1.5 hp when the input power is 1.5 kW. What is the efficiency of the motor under these operating conditions?

4-13. Assume that the cost of a kilowatt-hour is 9.5 cents. An appliance with an input power of 2 kW operates for 4 h/day. What is the cost of daily operation?

4-14. A load requires 390 W. The efficiency of the coupling system is 38%. How much power is required of the source?

5

Resistance

5-1. INTRODUCTION

Resistance is the circuit property that opposes current flow and dissipates power. There are other forms of opposition to current flow, but only resistance can dissipate electrical power.

Recall that current is the directed drift of free electrons. In the movement of electrons around a circuit, collisions take place between electrons and the atoms and molecules of the conductors and other components of a circuit. These collisions are a form of opposition to current that may be compared to the effects of friction in a mechanical system. The collisions produce heat, resulting in a loss of electrical power.

The unit of resistance is the *ohm*. The symbol for ohms is Ω. The symbol for resistance is R.

5-2. CONDUCTANCE

Conductance is defined as the reciprocal of resistance. The symbol is G and the International System unit is the siemen (S). Be aware that many persons still use the old unit, the mho (ohm spelled backward). Hence we have

$$G = \frac{1}{R} \quad \text{siemens} \qquad (5\text{-}1)$$

$$R = \frac{1}{G} \quad \text{ohms} \qquad (5\text{-}2)$$

It follows that a circuit with high resistance has a low amount of conductance. Conversely, a circuit with high conductance has a low amount of resistance.

EXAMPLE 5-1:
Solve for the conductance of each of the following resistances:
$$R_1 = 0.02 \, \Omega, \quad R_2 = 4.7 \, \Omega, \quad R_3 = 10 \, \Omega,$$
$$R_4 = 250 \, \Omega, \quad R_5 = 2.2 \, k\Omega, \quad R_6 = 33 \, k\Omega$$

Solution:

$$G_1 = \frac{1}{R_1} = \frac{1}{0.02} = 50 \text{ S}$$

$$G_2 = \frac{1}{R_2} = \frac{1}{4.7} = 0.213 \text{ S}$$

$$G_3 = \frac{1}{R_3} = \frac{1}{10} = 0.1 \text{ S}$$

$$G_4 = \frac{1}{R_4} = \frac{1}{250} = 4 \times 10^{-3} \text{ S} = 4 \text{ mS}$$

$$G_5 = \frac{1}{R_5} = \frac{1}{2200} = 0.455 \times 10^{-3} = 0.455 \text{ mS}$$

$$G_6 = \frac{1}{R_6} = \frac{1}{33 \times 10^3} = 0.0303 \times 10^{-3} = 30.3 \, \mu S$$

EXAMPLE 5-2:
Given the following conductances, solve for the resistances: $G_1 = 10$ mS, $G_2 = 5$ s, $G_3 = 2 \, \mu S$.

Solution:

$$R_1 = \frac{1}{G_1} = \frac{1}{10 \times 10^{-3}} = 0.1 \times 10^3 = 100 \, \Omega$$

$$R_2 = \frac{1}{G_2} = \frac{1}{5} = 0.2 \, \Omega$$

$$R_3 = \frac{1}{G_3} = \frac{1}{2 \times 10^{-6}} = 0.5 \times 10^6 = 500 \, k\Omega$$

It follows that for a given applied voltage a circuit with greater conductance will have more current than a circuit with a lesser amount of conductance. *As conductance increases, opposition to current decreases.* Conductance is a measure of the ease with which current may be established, while resistance is a measure of opposition to current.

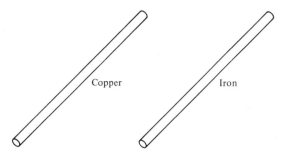

Figure 5-1 The effect of the material upon the resistance of a wire. The resistance of a wire depends upon many factors. Its diameter, length, temperature, and type of material all affect resistance. In this sketch, consider that the wires are identical except for the type of material used. The wires are also at the same temperature.

Iron does not have as many free electrons as copper. For a given applied voltage, the current in the copper wire is much greater than the current in the iron wire. The iron wire has a resistance that is nearly six times that of copper.

5-3. PHYSICAL CONSIDERATIONS

The amount of resistance found in a circuit depends upon many factors. The type of material used for conduction, the cross-sectional area of the connecting wires, the length of the wires, and the temperature of the materials all have effects upon the total resistance of a circuit.

Recall from our studies on atomic structure that various materials have differing amounts of free electrons available for the conduction of current. While good conductors like copper and silver have large quantities of free electrons, others like nichrome and carbon have far fewer conduction-band (free) electrons (Fig. 5-1).

The resistance of any conductor is inversely proportional to its cross-sectional area. The greater the area for conduction of a given current is, the fewer the collisions with the atoms of the material, and therefore a lesser amount of resistance. Also, the greater the cross-sectional area is, the greater the number of free electrons available for the conduction of current (Fig. 5-2).

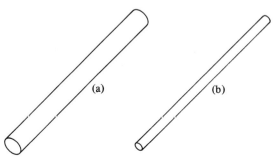

Figure 5-2 Conductor (a) has a diameter that is twice that of (b). Resistance is *inversely* proportional to area.

Increasing the length of a conductor increases its resistance. If we assume some average number of collisions per unit of length, it follows that increased length results in a greater total number of collisions and therefore higher resistance.

Finally, we must consider the temperature of the material. All atoms are in motion provided that they receive thermal energy. This is true for solids as well as gases. In a solid, the motion consists of vibration around a point in the structure, rather than random drift as in a gas. Vibration increases with temperature, because a greater amount of thermal energy is converted to mechanical energy. Since the numbers of collisions between electrons and atoms are affected by the vibrational motion of the atoms, the resistance of a substance is affected by the temperature of the substance.

5-4. PHYSICAL STANDARDS

Before we go any further with our studies, we must establish some standard units of measure as used with wires.

Area. Because most conductor wires have a circular cross section, the defined unit for area is the circular mil (CM).

The circular mil is simply the square of the diameter of the wire, when that diameter is in mils. *One mil equals one-thousandth of an inch.*

$$\text{CM} = d^2 \tag{5-3}$$

where d is the diameter in mils. Keep in mind that circular measure is significantly different from square measure of area as shown in Fig. 5-3.

To change square units of area to circular units, multiply the square units by 1.273:

$$\text{CM} = \text{square mils} \times 1.273 \tag{5-4}$$

To convert circular units of area into square units, divide the circular units by 1.273.

$$\text{area (square units)} = \frac{\text{area (circular units)}}{1.273} \tag{5-5}$$

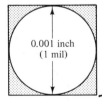

Figure 5-3 Square and circular measures of area. The circular mil is a smaller unit of measure than the square mil. The shaded area shows the additional area of a square mil.

EXAMPLE 5-3:
The diameter in mils of 12-gauge wire is 80.81 mils. What is the area in circular mils?

Solution:
$$\text{area (CM)} = \text{diameter}^2 \text{ (mils)}$$
$$= 80.81^2$$
$$= 6530 \text{ CM}$$

EXAMPLE 5-4:
A bus bar used for carrying very heavy currents has a cross section of 4 × 0.3 in. What is its area in circular mils?

Solution:
(a) Change the dimensions to mils.
$$4 \text{ in.} = 4000 \text{ mils}$$
$$0.3 \text{ in.} = 300 \text{ mils}$$

(b) Solve for area in square mils.
$$A = 4000 \times 300 = 1{,}200{,}000 \text{ square mils}$$

(c) Convert square mils to circular mils.
$$A \text{ (CM)} = 1{,}200{,}000 \times 1.273$$
$$A = 1{,}527{,}600 \text{ CM}$$

Length. The defined unit of length used in resistance computations is the *foot*.

Eventually it will be necessary to change to the International System, but for the present we do work with the foot.

Temperature. The selected standard is an approximation of "room temperature." *It is defined as 20°C.*

5-5. RESISTANCE OF A WIRE

The resistance of a wire is found by

$$R = \rho \frac{\ell}{A} \; \Omega \tag{5-6}$$

where ℓ = length in feet
A = area in CM
ρ = resistivity of the material

The resistivity (ρ) in Eq. (5-6) is normally given at 20°C. We shall develop formulas for finding resistance at other temperatures in Sec. 5-6. Resistivity is given as a quantity with units of ohms × circular mils per foot.

Note from Eq. (5-6) how the use of all the units results in ohms.

$$R = \frac{\Omega \times \text{CM}}{\ell} \times \frac{\ell}{\text{CM}}$$

Table 5-1 lists resistivity for a wide variety of conductors.

TABLE 5-1

Material	Resistivity (ohms·CM/ft at 20°C)
Aluminum	17.0
Antimony	251.0
Brass, annealed	42.0
Copper, annealed	10.37
Copper, hard drawn	10.7
Gold	16.7
Iron	66.4–81.4
Nickel	47.0
Platinum	60.2
Silver	9.9
Tungsten	33.2
Zinc	37.4
Nichrome	600.0

EXAMPLE 5-5:
An annealed copper wire is 250 ft long. If the diameter of the wire is 50.82 mils (American Wire Gauge #16), what is the resistance at 20°C?

Solution:

$$R = \frac{\rho \ell}{A} = \frac{10.37 \times 250}{50.82^2}$$

$$= 1.004 \; \Omega$$

EXAMPLE 5-6:
Assume the wire of Example 5-5 is aluminum instead of copper. What is the resistance at 20°C?

Solution:

$$R = \frac{17 \times 250}{50.82^2} = 1.646 \; \Omega$$

Equation (5-6) may be solved for length:

$$\ell = \frac{R \times A}{\rho} \qquad (5\text{-}7)$$

EXAMPLE 5-7:
How many feet of AWG #20 nichrome wire is needed to make a resistance of 20 ohms at 20°C?

Solution:
(a) Determine the CM area of 20-gauge wire. From the AWG table #20 wire has a diameter of 31.96 mils.

$$A = 31.96^2 = 1021.4 \text{ CM}$$

(b) Obtain the resistivity of nichrome at 20°C from Table 5-1.

$$\rho = 600 \text{ } \Omega \cdot \text{CM/ft}$$

(c) Use Eq. (5-7):

$$\ell = \frac{10 \times 1021.4}{600} = 17.02 \text{ ft}$$

5-6. TEMPERATURE EFFECTS

The temperature of a material has an effect upon its resistance. A change in temperature will cause resistance to change. *If resistance increases with increased temperature, the material has a positive temperature coefficient of resistance.* Metal conductors have positive temperature coefficients. Semiconductors and carbon have *negative* temperature coefficients of resistance. This means that as temperature increases resistance decreases. These materials have relatively few free electrons. When more thermal energy is supplied, there are much greater numbers of free electrons to contribute to conduction of current. Therefore, resistance decreases despite increased vibration of atoms.

The temperature coefficient of resistance is given as the resistance change per degree Celsius per ohm at the standard temperature. Standard temperature is usually 0°C. The symbol for the temperature coefficient of resistance is the Greek letter alpha (α). When given at the standard temperature, the symbol is written as α_s.

The change in resistance as a result of temperature change is found by

$$\Delta R = \alpha_s (R_s)(T_1 - T_s) \tag{5-8}$$

where ΔR = change in resistance
α_s = temperature coefficient of resistivity at standard temperature
T_1 = any temperature
T_s = standard temperature (usually 0°C)

EXAMPLE 5-8:
Annealed copper has an α_s of 0.00426. The resistance of a length of this wire is 11 Ω at 0°C. What is the resistance of 40°C?

Solution:
(a) Solve for the resistance change by Eq. (5-8).

$$\Delta R = 0.00426 \times 11 \times (40 - 0)$$
$$= 1.87 \text{ } \Omega$$

(b) The resistance at 40°C is the sum of ΔR and R_s.
$$R = 1.87 + 11 = 12.87 \, \Omega$$

We may not know the resistance of a wire at 0°C. Provided that we know the resistance of the wire at some temperature, we may solve for resistance at any other temperature.

$$R_2 = \frac{(1 + \alpha_s T_2)(R_1)}{1 + \alpha_s T_1} \qquad (5\text{-}9)$$

where R_1 = known value of resistance
T_1 = temperature at known resistance
T_2 = temperature at unknown resistance
R_2 = unknown resistance
α_s = temperature coefficient of resistance at 0°C

EXAMPLE 5-9:
What is the resistance of 100 ft of 22-gauge tungsten wire at 50°C?

Solution:
(a) To use Eq. (5-9), we must know the resistance at some other temperature. We can find the resistance at 20°C by use of Eq. (5-6).

$$R = \frac{\rho \ell}{A}$$

1. From Table 5-1, $\rho = 33.2 \, \Omega \cdot \text{CM/ft}$.
2. From the AWG table we find that #22 wire has a diameter of 25.347 mils.

$$R = \frac{33.2 \times 100}{25.347^2} = 5.174 \, \Omega = R_1$$

(b) Look up α_s in Table 5-2.

$$\alpha_s \text{ for tungsten} = 0.0049$$

TABLE 5-2 TEMPERATURE COEFFICIENTS OF RESISTANCE AT 0°C

Material	α_s
Aluminum	0.0042
Antimony	0.00388
Brass	0.00208
Annealed copper	0.00426
Carbon	−0.005
Gold	0.00365
Iron	0.00618
Nickel	0.006
Platinum	0.0037
Silver	0.0041
Tungsten	0.0049
Zinc	0.004
Nichrome	0.00044

(c) Solve the problem by using Eq. (5-9).

$$R_2 = \frac{1 + \alpha_s T_2}{1 + \alpha_s T_1}(R_1)$$

$$= \frac{1 + (0.0049)(50)}{1 + (0.0049)(20)} \times 5.174$$

$$= \frac{1.245}{1.098} \times 5.174 = 5.867 \, \Omega$$

EXAMPLE 5-10:
Carbon has a temperature coefficient of resistance (α_s) of -0.005. If a carbon resistor is rated at 150 Ω at 25°C, what is its resistance at (a) 0°C; (b) 75°C?

Solution:
(a) At 0°C,

$$R_2 = \frac{1 + (-0.005)(0)}{1 + (-0.005)(25)} \times 150$$

$$= \frac{1}{0.875} \times 150$$

$$= 171.4 \, \Omega$$

(b) At 75°C,

$$R_2 = \frac{1 + (-0.005)(75)}{1 + (-0.005)(25)} \times 150$$

$$= \frac{0.625}{0.875} \times 150$$

$$= 107.1 \, \Omega$$

5-7. CURRENT CAPACITY OF A WIRE

Whenever current flows in a wire, heat is developed as a result of the resistance of the wire. The greater the diameter of the wire is, the lower the resistance of the wire, and therefore the greater the current capacity of the wire.

Wires are normally covered with insulation. It is this insulation that burns when excessive currents are carried. For this reason, safety standards have been established by the National Fire Protection Association. Table 5-3 lists the American Wire Gauge number, the diameter in mils, and the maximum allowable current in amperes.

Note that maximum allowable current ratings are not given for wires smaller than 14 gauge. Commercial wiring is not permitted for wire sizes less than #14. Applications for the smaller sizes are found in electronic circuits where currents are usually in milliamperes.

TABLE 5-3

AWG	Diameter (20°C)	Maximum Current (A)
0000	460.0 mils	230
000	409.6	200
00	364.8	175
0	324.9	150
1	289.3	130
2	257.6	115
3	229.4	100
4	204.3	85
5	181.9	—
6	162.0	65
7	144.3	—
8	128.5	45
9	114.4	—
10	101.9	30
11	90.74	—
12	80.81	20
13	71.96	—
14	64.08	15
15	57.07	
16	50.82	
17	45.26	
18	40.30	
19	35.89	
20	31.96	
22	25.35	
26	15.94	
30	10.03	
40	3.145	

5-8 RESISTORS

Resistance plays a very important part in the operation of electronic circuits. Resistance values range from as low as a few ohms to millions of ohms. *Resistors* are lumped quantities of resistance, made into convenient standard sizes according to resistance value and power dissipation. They may be fixed or variable.

We referred to resistors as "lumped" quantities. This reference is in comparison to the resistance of a wire. The resistance of a wire is considered to be a *distributed* quantity, because its resistance is distributed over its length. We shall find this concept of lumped and distributed quantities again as we learn about other circuit properties.

5-9. TYPES OF FIXED RESISTORS

The degree of precision required of a resistor is one factor that determines the method of manufacturing the resistor. Another factor is the amount of power that the resistor must dissipate in the form of heat.

The accuracy of an electrical instrument is affected by the accuracy of its parts. Therefore, we find that precision resistors are used in such equipment. These resistors are made to tolerances of 1% or less. For high values of resistance, deposited carbon or metal film types are used. For low values of resistance, the precision resistor may be wound of fine resistance wire on a porcelain or ceramic core. Precision resistors have low power ratings of usually less than 2 W.

For higher power dissipations, wire-wound resistors are used. In these applications, very high amounts of resistance are not needed. These resistors are wound onto a ceramic core and then coated with a temperature-resistant enamel. The variety of sizes and shapes in this category is quite large. The tolerance rating of high power, wire-wound resistors is about 5% to 10% (Fig. 5-4).

The most widely used variety of fixed resistors is the molded-carbon composition resistor. This resistor is commonly called the *carbon resistor*. Carbon resistors are made in wattage ratings as low as $\frac{1}{8}$ W. The maximum power rating for carbon resistors is 2 W. Power dissipation capability is a function of the physical size of the carbon resistor. Figure 5-5 clearly illustrates the variety of sizes and power ratings for carbon resistors.

Carbon resistors are made with different standards of tolerance. The least expensive are rated at ±20%. One may also purchase carbon resistors with ±10% and ±5% tolerance ratings.

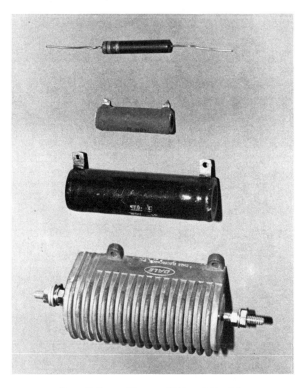

Figure 5-4

Sec. 5-9.   Types of Fixed Resistors

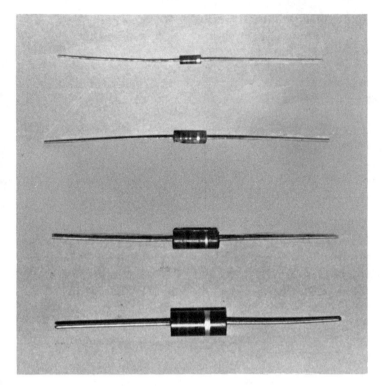

Figure 5-5

5-10. RESISTANCE COLOR CODE

Resistance and tolerance specifications for precision resistors are usually printed on the body of the resistor. Most wire-wound resistor specifications, including power rating, are also printed on the resistor. Molded-carbon composition resistors and some wire-wound resistors are identified by a three- to four-band color code as in Fig. 5-6. The color bands are used to identify the resistance and tolerance ratings of the resistor (see Table 5-4). The user is expected to determine power rating by the physical size of the resistor.

The band at the left edge of the body is the first digit of the resistance rating. The next band is the second digit. The third band is the multiplier. The fourth band is used to specify the tolerance rating. If there is no fourth band, the resistor has a tolerance rating of ±20%.

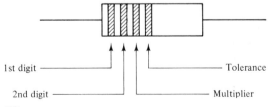

Figure 5-6 Color code for composition resistors.

Chap. 5 Resistance

TABLE 5-4 RESISTANCE COLOR CODE

Color	Digit Value	Multiplier	Tolerance
Black[a]	0	1	
Brown	1	10	
Red	2	100	
Orange	3	1,000	
Yellow	4	10,000	
Green	5	100,000	
Blue	6	1,000,000	
Violet	7	[b]	
Gray	8	[b]	
White	9	[b]	
Silver		0.01	10%
Gold		0.1	5%

[a] Cannot be used in the first band (first digit).
[b] Maximum resistance is 22 MΩ. Therefore, these colors are not used as multipliers.

EXAMPLE 5-11:
Determine the resistance and tolerance ratings for the following color codes:

	Band 1	Band 2	Band 3	Band 4
(a)	yellow	violet	red	silver
(b)	green	brown	black	gold
(c)	blue	gray	orange	silver
(d)	white	brown	green	gold
(e)	orange	white	silver	gold
(f)	brown	black	green	

Solution:
(a) 4 7 × 100 = 4.7 kΩ ±10%
(b) 5 1 × 1 = 51 Ω ±5%
(c) 6 8 × 1000 = 68 kΩ ±10%
(d) 9 1 × 100,000 = 9.1 MΩ ±5%
(e) 3 9 × 0.01 = 0.39 Ω ±5%
(f) 1 0 × 100,000 = 1 MΩ ±20%

EXAMPLE 5-12:
Specify the color code for the following resistors: 22 kΩ ±10%, 560 Ω ±20%, 10 Ω ±5%, 100 Ω ±5%, 22 MΩ ±10%.

Solution:
22 kΩ ±10%: red, red, orange, silver
560 Ω ±20%: green, blue, brown
10 Ω ±5%: brown, black, black, gold
100 Ω ±5%: brown, black, brown, gold
22 MΩ ±10%: red, red, blue, silver

5-11. VARIABLE RESISTORS

Certain applications require that variable or adjustable resistors be used rather than fixed resistors. By the use of these resistors it is possible to adjust current and voltage levels in a circuit.

When current is the controlled quantity, the variable resistor is called a *rheostat*. The *total* terminal resistance of the rheostat changes with various settings of the control. Rheostats are typically low-resistance units of rarely more than several thousand ohms. Commercial units are usually wire-wound types capable of high power dissipation. Some applications are in dc motor speed control, very precise adjustment of industrial control systems, and current control in small applicances. The typical wire-wound rheostat is a two-terminal device. It is possible to arrange other types of variable resistors as rheostats.

When voltage is the controlled quantity, the variable resistor is called a *potentiometer*. In this type of application, normally at very low power levels, the potentiometer is used as a voltage divider. The potentiometer is a three-terminal device, with the center terminal connected to a contact that may be rotated from one end of the resistor to the other. These devices may be of wire-wound construction, but the great majority of potentiometers are made with carbon composition as the resistor element.

Potentiometers are made in a wide range of resistance values up to several megohms. Most applications are in electronic circuits. One finds potentiometers used as volume controls, contrast and brightness controls, and in most applications where signal voltage levels must be controlled. Schematic symbols for potentiometers and rheostats are shown in Fig. 5-7.

Because of their ease of construction, potentiometers are less expensive than rheostats. Therefore, in those applications where current control at

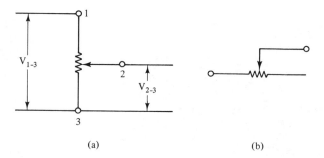

(a) (b)

Figure 5-7 Schematic symbols for potentiometer and rheostat. (a) Potentiometer. This variable device changes *resistance ratios*. As the wiper arm is moved away from terminal 1 towards terminal 3 the voltage V_{2-3} decreases. This is because the resistance between 2 and 3 becomes less and less of the total resistance between 1 and 3. The voltage ratio depends upon the resistance ratio and therefore the name, potentiometer. (b) Rheostat. This is truly a variable resistance. The main application of rheostats is in the control of current. The devices are usually wire-wound, frequently with very high wattage ratings. By comparison, the potentiometer is a low-current, low-wattage device.

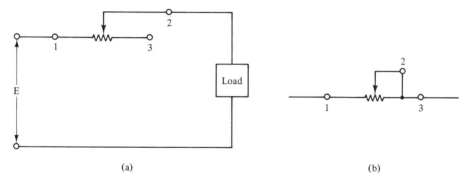

Figure 5-8 Potentiometer used as a rheostat. When used for this purpose, the current demand should be small; potentiometers are not usually designed to handle high currents.

low power levels is needed, potentiometers connected as rheostats are often used. Two ways of arranging a potentiometer as a rheostat are shown in Fig. 5-8.

Another category of variable resistors is a group that is not used for continuously variable control. Instead, these resistors are preset at a given value and held at that value for extended periods of operation. They are defined as *adjustable* resistors. In low-power applications, it is common to use potentiometers and rheostats as adjustable resistors. In these cases, the control shaft is usually not made available as a "front panel" adjustment.

5-12. POWER DISSIPATION RATINGS

We have noted that resistors are rated in terms of resistance, tolerance, and power dissipation.

Electrical power is dissipated in the form of heat. One must consider the effect that such heat has upon the resistor and upon other parts of a circuit located near the resistor. Frequently, the location of a resistor on a circuit board is based upon the amount of heat developed by the power to be dissipated.

Composition resistors normally will fail if operated at or near rated power values. The failure is gradual, taking many months. The resistor does not burn out, but changes gradually to a much lower amount of resistance. Such a change, particularly in electronic amplifier circuits, causes improper operation of the circuit. To extend the useful life of composition resistors, good practice calls for using carbon-composition resistors at much less power than rated values. A good rule of thumb is to use a dissipation rating that is approximately twice the actual power dissipation. *Simply limit power dissipation to about one-half of the power dissipation rating of the resistor.*

Wire-wound resistors may be operated at power levels equal to power dissipation rating. In fact, no harm is done to the resistor if actual power is slightly in excess of the power dissipation rating. However, we must be aware of the effects of heat upon other parts of the circuit.

SUMMARY

1. Resistance is the form of opposition to current that dissipates power in the form of heat.
2. Conductance (G) is the reciprocal of resistance.
3. The SI unit of conductance is the siemen (S).
4. The resistance of a conductor is affected by length, cross-sectional area, temperature, and the type of conducting material.
5. The cross-sectional area of a wire is measured in circular mils (CM): 1 CM = 0.7855 square mils.
6. Resistivity is the property of a conductor that depends upon the material.
7. A material with a positive temperature coefficient of resistance shows increased resistance with increased temperature. This is characteristic of metallic conductors.
8. Resistors are a way of obtaining controlled amounts of resistance in a "small package."
9. Fixed resistors may be carbon composition, deposited film, or wire wound.
10. Wire-wound resistors are used for power dissipations that are greater than a few watts. These types come in sizes up to 100 W.
11. Composition (carbon) resistors are available in a variety of resistance values, tolerances, and low power ratings.
12. A standard color code is used with carbon resistors to supply resistance and tolerance information.
13. Variable resistors are used to control voltage and current levels.
14. Variable resistors may be carbon or wire wound. Wire-wound types are used for higher power ratings and in instances where very high precision is required.
15. All resistors have power dissipation ratings. Wire-wound resistors are capable of extended operation at rated power; carbon resistors are not and should be used at about half of their power rating.

PROBLEMS

5-1. Solve for the conductance of the following resistances:
    10 Ω   100 Ω   1000 M
    10 kΩ   100 kΩ   1 MΩ
    150 Ω   47 kΩ   2.2 kΩ

5-2. Solve for the resistance of the following conductances:
    1.471 mS   1 mS   66.67 μS
    0.05 S   0.455 μS   2.128 mS

5-3. What is the area in square mils and circular mils of AWG #00 wire?

5-4. A copper wire has a resistance of 0.253 Ω at 20°C. What is the resistance of aluminum wire of the same length and gauge?

5-5. We wish to wind a coil of wire to make a 0.223 Ω precision resistor. Assuming #20 tungsten wire, what length of wire is needed?

5-6. Suppose that #20 nichrome is used in place of tungsten for Prob. 5-5. What length of wire is needed?

5-7. The resistance of a tungsten filament is 1.28 Ω at 150°C. What is its resistance at 20°C?

5-8. An electrical circuit is expected to carry a maximum current of 25 A. What gauge wire should be used?

5-9. A large motor draws a current of 70 A under maximum load conditions. What is the smallest gauge wire that may be used to connect the motor to a power source?

5-10. Write the color codes for the following resistors:

	Band 1	Band 2	Band 3	Band 4
2.7 kΩ ±10%				
56 kΩ ±20%				
0.47 Ω ±5%				
10 MΩ ±10%				
150 Ω ±5%				

5-11. Write the resistance values, given the color codes:

Band 1	Band 2	Band 3	Band 4	Resistance
brown	black	yellow	—	
yellow	violet	black	gold	
green	blue	brown	silver	
red	red	red	silver	
brown	gray	black	silver	
brown	black	silver	gold	

5-12. Wire a potentiometer into the circuit of Fig. 5-9 as a rheostat.

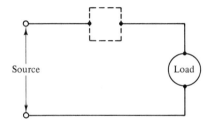

Figure 5-9

5-13. Wire a potentiometer into the circuit of Fig. 5-10.

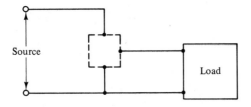

Figure 5-10

5-14. In the circuit of Fig. 5-11, $R_1 = 100\ \Omega\ \pm 5\%$, $R_2 = 3.3\ \text{k}\Omega\ \pm 10\%$, and $R_3 = 15\ \text{k}\Omega\ \pm 5\%$. Specify the color code for each resistor.

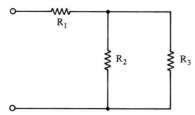

Figure 5-11

6

The Basic Electric Circuit

6-1. THE CLOSED CIRCUIT

All circuits, no matter how large and complex, have basic similarities. First, they all require a source of electrical energy, that is, input (applied) voltage. Second, all circuits must contain *loads*. Loading is a measure of the amount of current furnished by the source of voltage. The greater the load is, the greater amount of current drawn by the load from the source. Load and load resistance do not mean the same thing. Increasing load *resistance* results in *less* load current. Increasing *load* causes more current to be drawn from the source (Fig. 6-1).

A *closed* circuit is any circuit in which current is drawn from the source. For example, we may have a table lamp plugged into an outlet; the lamp will not light until we switch it on, *closing* the circuit. When the circuit is closed, current from the source to the load causes the lamp to light.

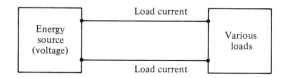

Figure 6-1 Basic electric circuit. Every circuit, regardless of size, must have (a) A source of electrical energy. (b) Every circuit must have at least one load. The load draws current from the source. A heavy load means high current and low resistance.

A single source of electrical energy may supply many loads. Each of these loads is a separate circuit. Each load represents a closed circuit only when connected to the source so that current is supplied to the load.

6-2. SOURCE VOLTAGE AND LOAD CURRENT

The current drawn by a closed circuit is called *load current*. Load current is *directly* proportional to source voltage.

EXAMPLE 6-1:
The load of Fig. 6-2 draws a current of 8 A from a 50-V source. What is the current if source voltage is (a) 100 V; (b) 125 V; (c) 40 V?

Solution: Because load current is proportional to source voltage for a given load, we may write

$$I_1 : V_1 = I_2 : V_2$$

$$\frac{I_1}{I_2} = \frac{V_1}{V_2} \quad \text{and} \quad I_1 = \frac{V_1}{V_2} \times I_2$$

Let I_2 be the 8-A current, and let V_2 be the 50-V source. Let I_1 and V_1 represent the new currents and voltages.

(a) $I_1 = \dfrac{100 \text{ V}}{50 \text{ V}} \times 8 \text{ A} = 16 \text{ A}$

(b) $I_1 = \dfrac{125}{50} \times 8 = 20 \text{ A}$

(c) $I_1 = \dfrac{40}{50} \times 8 = 6.4 \text{ A}$

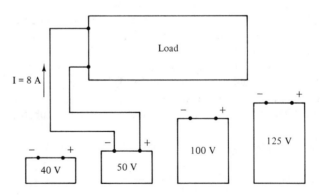

Figure 6-2

EXAMPLE 6-2:
The current drawn by a load from a 100-V source is 2 A. The source voltage is changed so that the load current is 1.5 A. What is the new source voltage?

Solution: Since current is proportional to voltage, we may find the ratio of the currents. This ratio must also be the ratio of the voltages.

$$\frac{I_1}{I_2} = \frac{2}{1.5} = 1.333$$

Therefore,

$$\frac{V_1}{V_2} = 1.333$$

This means that the original voltage is 1.333 times greater than the second voltage. We find the second voltage by *dividing* the original voltage by 1.333.

$$V_2 = \frac{100}{1.333} = 75 \text{ V}$$

6-3. EFFECT OF RESISTANCE ON CURRENT

In our discussion of the closed circuit, we covered the difference between load current and load resistance. In this section we shall reinforce this concept.

Assuming that source voltage is constant, *increasing* load resistance results in a *reduced* load current.

Current is inversely proportional to resistance.

EXAMPLE 6-3:

A load resistance of 150 Ω draws 1 A from a source. What is the current if the load resistance is (a) 300 Ω; (b) 75 Ω?

Solution:
(a) The load resistance is doubled in going from 150 to 300 Ω. Therefore, the load current is cut in half. The load current is one-half of 1 A.

$$I_L = 0.5 \text{ A}$$

(b) The load resistance is cut in half in going from 150 to 75 Ω. The load current therefore is doubled.

$$I_L = 2 \text{ A}$$

6-4. OPEN AND SHORT CIRCUITS

In an open circuit there is no path for current from the source through the load and back to the source. Therefore, the current is zero. A circuit may be opened on purpose by a switch, a relay, or some electronic device. On the other hand, the failure of a circuit component may also cause an open circuit. An open circuit has a load current of zero. It therefore represents an infinite amount of resistance. Examples of open circuits are shown in Fig. 6-3.

It is also an important practice to use the concept of an "open" when comparing a very high resistance to a much lower amount of resistance. If both resistors are connected to a source so that each is a separate circuit,

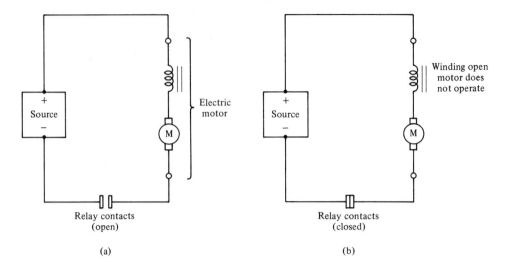

Figure 6-3 Two examples of open circuits. (a) The relay contacts open the circuit. (b) The open winding of the motor field prevents operation; the circuit is open.

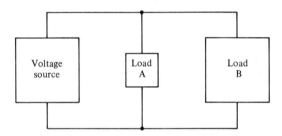

Figure 6-4 A comparison between very different loads. Load A is a night light and draws 42 mA. Load B is a small oven and draws 15 A. The total current drawn from the source is 15.042 A. This is nearly the same as the 15 A drawn by the oven. Therefore, while load A certainly performs an important job, when we compare its current with the current drawn by load B, it appears "open."

each draws load current. The very large resistance draws very much less load current than the small resistance. For all practical purposes, the load on the source is due to the small resistance only. We consider the other circuit "open" in comparison to the heavy load (Fig. 6-4).

EXAMPLE 6-4:
A 200-V source is working into two separate loads. One load resistance is 10 Ω and draws 20 A. The other is 10,000 Ω. Solve for (a) the current through the 10-kΩ resistor, and (b) the total load current supplied by the source.

Solution:
(a) The ratio of the load resistances is 1000 : 1. Therefore, the ratio of the load currents is 1 : 1000. The 10-kΩ resistor draws a load current that is $1/1000$ of 20 A, which is 0.02 A.

(b) Total source current is the sum of the load currents.

$$I_S = I_{L1} + I_{L2} = 20 + 0.02 = 20.02 \text{ A}$$

Note that for all practical purposes the total current is the same as the current in the 10-Ω load. Therefore, so far as the source is concerned, the 10 kΩ appears to be equivalent to an open circuit. However, the 20 mA drawn by the small load serves some useful purpose, perhaps a meter of some type.

We shall now move from open circuits to a discussion of *short circuits*. A short circuit represents *zero* ohms. Clearly, a short circuit across a voltage source would draw an extremely high load current. Short circuits are prevented from drawing very high, destructive currents by protective devices such as fuses (Fig. 6-5), circuit breakers and overload relays. Also, the current capacity of the voltage source is a limiting factor. If we short out a flashlight battery, we do not find dangerously high currents. We find, instead, that the battery furnishes a current that is limited by its internal resistance. This current quickly discharges the battery, which makes it useless for other applications.

Short circuits of the type we have been discussing are, of course, accidental. These types of shorts are due either to component failure or wiring errors. But, as we did with concepts of open circuits, we may treat certain parts of a working circuit as "short circuits" when compared to the rest of the circuit. For example, particularly in electronic circuits, we consider connecting wires to be "shorts" in comparison to other load resistances. Suppose, for example, that a load is 15 Ω and that the resistance of the connecting wires is 0.05 Ω. The load current is essentially the same whether or not we consider the resistance of the wires.

It is important to recognize that a circuit component may be treated as a short in comparison to another component, even when it represents a great deal of resistance. As an example, suppose that we have a 1-MΩ resistor connected to a 10-V source. A current of 10 μA is drawn from the source. If we connect a 1-kΩ resistor into the circuit so that the source current must flow through both resistors, the source furnishes current to a load resistance that is equal to the sum of the resistances. A 1-kΩ resistor is only 1/1000 of

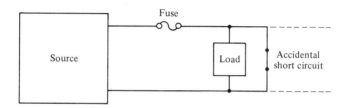

Figure 6-5 Fuse protection against accidental short-circuit. The fuse has a current rating that is slightly higher than the maximum load current. A short-circuit across the load raises current to very high levels and quickly melts the fuse link. The open fuse protects the *source* from dangerously high currents.

a 1-MΩ resistor. The total resistance is 1.001 MΩ. The current drawn from the source is 9.99 µA, a current that is certainly the same as 10 µA for all practical purposes. Therefore, we can consider the 1-kΩ resistor to be a "short" in comparison to the 1-MΩ resistor.

6-5. CURRENT AND POWER

Power is the rate of doing work. Recall that the unit of electrical power is the *watt* (W). Resistance is the *only* circuit property that is capable of dissipating electrical power.

The amount of power dissipated in a resistor is proportional to the square of its current.

Suppose that a 0.5-A current flows through a 10-Ω resistor. The power dissipated (as heat) by the resistor is 2.5 W. If the current is increased to 1A, the power dissipated in the resistor increases to 10 W. Note that we doubled the current and that the power was multiplied by four ($2^2 = 4$). If we were to triple the current, we would expect the power to increase by a factor of 9 ($3^2 = 9$). A formula for determining power that relates current and resistance to power dissipation is

$$P = I^2 R \qquad (6\text{-}1)$$

where P = power in watts
I = current in amperes
R = resistance in ohms

> **EXAMPLE 6-5:**
> Find the power dissipated in a 10-Ω resistor by a current of (a) 0.7 A, (b) 1.4 A, (c) 2.1 A.
>
> **Solution:** $P = I^2 R$.
> (a) $P = 0.7^2 \times 10$
> $= 0.49 \times 10 = 4.9$ W
> (b) $P = 1.4^2 \times 10$
> $= 1.4 \times 1.4 \times 10 = 19.6$ W
> (c) $P = 2.1^2 \times 10$
> $= 2.1 \times 2.1 \times 10 = 44.1$ W
> Note that the power in (c) is nine times the power in (a) because the current in (c) is three times the current in (a).

We observe from Eq. (6-1) that power is *directly* proportional to resistance. Doubling resistance, while maintaining the *same current*, causes power to double.

EXAMPLE 6-6:

Repeat Example 6-5 but with $R = 20\ \Omega$.

Solution:
(a) $P = 0.7^2 \times 20$
$= 0.49 \times 20 = 9.8$ W
(b) $P = 1.4^2 \times 20$
$= 1.4 \times 1.4 \times 20 = 39.2$ W
(c) $P = 2.1^2 \times 20$
$= 2.1 \times 2.1 \times 20 = 88.2$ W

6-6. CURRENT MEASUREMENT

Current is measured by an instrument called an *ammeter*. An ammeter may also be part of a multifunction instrument called a *multimeter*. Recall that current is the *rate* of drift of charge (current flow). Therefore, the ammeter must be connected so that the current to be measured passes through the meter. This is done by first removing power from the circuit, then opening the circuit in that portion where current is to be measured, and finally connecting the ammeter as in Fig. 6-6 so that it *closes* the circuit. Then power is supplied to the circuit and the meter is read. Because the ammeter becomes part of the circuit, the ammeter must have a very low resistance so that it causes minimum error. In effect, the ammeter must have so little resistance that we consider it to be a short circuit. Examine Fig. 6-7 very carefully so that you understand how current measurements are made. Figure 6-8 shows current measurement in part of a circuit.

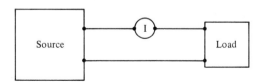

Figure 6-6 Basic current measurement.

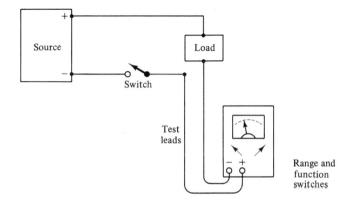

Figure 6-7 Current measurement with a multimeter.

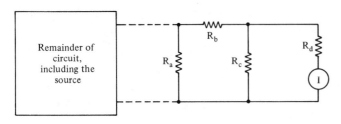

Figure 6-8 Current measurement in part of a circuit. In this case, we are measuring the current flow in R_d.

6-7. VOLTAGE MEASUREMENT

Recall that voltage is a *difference of potential*. Voltage measurements may be made with instruments called *voltmeters*. As with ammeters, these are frequently part of a multimeter.

Voltage measurements are made by connecting the voltmeter across the parts of a circuit where voltage is to be determined. Thus, if applied voltage is to be measured, the voltmeter is connected across (between) the source terminals. If we wish to measure the voltage across a part of a circuit, the voltmeter is connected across that part of the circuit.

A voltmeter must have a very high resistance so that, when it is connected across a part of a circuit, it does not change the resistance of that part of the circuit. An ideal voltmeter acts like an open circuit in comparison to the resistance of the circuit under test. In practice, electronic voltmeters and multimeters do approach this ideal. Figure 6-9 shows how to use a voltmeter.

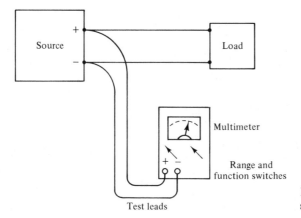

Figure 6-9 Using a multimeter to measure source voltage.

6-8. VOLTAGE DROP

We recognize that a voltage source is needed in order for a closed circuit to have current flow. We will now consider the voltages developed across each part of an operating circuit.

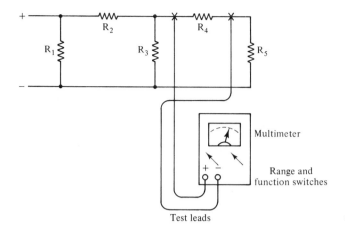

Figure 6-10 Multimeter measurement of voltage drop.

Whenever current flows in a resistor, there is a voltage developed across this resistor. This voltage, called a *voltage drop*, is the amount of potential difference needed to sustain the current.

We calculate the voltage drop as the product of current and resistance.

$$V = IR \qquad (6\text{-}2)$$

where V = voltage in volts
I = current in amperes
R = resistance in ohms

Many circuits, particularly in electronics, consist of more than one resistor or other types of circuit elements. When these parts are connected so that the *same* current flows through each part, the circuit is called a *series* circuit. We shall study series circuits in detail in Chapter 8.

The current in a series circuit produces voltage drops around the circuit. Each drop is the amount of voltage needed to cause the required current in that part of the circuit. We will demonstrate in Chapter 8 that, because the current is the same throughout a series circuit, the sum of the voltage drops around the circuit is equal to the applied voltage. Figure 6-10 is an example of voltage drop measurement.

EXAMPLE 6-7:
In Fig. 6-11, find the voltage drop across a 100-Ω resistor when current is (a) 10 μA, (b) 20 mA, (c) 5 A.

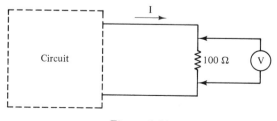

Figure 6-11

Sec. 6-8.   Voltage Drop

Solution: $V = IR$.

(a) $V = 10 \times 10^{-6} \times 100 = 1000 \ \mu V = 1 \ mV$
(b) $V = 20 \times 10^{-3} \times 100 = 2000 \ mV = 2 \ V$
(c) $V = 5 \times 100 = 500 \ V$

SUMMARY

1. Every closed circuit requires a power source and a load.
2. Load is a measure of current drawn from a source. The greater the load is, the greater the current.
3. The current drawn by a load is directly proportional to source voltage.
4. The lower the load resistance is, the greater the load current.
5. An open circuit represents "infinite" resistance and zero current.
6. A short circuit represents zero resistance and "infinite" current, provided we could supply such a current! In practice, fuses or circuit breakers would disconnect the short circuit.
7. Power dissipation is proportional to the square of the current in a resistor.
8. Power dissipation is proportional to resistance for a given current.
9. Current is measured by a meter called an *ammeter*. The meter must look like a short to the rest of the circuit. The meter must be made part of the circuit for the measurement.
10. Voltage is measured by a voltmeter. The meter should look like an open to the circuit. Measurement is made by connecting across points in the circuit.
11. Voltage drop is the product of current and resistance.
12. Voltage drop represents the voltage needed across a component in order to maintain current.

PROBLEMS

6-1. In the circuit of Fig. 6-12, the current is 3 A when the source voltage is 100 V. What is the source voltage when current is 4.5 A?

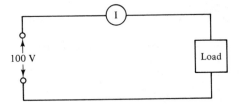

Figure 6-12

6-2. Given the circuit of Fig. 6-12, what is the load current when source voltage is 50 V?

6-3. Suppose that the current drawn by a load changes to half of its previous value. What are the two possible causes for this change?

6-4. A load draws a current of 1.5 A from a source. Suppose that loading increases to 1.9 A and that source voltage has not been changed. Does this increased load represent more or less load resistance?

6-5. When connected to the 50-Ω resistor of Fig. 6-13, current is 2 A. What is current when the 50-Ω resistor is replaced by a 100-Ω resistor? A 25-Ω resistor?

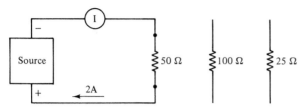

Figure 6-13

6-6. We have a circuit with a load current of 125 mA. If we wish to reduce the load current to 100 mA, do we increase or decrease load resistance?

6-7. In the circuit of Fig. 6-14, $R_1 = 5000$ Ω and $R_2 = 100$ Ω. Which resistor has the greatest effect upon current drawn from the source?

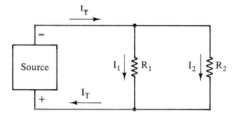

Figure 6-14

6-8. In the circuit of Fig. 6-14, suppose that the current in R_2 is 0.5 A. What is the current in R_1?

6-9. Given the conditions of Prob. 6-8, what is the total current drawn from the load?

6-10. In the circuit of Fig. 6-14, what would be the current drawn from the load if R_1 is disconnected?

6-11. In the circuit of Fig. 6-14, can we treat R_1 as an open circuit in comparison to R_2? Explain.

6-12. In the circuit of Fig. 6-15, the current drawn from the source is 50 mA. What would the current be if we should short out R_2?

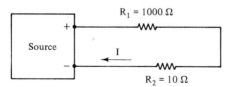

Figure 6-15

6-13. Why can we treat R_2 as a short circuit in comparison to R_1 in the circuit of Fig. 6-15?

6-14. The current in a 10-Ω resistor is 1 A. What is the power dissipated in the resistor?

6-15. Suppose that the current in Prob. 6-14 is reduced to 0.5 A. What is the power dissipated in the 10-Ω resistor?

6-16. Calculate the power dissipation in each resistor in the circuit of Fig. 6-16.

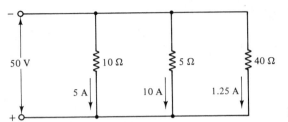

Figure 6-16

6-17. Calculate the power dissipated in each resistor in the circuit of Fig. 6-17.

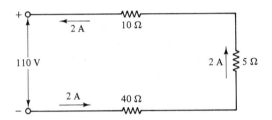

Figure 6-17

6-18. Given the circuit of Fig. 6-18:
    (a) Redraw the circuit adding an ammeter to measure I_1.
    (b) Repeat, but for I_2.
    (c) Repeat, but for I_3.

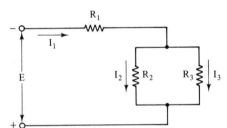

Figure 6-18

6-19. A technician makes current measurements in a circuit similar to that of Fig. 6-18. He finds that I_1 is less than expected, that I_2 is greater than the proper value and is equal to I_1, and that I_3 is zero. What is the likely cause?

Use the circuit of Fig. 6-19 for Probs. 6-20 through 6-22.

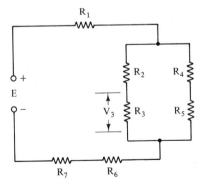

Figure 6-19

6-20. Sketch the circuit to show how a voltmeter must be connected to the measure V_1, the drop across R_1.

6-21. How would a voltmeter be connected to measure V_3? Sketch the circuit.

6-22. Show how you would connect a voltmeter to measure the sum of the voltage drops across R_6 and R_7.

6-23. In the circuit of Fig. 6-20, a technician has attempted to measure V_3. Given the way the meter has been connected, what would you expect the meter reading to be?

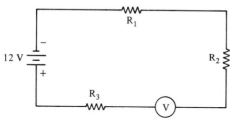

Figure 6-20

6-24. Voltage drops in the circuit of Fig. 6-21 are measured by an experienced technician. He finds that $V_1 = 24$ V, $V_2 = 0$ V, and $V_3 = 0$ V. What is the likely cause of the trouble if $R_1 = 18\ \Omega$, $R_2 = 22\ \Omega$, and $R_3 = 39\ \Omega$?

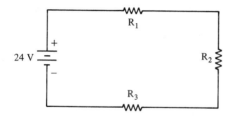

Figure 6-21

6-25. Solve for the voltage drop across R_2 in the circuit of Fig. 6-22.

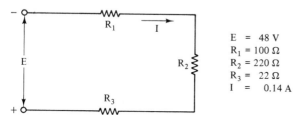

E = 48 V
R_1 = 100 Ω
R_2 = 220 Ω
R_3 = 22 Ω
I = 0.14 A

Figure 6-22

6-26. Solve for the voltage drop across R_3 in the circuit of Fig. 6-22.

6-27. Given the circuit of Fig. 6-23, what should the voltmeter reading be?

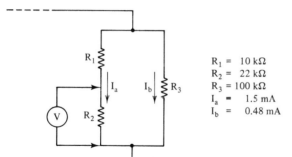

R_1 = 10 kΩ
R_2 = 22 kΩ
R_3 = 100 kΩ
I_a = 1.5 mA
I_b = 0.48 mA

Figure 6-23

6-28. In the circuit of Fig. 6-23, what is the voltage drop across R_3?

6-29. Given the circuit of Fig. 6-24, solve for the voltage drop across each resistor.

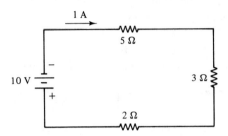

Figure 6-24

6-30. Compare the sum of the voltage drops calculated in Prob. 6-29 to the source voltage.

7

Ohm's Law

7-1. A BASIC PHYSICAL LAW

All physical systems obey a fundamental law: *The response of a system to a force is in direct proportion to the force and is inversely proportional to the opposition of the system.*

Thus the rate of flow of water in a plumbing system depends upon pressure and inversely upon the friction of the pipes. If we increase pressure, the rate of flow increases. If friction is increased by using smaller pipes or by making the water path longer, the rate of water flow decreases.

Electrical circuits also obey this fundamental relationship between rate, pressure, and opposition. It was first published by Georg Simon Ohm in 1826 and is known as Ohm's law.

> The current in a circuit is directly proportional to source voltage and inversely proportional to the resistance of the circuit.

Therefore, an *increase* in resistance *reduces* current; a *reduction* in resistance *increases* current. Also, an *increase* in voltage *increases* current; a *decrease* in voltage *decreases* current. These concepts are a restatement of the principles on current discussed in Chapter 6.

7-2. CURRENT FORMULA

The current formula is simply Ohm's original conclusion stated in mathematical form.

$$I = \frac{E}{R_T} \qquad (7\text{-}1)$$

$$I = \frac{V}{R} \qquad (7\text{-}2)$$

where I is in amperes, V and E are in volts, and R is in ohms.

Note that we wrote two formulas that are basically the same, except for the symbol used for voltage. We shall use E to indicate source (applied) voltage and V to indicate voltage drop. This notation will be used throughout the text. Whenever the symbol E is used, remember that the voltage is source voltage. All other voltages will be identified by the symbol V.

EXAMPLE 7-1:

A circuit with a resistance of 200 Ω has an applied voltage of (a) 100 V, (b) 20 V, and (c) 10 mV. Solve for current at each step.

Solution:

$$I = \frac{E}{R_T}$$

(a) $I = \dfrac{100}{200} = 0.5$ A

(b) $I = \dfrac{20}{200} = 0.1$ A

(c) $I = \dfrac{10 \times 10^{-3}}{200}$
$= 0.05 \times 10^{-3}$
$= 0.05$ mA $= 50$ μA

EXAMPLE 7-2:

We measure a voltage drop of 50 V across a 100-Ω resistor (Fig. 7-1). What is the current flow in the resistor?

Solution:

$$I = \frac{V}{R} = \frac{50}{100} = 0.5 \text{ A}$$

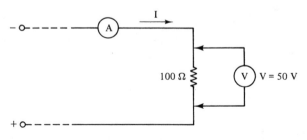

Figure 7-1

7-3. RESISTANCE FORMULA

Ohm's law is used to solve for resistance when current and voltage are known.

$$R = \frac{V}{I} \tag{7-3}$$

Formula 7-3 is used to find the resistance of *part* of a circuit. The voltage drop across that part of the circuit is given by V. The current through that part of the circuit is given by I.

We may write a similar formula for the *total* resistance of a circuit. In this case, the voltage must be the applied (source) voltage and the current, the *total* current drawn from the source by the circuit.

$$R_T = \frac{E}{I_T} \tag{7-4}$$

where R_T = total resistance in ohms
E = source voltage in volts
I_T = source current in amperes

EXAMPLE 7-3:

The voltage drop across a resistor (Fig. 7-2) is 150 mV. We measure a current of 75 µA in the resistor. What is the resistance of the resistor?

Solution:

$$R = \frac{V}{I}$$
$$= \frac{150 \times 10^{-3}}{75 \times 10^{-6}}$$
$$= 2 \times 10^3 = 2 \text{ k}\Omega$$

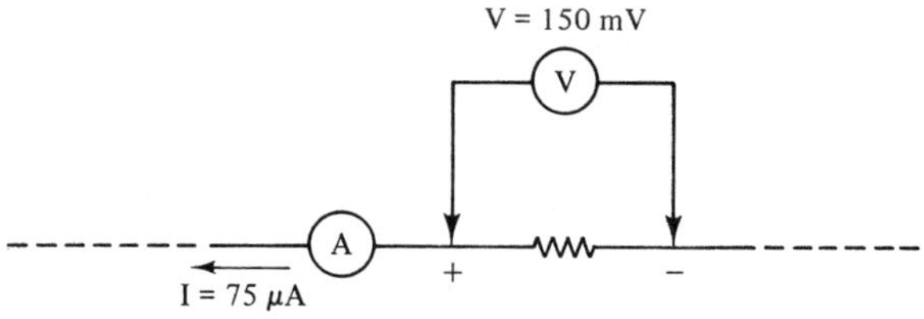

Figure 7-2

EXAMPLE 7-4:

A circuit draws 0.5 mA from a 30-V source. What is the total resistance of the circuit?

Solution:

$$R_T = \frac{E}{I_T}$$
$$= \frac{30}{0.5 \times 10^{-3}}$$
$$= 60 \times 10^3 = 60 \text{ k}\Omega$$

EXAMPLE 7-5:
In Fig. 7-3 two resistors are connected so that the same current must flow through each. The current is 10 mA. A voltage measurement is made across the two resistors and is found to be 25 V. What is the resistance represented by the two resistors?

Solution:

$$R_1 + R_2 = \frac{V}{I}$$

$$= \frac{25}{10 \times 10^{-3}}$$

$$= 2.5 \times 10^3 = 2.5 \text{ k}\Omega$$

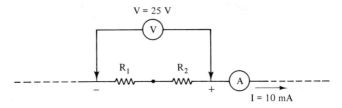

Figure 7-3

7-4. VOLTAGE FORMULA

We can use Ohm's law to solve for voltage when current and resistance are known. Once again, Ohm's law formulas apply to parts of a circuit as well as to the entire circuit.

Voltage drop, the voltage across a part of a circuit, is the product of current times resistance. The resistance used in the calculation is the resistance across which the drop is to be calculated. The current is the current through that resistance.

$$V = IR \qquad (7\text{-}5)$$

Source voltage is calculated by a similar formula. However, the resistance must be the *total* resistance and the current must be the source current (*total current*).

$$E = I_T R_T \qquad (7\text{-}6)$$

EXAMPLE 7-6:
A load resistance of 500 Ω draws 0.1 A from a source. What is the source voltage?

Solution:

$$E = I_T R_T$$
$$= 0.1 \times 500 = 50 \text{ V}$$

EXAMPLE 7-7:
A source, operating into a 10-kΩ load, supplies a current of 10 mA. What is the source voltage?

Solution:
$$E = I_T R_T$$
$$= (10 \times 10^{-3})(10 \times 10^3)$$
$$= 100 \text{ V}$$

EXAMPLE 7-8:
What is the voltage drop across a 27-kΩ resistor if the current flow is (a) 0.2 mA, (b) 0.8 mA, (c) 3 mA?

Solution: $V = IR$.
(a) $V = 0.2 \times 10^{-3} \times 27 \times 10^3$
  $= 5.4$ V
(b) $V = 0.8 \times 10^{-3} \times 27 \times 10^3$
  $= 21.6$ V
(c) $V = 3 \times 10^{-3} \times 27 \times 10^3$
  $= 81$ V

7-5. LEARNING THE FORMULAS

All the Ohm's law formulas are the foundation upon which we shall build your knowledge of electric circuits. Because they are so important, we make use of the memory aid given in Fig. 7-4.

Any one of the basic quantities is found by use of the other two in the relationship shown.

If we wish to calculate voltage, we see that current and resistance are side by side as in multiplication.

voltage = current $\times$ resistance

To find resistance, we note that voltage is over current, as in division.

resistance = voltage $\div$ current

To find current, we see from the sketch that voltage is over resistance, as in division.

current = voltage $\div$ resistance

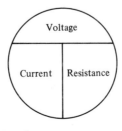

Figure 7-4

7-6. OHM'S LAW AND ENTIRE CIRCUITS

We may use Ohm's law with entire circuits or with parts of a circuit. Formulas 7-1, 7-4, and 7-6 have already been discussed in dealing with current, resistance, and voltage relationships in entire circuits. In this section we shall restate important concepts on the use of Ohm's law.

Whenever we use Ohm's law to solve an entire circuit, we must work with the voltage, resistance, and current that apply to the entire circuit.

1. Total current must be found by using source voltage and total resistance.

$$I_T = \frac{E}{R_T} \tag{7-1}$$

2. Total resistance can be found only by use of total current and source voltage.

$$R_T = \frac{E}{I_T} \tag{7-4}$$

3. Source voltage is calculated by use of total current and total resistance.

$$E = I_T R_T \tag{7-6}$$

EXAMPLE 7-9:
The total resistance of a circuit is 39 kΩ. If 10 V is applied to the circuit, what is the current (I_T) drawn from the source?

Solution:

$$I = \frac{E}{R_T}$$
$$= \frac{10}{39 \times 10^3}$$
$$= 0.256 \times 10^{-3}$$
$$= 0.256 \text{ mA} = 256 \text{ }\mu\text{A}$$

EXAMPLE 7-10:
The circuit of Fig. 7-5 has a total resistance of 2.7 kΩ. What is the total current if the source voltage is 12 V?

Solution:

$$I = \frac{E}{R_T}$$
$$= \frac{12}{2.7 \times 10^3}$$
$$= 4.4 \times 10^{-3} = 4.4 \text{ mA}$$

Figure 7-5

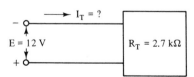

EXAMPLE 7-11:

A current draws 5 mA from a 20-V source. What is the total resistance of the circuit of Fig. 7-6?

Solution:

$$R_T = \frac{E}{I_T}$$
$$= \frac{20}{5 \times 10^{-3}}$$
$$= 4 \times 10^3 = 4 \text{ k}\Omega$$

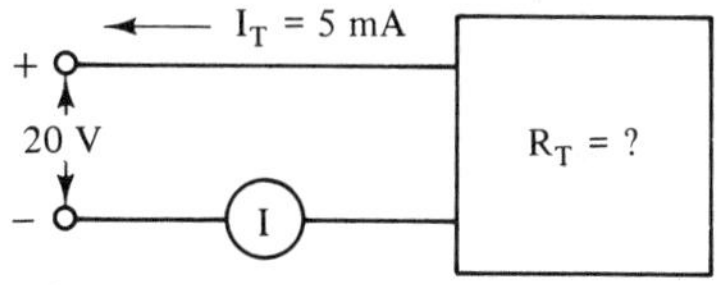

Figure 7-6

EXAMPLE 7-12:

When 45 V dc is applied to the power terminals of an amplifier, the load current drawn is 200 mA. What is the total resistance of the load?

Solution:

$$R_T = \frac{E}{I_T}$$
$$= \frac{45}{0.2} = 225 \text{ }\Omega$$

EXAMPLE 7-13:

A current of 20 mA is drawn from a source by a load of 10 kΩ. What is the applied voltage?

Solution:

$$E = I_T R_T$$
$$= 20 \times 10^{-3} \times 10 \times 10^3$$
$$= 200 \text{ V}$$

EXAMPLE 7-14:

A 300-Ω load (Fig. 7-7) requires a current of 150 mA. What is the applied voltage needed in order for the load to draw the specified current?

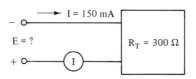

Figure 7-7

Solution:

$$E = I_T R_T$$
$$= 0.15 \times 300 = 45 \text{ V}$$

7-7. OHM'S LAW APPLIED TO PART OF A CIRCUIT

In actual practice, we will find that we frequently make calculations on parts of circuits, rather than entire circuits. We can and do use Ohm's law for many of these calculations.

When applying Ohm's law to part of a circuit, we must use the voltage, current, and resistance relating to only that part of the circuit.

All the examples that follow make use of this rule. The "secret" to success in the solution of Ohm's law problems is simply to use the correct values of voltage, current, and resistance.

EXAMPLE 7-15:
A voltage drop of 26 V is measured across a 10-kΩ resistor. What is the current flow in the resistor?

Solution:

$$I = \frac{V}{R}$$
$$= \frac{26}{10 \times 10^3}$$
$$= 2.6 \times 10^{-3} = 2.6 \text{ mA}$$

EXAMPLE 7-16:
The voltage drop across a 27-kΩ resistor is measured as 40 V (Fig. 7-8). Calculate the current flow in the resistor.

Solution:

$$I = \frac{V}{R}$$
$$= \frac{40}{27 \times 10^3}$$
$$= 1.48 \times 10^{-3} = 1.48 \text{ mA}$$

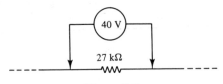

Figure 7-8

EXAMPLE 7-17:

Voltage and current measurements are made on an unknown resistor (Fig. 7-9). The voltage drop is 30 mV and the current is 150 μA. What is the resistance of the resistor?

Solution:

$$R = \frac{V}{I}$$
$$= \frac{30 \times 10^{-3}}{150 \times 10^{-6}}$$
$$= 0.2 \times 10^3 = 200 \: \Omega$$

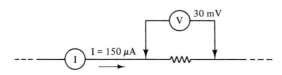

Figure 7-9

EXAMPLE 7-18:

In Fig. 7-10, two resistors are connected so that the same current must flow through each (series connection). The voltage drop across one is 15 V. The resistance of that resistor is 10 kΩ. The resistance of the other resistor is 12 kΩ. What is the voltage drop across the 12-kΩ resistor?

Solution:
(a) Solve for current.

$$I = \frac{V}{R}$$
$$= \frac{15}{10} \times 10^3$$
$$= 1.5 \times 10^{-3} = 1.5 \text{ mA}$$

(b) Once the current is known, we may solve for the voltage drop across the 12-kΩ resistor.

$$V = IR$$
$$= (1.5 \times 10^{-3})(12 \times 10^3)$$
$$= 18 \text{ V}$$

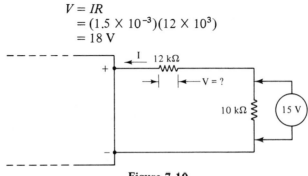

Figure 7-10

7-8. COMPUTATIONS

The use of engineering units allows us to take some arithmetic shortcuts with Ohm's law problems. We did not use these shortcuts in any of the examples in order to concentrate upon *concepts* and *applications*.

Current Calculations

1. When resistance is in kilohms and the voltage is in volts, the resultant current is always in milliamperes.

$$\frac{V}{k\Omega} = mA$$

This is true because the resistance unit in the denominator (10^3) appears in the result as 10^{-3}, the power of 10 corresponding to milli, as in mA.

2. When the resistance is in megohms and the voltage is in volts, the resultant current is in microamperes. The 10^6 in the denominator appears as 10^{-6} in the result; therefore, we have microamperes.

$$\frac{V}{M\Omega} = \mu A$$

3. Voltage in millivolts and resistance in kilohms results in microamperes.

$$\frac{mV}{k\Omega} = \frac{10^{-3}}{10^3} = 10^{-6} = \mu A$$

4. Voltage in millivolts and resistance in megohms results in nanoamperes.

$$\frac{mV}{M\Omega} = \frac{10^{-3}}{10^6} = 10^{-9} = nA$$

5. Voltage in microvolts and resistance in kilohms results in nanoamperes.

$$\frac{\mu V}{k\Omega} = \frac{10^{-6}}{10^3} = 10^{-9} = nA$$

Voltage Calculations

1. When resistance is in kilohms and the current is in milliamperes, the voltage is in the basic unit, volts. The powers of 10 of the units cancel.

$$mA \times k\Omega = V$$

2. Current in milliamperes and resistance in megohms results in kilovolts. The product of 10^{-3} and 10^6 results in 10^3, the unit for kilo.

$$mA \times M\Omega = kV$$

3. Current in microamperes and resistance in megohms results in volts, because the powers of 10 for the units cancel.

$$\mu A \times M\Omega = V$$

4. Current in microamperes and resistance in kilohms results in millivolts ($10^{-6} \times 10^3 = 10^{-3}$).

$$\mu A \times k\Omega = mV$$

5. Current in microamperes and resistance in ohms results in microvolts.

$$\mu A \times \Omega = \mu V$$

6. Current in milliamperes and resistance in ohms results in millivolts.

$$mA \times \Omega = mV$$

Resistance Calculations

1. When voltage is in the basic unit, volts, and current is in milliamperes, the resistance is in kilohms ($10^0/10^{-3} = 10^3$).

$$\frac{V}{mA} = k\Omega$$

2. Voltage in volts and current in microamperes results in megohms ($10^0/10^{-6} = 10^6$).

$$\frac{V}{\mu A} = M\Omega$$

3. Voltage in millivolts and current in milliamperes results in ohms. The units cancel.

$$\frac{mV}{mA} = \Omega$$

4. Voltage in millivolts and current in microamperes results in kilohms ($10^{-3}/10^{-6} = 10^3$).

$$\frac{mV}{\mu A} = k\Omega$$

All scientific calculators accept power of 10 notation. The computational shortcuts we have just discussed are very valuable tips despite the capability of electronic calculators.

1. Our fingers make mistakes. A wrong [EE] entry into the calculator may not be noticed. The answer will simply be wrong!
2. We must develop a proficiency with the units, so that we *know* what to expect. This is true whether as a student or as a working electronic technician.

SUMMARY

1. Ohm's law calculations may be made on parts of a circuit or on an entire circuit.
2. Current is directly proportional to voltage. $I = V/R$.
3. Current is inversely proportional to resistance.
4. Voltage drop is the product of current and resistance and is directly proportional to these quantities. $V = IR$.
5. When using Ohm's law with an entire circuit, we must work with total voltage, total current, and total resistance.
6. Ohm's law calculations on a part of a circuit must use the voltage, current, and resistance that relate to that part of the circuit.

PROBLEMS

7-1. Solve for current in the circuit of Fig. 7-11. Each resistor, one at a time, is connected to the 40-V source.

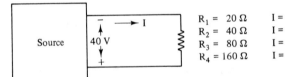

Figure 7-11

7-2. In the circuit of Fig. 7-12, the source voltage applied to R_L depends upon the switch setting. Calculate load current, given the voltage at each switch setting.
  (a) $E_a = 25$ V, $I =$
  (b) $E_b = 50$ V, $I =$
  (c) $E_c = 100$ V, $I =$
  (d) $E_d = 150$ V, $I =$
  (e) $E_e = 200$ V, $I =$

Figure 7-12

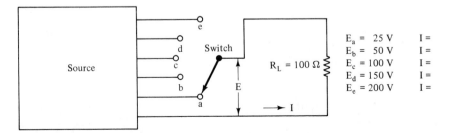

7-3. The total resistance of a circuit is 1.2 kΩ. If 36 V is applied, what is the total current?

7-4. The voltage measured across a 56-Ω resistor is 11.2 V. What is the current flow in milliamperes?

7-5. We measure 66 V across a 3.3-kΩ resistor. What is the current in the resistor?

7-6. Calculate the current for each set of conditions:
(a) $V = 24$ V, $R = 1.2$ kΩ, $I =$
(b) $V = 6$ V, $R = 1$ MΩ, $I =$
(c) $V = 4.8$ kV, $R = 300$ Ω, $I =$
(d) $V = 15$ V, $R = 10$ Ω, $I =$

7-7. Given the circuit of Fig. 7-13, solve for R for each set of measurements.
(a) $V = 13.6$ V, $I = 2.43$ mA, $R =$
(b) $V = 275$ mV, $I = 15.3$ mA, $R =$
(c) $V = 1.2$ kV, $I = 40$ A, $R =$
(d) $V = 75$ V, $I = 192$ μA, $R =$

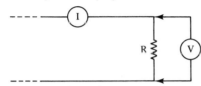

Figure 7-13

7-8. A power supply is used with a variety of loads. Assuming a constant supply voltage of 60 V, calculate R_T for each value of I_T.
(a) $I_T = 60$ mA, $R_T =$
(b) $I_T = 128$ mA, $R_T =$
(c) $I_T = 900$ mA, $R_T =$
(d) $I_T = 0.45$ mA, $R_T =$

7-9. Calculate the total resistance of a circuit if it draws 0.75 A from a 12-V source.

7-10. What is the total resistance of a certain research instrument if it draws 25 μA from a 42-kV source?

7-11. The total resistance of a circuit is 3 kΩ. What is source voltage if total current is 40 mA?

7-12. For each set of data, solve for voltage drop.
(a) $I = 25$ μA, $R = 10$ kΩ, $V =$
(b) $I = 4.2$ mA, $R = 390$ Ω, $V =$
(c) $I = 2.1$ A, $R = 75$ Ω, $V =$
(d) $I = 0.35$ A, $R = 470$ Ω, $V =$

7-13. A circuit has a total resistance of 125 Ω. When connected to a source, it draws 384 mA. What is the source voltage?

7-14. The current through a 10-kΩ resistor is 2.4 mA. Solve for the voltage drop across the resistor.

7-15. A voltmeter draws 20 μA when connected across 10 V. What is the resistance of the instrument?

7-16. The input resistance of an instrument is 9.6 MΩ. When 5 V is applied, what is the input current?

7-17. You have several unknown resistors. You also have a fixed source of 10 V and an accurate ammeter. Each resistor is connected to the source and current is measured. Based upon the following data, calculate the resistance of each resistor.

(a) $I_1 = 5$ mA, $R_1 =$
(b) $I_2 = 1$ A, $R_2 =$
(c) $I_3 = 3.7$ mA, $R_3 =$
(d) $I_4 = 2.13$ μA, $R_4 =$

7-18. The voltage drop across a 6.8-kΩ resistor is 27 V. Solve for the current flow in the resistor.

7-19. Solve for current if the voltage drop across a 15-kΩ resistor is 30 mV.

7-20. The current through a 22-kΩ resistor is 400 mA. What is the voltage drop across the resistor?

7-21. In the circuit of Fig. 7-14, the voltage across R_2 is 22 V. (a) Solve for the current in R_2. (b) Given that the voltage across R_1 is measured as approximately 15 V, what is the current in R_1?

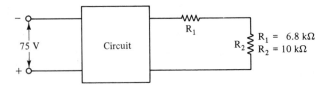

Figure 7-14

7-22. In the circuit of Fig. 7-15, we must determine the resistance of R_3. Redraw the circuit, and show how an ammeter and voltmeter would be connected to measure V_3 and I_3.

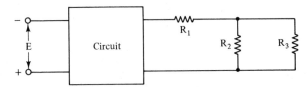

Figure 7-15

7-23. Use the graph of Fig. 7-16 to plot graphs of current and voltage for $R_1 = 10$ Ω and $R_2 = 15$ Ω. Use the voltages given on the horizontal axis.

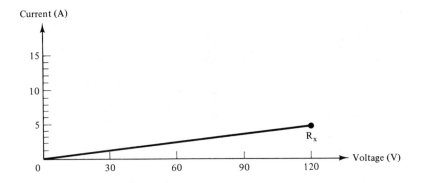

Figure 7-16

7-24. In the graph of Prob. 7.23, a graph of current versus voltage for an unknown resistor (R_x) is plotted. Is R_x greater or less resistance than R_2?

7-25. Do the following computations *mentally*.
 (a) $I = 4$ mA, $R = 12$ kΩ, $V =$
 (b) $R = 1$ kΩ, $V = 15$ V, $I =$
 (c) $V = 20$ mV, $I = 5$ mA, $R =$
 (d) $V = 40$ kV, $R = 100$ kΩ, $I =$

Series Circuits

8-1. INTRODUCTION

Whenever parts of a circuit are connected so that there is *only one path for current*, the parts are connected in *series*. For example, if we connect a wire from one resistor to another, without connecting any other parts to the wire, there is a single current path, and the resistors are in series.

The point in a circuit where parts are connected is called a *node*. In the series connection shown in Fig. 8-1, there are only two wires at a node. One carries the current entering the node and the other carries the current from the node.

When all the parts of a circuit including the source are connected in series, we have a series circuit.

8-2. CURRENT IN A SERIES CIRCUIT

One of the fundamental properties of series circuits relates to current.

The current in a series circuit is the same everywhere in the circuit.

This concept should be clear from the fact that there is just one current path. Therefore, current entering the circuit, flowing throughout the circuit, leaving the circuit, entering the source, and leaving the source must be the same. We always measure the same amount of current at any point or node in the circuit.

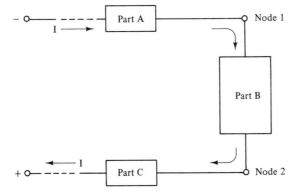

Figure 8-1 The series connection for parts of a circuit. Because there is just one path for current flow, the current in parts A, B, and C is the same. These parts are in series with each other, no matter what the rest of the circuit may look like!

8-3. RESISTORS IN SERIES

Resistors are connected in series simply by wiring them end to end. We may visualize them as being connected in a line, but the actual geometry of the circuit is unimportant. What is important is that we have only one path for current. You will note several examples of series resistive circuits in the illustrations. In each case, the "end" of one resistor is connected to the "start" of another resistor. The entire "string" of resistors is then connected to a source (Fig. 8-2).

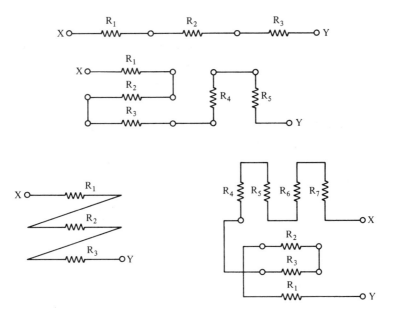

Figure 8-2 Various series resistive circuits. Each of these examples of series resistive circuits adheres to the principle that there is just one path for current between terminals X and Y.

Sec. 8.3   Resistors in Series

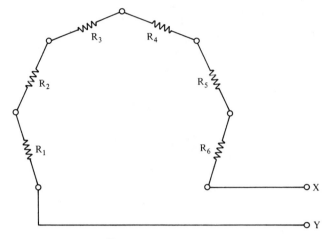

Figure 8-2 *(continued)*

8-4. TOTAL RESISTANCE

Resistance is opposition to current flow. When resistors are series connected, each adds to the total opposition. We state this concept as one of the fundamental properties of series circuits.

The total resistance of a series circuit is the sum of the individual resistances.

$$R_T = R_1 + R_2 + R_3 + \cdots + R_n \qquad (8\text{-}1)$$

The notation n in R_n represents the number of resistors in the circuit. For example, in a four-resistor circuit, R_n would be R_4. In a seven-resistor circuit R_n would be R_7. The total resistance (in this case) would be

$$R_T = R_1 + R_2 + R_3 + R_4 + R_5 + R_6 + R_7$$

EXAMPLE 8-1:

In the circuit of Fig. 8-3, $R_1 = 4.7$ kΩ, $R_2 = 5.6$ kΩ, and $R_3 = 910$ Ω. What is the total resistance of the series circuit?

Solution:

$$\begin{aligned} R_T &= R_1 + R_2 + R_3 \\ &= 4700 + 5600 + 910 \\ &= 11{,}210 \ \Omega \\ &= 11.21 \ \text{k}\Omega \end{aligned}$$

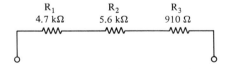

Figure 8-3

110 Chap. 8 Series Circuits

EXAMPLE 8-2:
A series circuit consists of four resistors: $R_1 = 220\ \Omega$, $R_2 = 470\ \Omega$, $R_3 = 560\ \Omega$, and $R_T = 2250\ \Omega$. What is the resistance of R_4?

Solution:
$$R_T = R_1 + R_2 + R_3 + R_4$$

Therefore,
$$\begin{aligned} R_4 &= R_T - (R_1 + R_2 + R_3) \\ &= 2250 - (220 + 470 + 560) \\ &= 2250 - 1250 \\ &= 1000\ \Omega \\ &= 1\ k\Omega \end{aligned}$$

EXAMPLE 8-3:
A series circuit has a total resistance of 7 kΩ. A 10-kΩ resistor is connected in series with the circuit. What is the total resistance of the system?

Solution:
$$\begin{aligned} R_T &= 7\ k\Omega + 10\ k\Omega \\ &= 17\ k\Omega \end{aligned}$$

8-5. VOLTAGE DROPS

When there is current flow in a series circuit, each resistor will have a voltage drop across it. The amount of the drop depends upon the current and the amount of resistance. We have already developed this idea in Chapter 7, Eq. (7-5).

$$V = IR$$

Because the current is the same amount in all parts of a series circuit, the voltage drops around the circuit will be proportional to the individual resistances. A resistor that has twice the resistance of another resistor will have twice the voltage drop of the smaller resistance. If a resistance is one third of another, the larger resistance will have three times as much voltage drop as the smaller one, and so on. This proportional relationship between voltage drops and resistance in a series circuit results in some applications we shall find very useful.

EXAMPLE 8-4:
Three resistors are connected in series in the circuit of Fig. 8-4: $R_1 = 2.2\ k\Omega$, $R_2 = 1\ k\Omega$, and $R_3 = 3.3\ k\Omega$. The voltage drop across R_2 is 6V. What are the voltage drops across the other resistors?

Solution:
(a) We may solve the problem by first solving for current, and then for the remaining voltage drops.

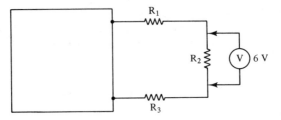

Figure 8-4

1. $I = \dfrac{V_2}{R_2} = \dfrac{6}{1} k\Omega = 6\ mA$
2. $V_1 = IR_1 = 6\ mA \times 2.2\ k\Omega$
   $= 13.2\ V$
3. $V_3 = IR_3 = 6\ mA \times 3.3\ k\Omega$
   $= 19.8\ V$

(b) The problem may be solved by recognizing that R_1 is 2.2 times greater than R_2, and that R_3 is 3.3 times greater. Therefore,

$$V_1 = 2.2 \times V_2 = 2.2 \times 6\ V = 13.2\ V$$
$$V_3 = 3.3 \times V_2 = 3.3 \times 6\ V = 19.8\ V$$

EXAMPLE 8-5:

The total resistance of the series circuit of Fig. 8-5 is 12.6 kΩ. One part of the circuit is a 3.9-kΩ resistor. If 20 V is applied to the circuit, what is the voltage drop across the 3.9-kΩ resistor?

Solution: Once again, we have two ways to solve the problem.
(a) Solve for current and then for the voltage drop.

$$I = \frac{E}{R_T} = \frac{20}{12.6 \times 10^3} = 1.59\ mA$$

$$V = IR = 1.59\ mA \times 3.9\ k\Omega = 6.19\ V$$

(b) Multiply the applied voltage by the ratio of 3.9 to 12.6 kΩ. This is an example of the use of resistance ratios to find a voltage drop. In Chapter 11 we shall call this the voltage division rule.

$$V = 20\ \frac{3900}{12,600} = 6.19\ V$$

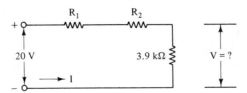

Figure 8-5

8-6. SUM OF THE VOLTAGE DROPS

We come now to the last of the basic properties of series circuits.

The sum of all the voltage drops is equal to the source voltage.

This property of series circuits should be apparent. Voltage is a form of force. Certainly, the force applied to a circuit must be exactly equal to the forces used in the circuit. We may also derive this principle.

$$I = \frac{E}{R_T} \quad \text{and} \quad E = IR_T$$

But

$$R_T = R_1 + R_2 + R_3 + \cdots + R_n$$

Therefore,

$$\begin{aligned} E &= I(R_1 + R_2 + R_3 + \cdots + R_n) \\ &= IR_1 + IR_2 + IR_3 + \cdots + IR_n \\ &= V_1 + V_2 + V_3 + \cdots + V_n \end{aligned} \tag{8-2}$$

EXAMPLE 8-6:
The source voltage for a series circuit is 45 V. The circuit consists of four resistors. The voltage drops across three of the resistors are $V_1 = 6$ V, $V_2 = 15$ V, and $V_3 = 10$ V. What is the voltage drop across R_4?

Solution:

$$\begin{aligned} E &= V_1 + V_2 + V_3 + V_4 \\ V_4 &= E - (V_1 + V_2 + V_3) \\ &= 45 - (6 + 15 + 10) \\ &= 45 - 31 = 14 \text{ V} \end{aligned}$$

EXAMPLE 8-7:
Three resistors are in series across a source of 100 V. $R_1 = 25\,\Omega, R_3 = 30\,\Omega$, and R_2 is unknown. The voltage drop across R_3 is measured and is 45 V. What is the resistance of R_2?

Solution:
(a) Find the current.

$$I = \frac{V_3}{R_3} = \frac{45}{30} = 1.5 \text{ A}$$

(b) Solve for the voltage drop across R_1.

$$\begin{aligned} V_1 &= IR_1 \\ &= 1.5 \times 25 = 37.5 \text{ V} \end{aligned}$$

(c) The voltage drop across R_2 is the difference between the applied voltage and the sum of the other voltage drops.

$$V_2 = E - (V_1 + V_3)$$
$$= 100 - (37.5 + 45)$$

(d) Solve for R_2.

$$R_2 = \frac{V_2}{I} = \frac{17.5}{1.5} = 11.67 \, \Omega$$

Check: $I = E/R_T$ and $R_T = 25 + 11.67 + 30 = 66.67 \, \Omega$
$I = 100/66.67 = 1.5$ A

An alternate method of solving for R_2 is to find current first, then the total resistance, and then R_2.

(a) $I = \dfrac{V_3}{R_3} = \dfrac{45}{30} = 1.5$ A

(b) $R_T = \dfrac{100}{1.5} = 66.67 \, \Omega$

(c) $R_2 = R_T - (R_1 + R_3)$
$= 66.67 - (25 + 30)$
$= 11.67 \, \Omega$

8-7. VOLTAGE SOURCES IN SERIES

Sources may be connected so that the total source voltage is the sum of the source voltages. This is known as the *series-aiding* connection. In this connection, the *positive* terminal of one source is connected to the *negative* terminal of the next source. The net effect is a single source with a greater voltage.

EXAMPLE 8-8:
Solve for the current in the circuit of Fig. 8-6, given that $E_1 = 30$ V, $E_2 = 20$ V, and $R_T = 200 \, \Omega$.

Solution:

$$E_T = E_1 + E_2 = 30 + 20 = 50 \text{ V}$$
$$I = \frac{E_T}{R_T} = \frac{50}{200} = 0.25 \text{ A}$$

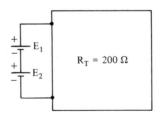

Figure 8-6

Sources may also be connected *series-opposing*. In this arrangement a terminal of one source is connected to a terminal of the *same polarity* on the other source. The net voltage available is the difference between the sources.

EXAMPLE 8-9:
Solve for net source voltage in the circuit of Fig. 8-7 and indicate the polarity of the net voltage.

Solution:
$$E_T = 50 - 10 = 40 \text{ V}$$

Polarity:

$$- \text{ at A } + \text{ at B}$$

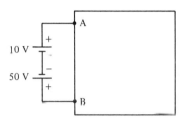

Figure 8-7

EXAMPLE 8-10:
Solve the circuit of Fig. 8-8 for E_T and indicate the polarity of the voltage.

Solution:
$$E_T = 30 + 20 - 10 = 40 \text{ V}$$

Polarity:

$$+ \text{ at A } - \text{ at B}$$

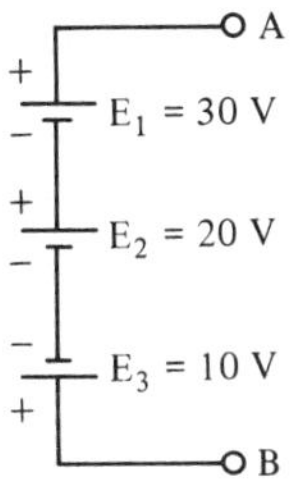

Figure 8-8

We may summarize the properties of series circuits as follows:

1. The current is the same amount in all parts of the circuit, including sources.

Sec. 8-7. Voltage Sources in Series 115

2. The total resistance is the sum of all the resistances.
3. The sum of all the voltage drops is equal to the source voltage.

8-8. CIRCUIT ANALYSIS

The electrical/electronics technician has a great need for a high level of skill in the analysis of circuits. We must analyze circuits in order to understand how they function and how to troubleshoot them. In your studies, as in this book, you will begin with elementary circuits and will progress to more complex circuits by building your knowledge from elementary circuits. You will find complex circuits quite easy to work with if you base your work upon a strong foundation in circuit analysis. The series circuit is an ideal starting point for learning how to apply the basic principles and laws of electric circuits.

In this section, we shall use a great many examples of how to apply Ohm's law and the rules of series circuits.

EXAMPLE 8-11:
Given the circuit of Fig. 8-9 and the source (applied) voltage, solve for (a) total resistance, (b) current, and (c) each voltage drop. Demonstrate that the sum of the voltage drops is equal to the applied voltage.

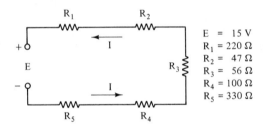

$E = 15$ V
$R_1 = 220\ \Omega$
$R_2 = 47\ \Omega$
$R_3 = 56\ \Omega$
$R_4 = 100\ \Omega$
$R_5 = 330\ \Omega$

Figure 8-9

Solution:
(a) Total resistance is the sum of the resistances.

$$R_T = R_1 + R_2 + R_3 + R_4 + R_5$$
$$= 220 + 47 + 56 + 100 + 330$$
$$= 753\ \Omega$$

(b) Current is found by Ohm's law. Note that in this case we shall use the *total* resistance and the applied voltage. The applied voltage is connected to the *entire* circuit and not to just part of the circuit. The current found is the current in all parts of the circuit.

$$I = \frac{E}{R_T} = \frac{15}{753} = 0.01992\ \text{A} = 19.92\ \text{mA}$$

(c) Each voltage drop is found by use of Ohm's law.

$$V = IR$$
$$V_1 = IR_1 = 0.01992 \times 220 = 4.382\ \text{V}$$

$$V_2 = IR_2 = 0.01992 \times 47 = 0.936 \text{ V}$$
$$V_3 = IR_3 = 0.01992 \times 56 = 1.116 \text{ V}$$
$$V_4 = IR_4 = 0.01992 \times 100 = 1.992 \text{ V}$$
$$V_5 = IR_5 = 0.01992 \times 330 = 6.574 \text{ V}$$

(d) The sum of the voltage drops is equal to the applied voltage.

$$4.832 + 0.936 + 1.116 + 1.992 + 6.574 = 15 \text{ V}$$

EXAMPLE 8-12:

Given a complete circuit and the current, we can solve for the *required* source voltage: $R_1 = 4.7 \text{ k}\Omega, R_2 = 6.8 \text{ k}\Omega, R_3 = 10 \text{ k}\Omega$, and $I = 1.4$ mA.

Solution: There are two ways we can solve for the applied voltage.
(a) Compute the total resistance and then use Ohm's law to find E.

$$R_T = 4.7 \text{ k}\Omega + 6.8 \text{ k}\Omega + 10 \text{ k}\Omega$$
$$= 21.5 \text{ k}\Omega$$
$$E = IR_T$$
$$= 1.4 \text{ mA} \times 21.5 \text{ k}\Omega$$
$$= 30.1 \text{ V}$$

(b) Solve for each voltage drop. The sum of the voltage drops is equal to the applied voltage.

$$V_1 = IR_1 = 1.4 \text{ mA} \times 4.7 \text{ k}\Omega = 6.58 \text{ V}$$
$$V_2 = IR_2 = 1.4 \text{ mA} \times 6.8 \text{ k}\Omega = 9.52 \text{ V}$$
$$V_3 = IR_3 = 1.4 \text{ mA} \times 10 \text{ k}\Omega = 14 \text{ V}$$
$$E = V_1 + V_2 + V_3$$
$$= 6.58 + 9.52 + 14 = 30.1 \text{ V}$$

EXAMPLE 8-13:

Given the applied voltage and current flow in Fig. 8-10, solve for the unknown resistance.

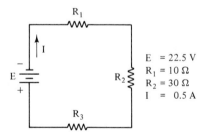

$E = 22.5$ V
$R_1 = 10 \ \Omega$
$R_2 = 30 \ \Omega$
$I = 0.5$ A

Figure 8-10

Solution: There are two ways to solve for R_3.
(a) Use Ohm's law to solve for total resistance. R_3 is the difference between R_T and the sum of R_1 and R_2.

$$R_T = \frac{E}{I} = \frac{22.5}{0.5} = 45 \ \Omega$$
$$R_3 = R_T - (R_1 + R_2)$$
$$= 45 - (10 + 30) = 5 \ \Omega$$

(b) We may solve for R_3 by first finding the voltage across R_3 then solve by use of Ohm's law.

$$V_3 = E - V_1 - V_2$$
$$= 22.5 - 0.5 \times 10 - 0.5 \times 30$$
$$= 22.5 - 5 - 15 = 2.5 \text{ V}$$
$$R_3 = \frac{V_3}{I} = \frac{2.5}{0.5} = 5 \text{ }\Omega$$

EXAMPLE 8-14:
Given the applied voltage and a voltage drop as in Fig. 8-11, solve for the unknown resistance.

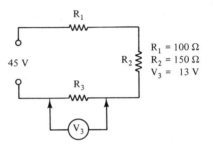

$R_1 = 100 \text{ }\Omega$
$R_2 = 150 \text{ }\Omega$
$V_3 = 13 \text{ V}$

Figure 8-11

Solution:
Current is found by Ohm's law:

$$I = \frac{V_3}{R_3} = \frac{13}{150} = 0.0867 \text{ A}$$

Total resistance is then calculated.

$$R_T = \frac{E}{I} = \frac{45}{0.0867} = 519 \text{ }\Omega$$

R_2 is then found.

$$R_2 = R_T - (R_1 + R_2)$$
$$= 519 - (100 + 150)$$
$$= 269 \text{ }\Omega \quad \text{(in practice, we would use a 270-}\Omega \text{ resistor)}$$

EXAMPLE 8-15:
Given the applied voltage, voltage drop across an unknown resistance, and the remainder of the circuit of Fig. 8-12, solve for the unknown resistance.

Solution:
With 38.25 V across R_2, the remainder of the applied voltage must be dropped across R_1 and R_3.

$$V_1 + V_3 = E - V_2 = 70 - 38.25 = 31.75 \text{ V}$$
$$IR_1 + IR_3 = 31.75 \text{ V}$$
$$I(R_1 + R_3) = 31.75$$

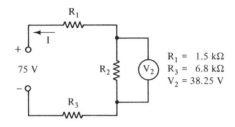

Figure 8-12

The current in the series circuit is then found by use of Ohm's law.

$$I = \frac{31.75}{1.5 \text{ k}\Omega + 6.8 \text{ k}\Omega} = 3.825 \text{ mA}$$

Once current is calculated, we can calculate the total resistance.

$$R_T = \frac{E}{I} = \frac{70}{3.825 \text{ mA}} = 18.3 \text{ k}\Omega$$

Then

$$\begin{aligned} R_2 &= R_T - (R_1 + R_3) \\ &= 18.3 \text{ k}\Omega - (1.5 \text{ k}\Omega + 6.8 \text{ k}\Omega) \\ &= 10 \text{ k}\Omega \end{aligned}$$

8-9. MEASUREMENTS AND TROUBLESHOOTING

Current Measurement. Because the current is the same in all parts of the series circuit, only one measurement is needed. The multimeter leads may be connected into the circuit at any convenient point. Remember, the ammeter must be placed in *series* into the circuit.

Voltage Measurement. Applied voltage should always be measured at the source terminals with all loads connected. Sources have enough internal resistance to cause the terminal voltage to change in going from no load to full load, even with fairly low load currents.

Voltage drops are measured by connecting meter leads between points in the circuit.

EXAMPLE 8-16:
We shall make use of voltage measurements to explain fully the relationship between voltage drops and applied voltage. Given the circuit of Fig. 8-13:
(a) Measure the applied voltage by connecting the voltmeter (VM) leads to nodes y and x.

$$V_{y-x} = 115 \text{ V} = V_1 + V_2 + V_3 + V_4$$

(b) Measure the voltage drop across R_1. The VM leads are connected to nodes a and x.

$$V_{a-x} = 18 \text{ V} = E - (V_2 + V_3 + V_4)$$

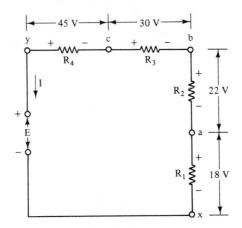

Figure 8-13

(c) Measure the voltage drop between nodes b and x. The VM leads are connected to nodes b and x.

$$V_{b-x} = 40 \text{ V} = V_1 + V_2$$

Note also, that $V_{b-x} = E - (V_3 + V_4)$.

(d) Measure V_{y-c}

$$V_{y-c} = 45 \text{ V} = V_4$$

You should note that in each measurement we found *series-aiding* and *series-opposing* voltages. Voltage drops as a result of a single source are always aiding each other and *opposing* the source. This is why V_{b-x} is equal to the sum of V_1 and V_2 and is also equal to the difference between the supply voltage and drops of V_3 and V_4.

(e) Assume that load current (I_L) is 300 mA and the internal resistance of the source is 10 Ω. What is the terminal voltage (V_{y-x}) if the load circuit is opened ($I_L = 0$)?

Solution:
When $I_L = 300$ mA, the voltage $V_{y-x} = 115$ V. This voltage was equal to the emf of the source *minus* the internal voltage drop.

$$115 = \text{emf} - 0.3 \times 10$$
$$115 = \text{emf} - 3$$
$$\text{emf} = 115 + 3 = 118 \text{ V}$$

Troubleshooting. The nature of an actual problem is very often determined by the kind of circuit parts used. So far, we have studied series resistive circuits. Resistors are either carbon composition, wirewound, or deposited carbon. The most widely used resistor is the carbon-compound type. The most common fault with these resistors is a change of resistance value, almost always to a very much *lower* value of resistance. This type of fault occurs very slowly. It can take months for the resistance change to be great enough to affect the performance of a circuit. We find these defective

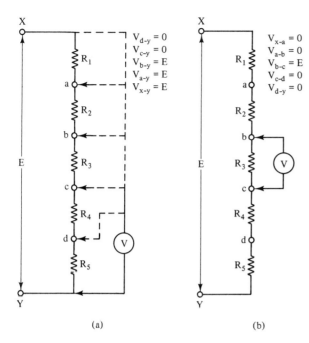

Figure 8-14 Using a voltmeter to locate an open resistor. In the method shown in (a) one lead of the voltmeter is connected to one side of the source. As we move the other lead from point to point in the circuit, we get no change until we measure just past the open resistor. In the method shown in (b) we make measurements directly across each resistor. Because the circuit is open, there are no voltage drops; and source voltage is equal to the voltage across the open resistor.

resistors by simply measuring voltage drop and comparing the value to the voltage listed in the manufacturer's service manual.

Another, less common, fault is an open resistor. If the part is a carbon resistor, we will spot it without voltage measurements. Carbon resistors burn across, resulting in a completely open unit with an easily recognized charred area in the center.

Open resistors that cannot be spotted by a visual check are rapidly located with a *voltmeter*. The open resistor opens the circuit, causing the current to drop to zero. Therefore, there are no voltage drops across any of the "good" resistors in the circuit. The open resistor will have all the applied voltage across it. When we come across such a resistor, we have found an open resistor. Figure 8-14 shows two ways we can find an open resistor in a series string.

SUMMARY

1. Current is the same in all parts of a series circuit.
2. Resistors in series add. Total resistance is the sum of the individual resistances.
3. The sum of the voltage drops around a series circuit is equal to the applied voltage.
4. Voltage sources may be connected in series aiding. This is $+-, +-$ connection. Total voltage is the sum of the source voltages.
5. Voltage sources may be connected in series-opposition, resulting in a total voltage that is the difference between source voltages. This $+-, -+$ connection.

PROBLEMS

8-1. Given the series circuit of Fig. 8-15, identify each *node* as A, B, C, and so on. Node A is indicated in the figure.

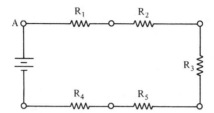

Figure 8-15

8-2. Redraw the circuit of Fig. 8-15 by inserting an ammeter between R_4 and R_5. Suppose that the meter indicates a current of 220 mA; what is the current in R_1? Why?

8-3. In the circuit of Fig. 8-16, the ammeter reads 100 mA. Solve for the voltage drop across each resistor.

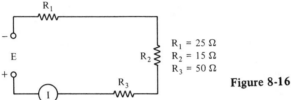

$R_1 = 25\ \Omega$
$R_2 = 15\ \Omega$
$R_3 = 50\ \Omega$

Figure 8-16

8-4. In the circuit of Fig. 8-17, the current in R_1 is 25 mA. The voltage across R_2 is 25 V. What is the resistance of R_2?

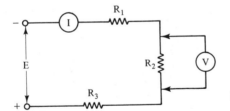

Figure 8-17

Figure 8-18 is a circuit board with a variety of separate resistors. We may interconnect resistors to obtain a variety of total resistances. Use the figure for Probs. 8-5 through 8-7.

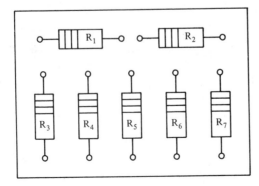

$R_1 = 10\ \Omega$
$R_2 = 18\ \Omega$
$R_3 = 47\ \Omega$
$R_4 = 68\ \Omega$
$R_5 = 100\ \Omega$
$R_6 = 150\ \Omega$
$R_7 = 220\ \Omega$

Figure 8-18

8-5. What resistors are connected in series to obtain $R_T = 168\ \Omega$? Sketch the circuit.

8-6. Sketch a series circuit that would result in $R_T = 75\ \Omega$.

8-7. Connect the resistors on the circuit for maximum total resistance. What is R_T? Sketch the circuit.

8-8. A series circuit consists of the following resistors: $R_1 = 4.7\ \text{k}\Omega, R_2 = 2.2\ \text{k}\Omega$, and $R_3 = 10\ \text{k}\Omega$. What is the total resistance of the circuit?

8-9. A three-resistor series circuit draws 0.5 A when connected to a 100-V source. Solve for R_1, given that $R_2 = 100\ \Omega$ and $R_3 = 75\ \Omega$.

8-10. Solve for current in a series circuit given that $E = 48\ \text{V}, R_1 = 10\Omega, R_2 = 96\ \Omega$, and $R_3 = R_4 = 47\ \Omega$. Sketch the circuit.

8-11. Solve for total resistance in the circuit of Fig. 8-19. Each resistor has $R = 100\ \Omega$.

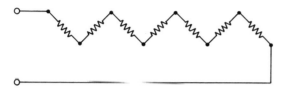

Figure 8-19

8-12. Four equal resistors are connected in series. If the resistance of one is 220 Ω, what is the total resistance? Write a formula for the total resistance of equal resistances in series.

8-13. A series circuit consists of two resistors and a 24-V source. $R_1 = 4.7\ \text{k}\Omega, R_2 = 5.6\ \text{k}\Omega$. What is the voltage drop across each resistor?

8-14. The source in Prob. 8-13 is changed so that the current is 290 μA. What is the source voltage?

8-15. A three-resistor series circiut has the following voltage drops: $V_1 = 12\ \text{V}, V_2 = 8\ \text{V}$, and $V_3 = 15\ \text{V}$. What is the source voltage?

8-16. Given a three-resistor series circuit, what is source voltage if the voltage drop across $R_2 = 2.37\ \text{V}$ and $R_1 = 6.8\ \text{k}\Omega, R_2 = 10\ \text{k}\Omega$, and $R_3 = 2.2\ \text{k}\Omega$?

8-17. Given the circuit of Fig. 8-20, solve for current, each voltage drop, and source voltage (E).

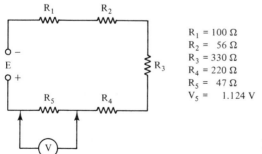

$R_1 = 100\ \Omega$
$R_2 = 56\ \Omega$
$R_3 = 330\ \Omega$
$R_4 = 220\ \Omega$
$R_5 = 47\ \Omega$
$V_5 = 1.124\ \text{V}$

Figure 8-20

8-18. In the circuit of Fig. 8-21, $E = 15\ \text{V}$ and $V_1 = 4.69\ \text{V}$. Solve for R_2, given that $R_1 = 1.5\ \text{k}\Omega$.

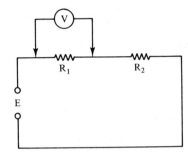

Figure 8-21

8-19. A series circuit consists of three resistors. A current of 1.43 mA is measured when 15 V is applied to the circuit. The voltage drops are found to be $V_1 = 2V_2$ and $V_2 = 2V_3$. Solve for $R_3, R_2,$ and R_1.

8-20. In the circuit of Fig. 8-22, solve for current and indicate its direction.

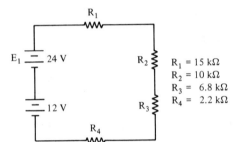

Figure 8-22

8-21. Refer to the circuit of Fig. 8-22. Reverse source E_1 so that its negative terminal is connected to point a. Solve for current and indicate its direction.

8-22. In the circuit of Fig. 8-23, solve for the voltage drop across R_3, given that $V_1 = 2.16$ V.

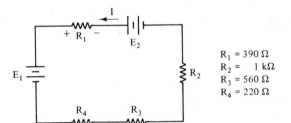

Figure 8-23

8-23. Given that $E_1 = 15$ V in the circuit of Fig. 8-23, solve for E_2, given that $V_1 = 3.77$ V.

8-24. Suppose that the terminals of E_2 in Fig. 8-23 are reversed so that the negative terminal is at R_1 and the positive at R_2. Using the value of E_2 found in Prob. 8-23, solve for each voltage drop.

8-25. Sketch a two-resistor series circuit with a source and ammeter. Indicate meter polarity as it relates to the power supply polarity.

8-26. In the circuit of Fig. 8-24, voltage measurements are made using the negative terminal of the source as the reference. We progress with the positive lead of the voltmeter around the circuit from points A through D. Solve for V_A, V_B, V_C, and V_D.

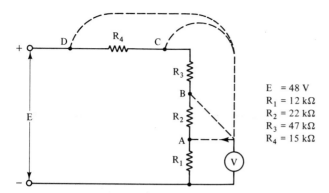

Figure 8-24

8-27. The circuit of Figure 8-24 develops a fault. Using a voltmeter, the technician makes the following measurements with respect to the negative terminal of the source: $V_A = 0$ V, $V_B = 0$ V, $V_C = 48$ V, and $V_D = 48$ V. What is the probable circuit fault?

8-28. The voltage drops are measured in the circuit of Prob. 8-27. $V_1 = 0$ V, $V_2 = 0$ V, $V_3 = 48$ V, and $V_4 = 0$ V. Do these results confirm the diagnosis made in Prob. 8-27?

8-29. Calculate each voltage drop in a three-resistor series circuit, given that $E = 18$ V, $R_1 = 10$ kΩ, $R_2 = 4.7$ kΩ and $R_3 = 3.3$ kΩ.

8-30. Voltage drops are measured in the circuit of Prob. 8-29. $V_1 = 3.6$ V, $V_2 = 8.46$ V, and $V_3 = 5.94$ V. The resistors are carbon composition types. What is the probable cause of the circuit problem?

9

Parallel Circuits

9-1. INTRODUCTION

In the series circuit, and in series-connected parts, there is only one path for current. The voltage drops around a series circuit can be different from each other. Also, applied voltage equals the sum of the voltage drops. The total resistance is the sum of the resistances.

When parts are connected in parallel, we find that there are *different paths* for current. These paths are called *branches*. Branch currents need not be the same amount. In the parallel connection, voltage is the only quantity that is the same amount for each branch of the circuit. We shall find that the total resistance of the circuit is not the sum of the resistances, but is less than the resistance of any branch. Various sketches of parallel circuits are shown in Fig. 9-1.

9-2. PARALLEL RESISTANCE CIRCUIT

In a parallel circuit load resistances are connected separately to a source. If we assume that connecting wires are "short circuits" in comparison to the resistance of the loads, then each load resistor is connected directly across the source. House wiring is a classic example of a parallel circuit. Each electrical appliance is separately connected to line voltage, even when they may share the same electrical outlet.

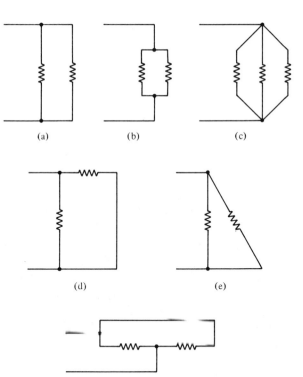

Figure 9-1 Various ways to show resistors in parallel. Of all the various sketches shown in the figure, only two are widely used. These are (a) and (b). The others are shown in order to make you aware that parallel circuits don't have to "look like parallel circuits."

There are three basic rules for parallel circuits:

1. *The voltage across each branch is the same.* Since the branches are connected to the same voltage source (or to the same pair of circuit nodes), it is not possible for the branch voltages to be different.
2. *The total current drawn by a parallel circuit is the sum of the branch currents.* Each branch draws a separate load current from the source. Clearly, the total load current supplied by the source must equal the current drawn by the branches, as shown in Fig. 9-2.

$$I_T = I_1 + I_2 + I_3 + \cdots + I_n \tag{9-1}$$

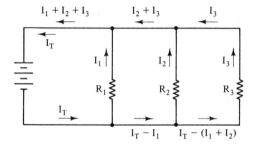

Figure 9-2 A three branch parallel circuit.

Sec. 9-2. Parallel Resistance Circuit 127

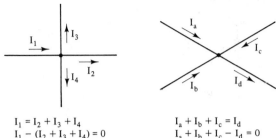

$I_1 = I_2 + I_3 + I_4$
$I_1 - (I_2 + I_3 + I_4) = 0$

$I_a + I_b + I_c = I_d$
$I_a + I_b + I_c - I_d = 0$

Figure 9-3 Currents at a circuit node.

where I_n represents the current of the nth branch. Another way to understand this current rule is by the use of one of the basic physical laws. *The amount of current entering a node must be exactly equal to the amount of current leaving that node* (Fig. 9-3).

3. *The total resistance of a parallel circuit is less than any branch resistance.* This may be seen easily if we remember that the total current of a parallel circuit is greater than any branch current. Since the voltage across a parallel circuit is the same, it follows that the ratio of voltage to total current, which is total resistance, must be less than the resistance of any branch.

EXAMPLE 9-1:
A three-branch parallel circuit is connected to a 100-V source. The branch currents are 2 A, 1 A, and 0.5 A. (a) What is the voltage across each branch? (b) What is the total current?

Solution:
(a) Each branch voltage is 100 V.
(b) Total current is 3.5 A.

$$I_T = 2 + 1 + 0.5 = 3.5 \text{ A}$$

9-3. EQUIVALENT RESISTANCE

The total resistance of a parallel circuit can be represented by a single resistance. This single resistance will draw exactly as much current as the total current drawn by the parallel circuit. Because this resistance represents the same amount of load as the parallel circuit, we refer to the resistance as the *equivalent resistance* (R_{eq}) of a parallel circuit.

$$R_{eq} = \frac{E}{I_T} \qquad (9-2)$$

EXAMPLE 9-2:
The parallel circuit of Fig. 9-4 draws 5 A from a 120-V source. What is the equivalent resistance of the parallel circuit?

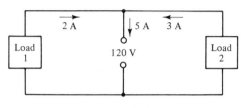

Figure 9-4

Solution:

$$R_{eq} = \frac{E}{I_1 + I_2} = \frac{120}{5} = 24 \, \Omega$$

The concept of equivalent resistance is very valuable in our future work with network analysis. It is helpful for us to begin using it with parallel circuits. The equivalent resistance of a parallel circuit is always less than any branch resistance. We shall use R_{eq} to represent the total resistance of a parallel circuit.

9-4. EQUIVALENT RESISTANCE: MANY-BRANCHED CIRCUIT

The equivalent (total) resistance of any parallel circuit may be found by Ohm's law, as in Eq. (9-2).

$$R_{eq} = \frac{E}{I_T} = \frac{E}{I_1 + I_2 + I_3 + \cdots + I_n}$$

Each branch is considered to be directly across the source. Therefore, each branch current equals source voltage divided by branch resistance.

$$R_{eq} = \frac{E}{E/R_1 + E/R_2 + E/R_3 + \cdots + E/R_n}$$

Factoring,

$$R_{eq} = \frac{E}{E(1/R_1 + 1/R_2 + 1/R_3 + \cdots + 1/R_n)}$$

Therefore,

$$R_{eq} = \frac{1}{1/R_1 + 1/R_2 + 1/R_3 + \cdots + 1/R_n} \qquad (9\text{-}3)$$

EXAMPLE 9-3:

What is the equivalent (total) resistance of a three-branched parallel circuit? $R_1 = 50 \, \Omega$, $R_2 = 100 \, \Omega$, and $R_3 = 40 \, \Omega$.

Solution:

$$R_{eq} = \frac{1}{1/R_1 + 1/R_2 + 1/R_3}$$

$$= \frac{1}{1/50 + 1/100 + 1/40}$$

$$= \frac{1}{0.02 + 0.01 + 0.025}$$

$$= \frac{1}{0.055} = 18.18 \ \Omega$$

Formula 9-3 may also be found by considering conductance concepts. Each branch is a conduction path between the terminals of the circuit. The total conductance of a parallel resistance circuit is the sum of the conductances of the branches.

$$G_T = G_1 + G_2 + G_3 + \cdots + G_n \qquad (9\text{-}4)$$

where G is conductance in siemens (S).

If we recall that resistance and conductance are reciprocals, that is,

$$R = \frac{1}{G} \quad \text{and} \quad G = \frac{1}{R}$$

then

$$G_T = \frac{1}{R_1} + \frac{1}{R_2} + \frac{1}{R_3} + \cdots + \frac{1}{R_n}$$

and

$$R_{eq} = \frac{1}{G_T} \qquad (9\text{-}5)$$

Therefore,

$$R_{eq} = \frac{1}{1/R_1 + 1/R_2 + 1/R_3 + \cdots + 1/R_n} \qquad (9\text{-}3)$$

EXAMPLE 9-4:
Given a parallel circuit with four branches, and $R_1 = 2.7$ kΩ, $R_2 = 3.3$ kΩ, $R_3 = 10$ kΩ, and $R_4 = 5.6$ kΩ. Solve for R_{eq} using a scientific calculator.

Solution: Equation (9-3) is readily used with the calculator. Calculation for R_{eq} requires that we take the reciprocal of a sum of reciprocals. This all sounds much more difficult than it really is! All scientific calculators have a reciprocal function, the $\boxed{1/x}$ key. Let us proceed to a solution by listing each step and the calculator display.

Step	Display
1. Enter 2700	2700
2. Press $\boxed{1/x}$	.00037037
3. Press $\boxed{+}$	.00037037
4. Enter 3300	3300

5. Press $\boxed{1/x}$		.0003030
6. Press $\boxed{+}$		.0006734
7. Enter 10,000		10000
8. Press $\boxed{1/x}$		.0001
9. Press $\boxed{+}$		.0007734
10. Enter 5600		5600
11. Press $\boxed{1/x}$		.00017857
12. Press $\boxed{=}$		.00095197
13. Press $\boxed{1/x}$		1050.45

$R_{eq} = 1050 \, \Omega$

EXAMPLE 9-5:

When all branches of a parallel circuit are in the same units, such as kilohms, we need not enter the unit into the calculator. We must, however, make the mental note to read the result as if it were in kilohms.

Calculate the equivalent resistance of 47 kΩ in parallel with 56 kΩ, in parallel with 82 kΩ.

Solution:

Enter 47		47
Press $\boxed{1/x}$		.0212765957
Press $\boxed{+}$		.0212765957
Enter 56		56
Press $\boxed{1/x}$		.0178571429
Press $\boxed{+}$		.0391337386
Enter 82		82
Press $\boxed{1/x}$		.012195122
Press $\boxed{=}$		.0513288606
Press $\boxed{1/x}$		19.48221701

$R_{eq} = 19.5 \, k\Omega$

EXAMPLE 9-6:

Given the circuit of Fig. 9-5, $R_1 = 390 \, \Omega$, $R_2 = 1.2 \, k\Omega$, $R_3 = 1.8 \, k\Omega$, and $R_4 = 2.2 \, k\Omega$. Solve for (a) total conductance and (b) R_{eq}.

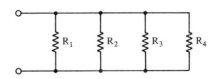

Figure 9-5

Solution:

(a) $G_T = \dfrac{1}{R_1} + \dfrac{1}{R_2} + \dfrac{1}{R_3} + \dfrac{1}{R_4}$

$= \dfrac{1}{390} + \dfrac{1}{1200} + \dfrac{1}{1800} + \dfrac{1}{2200}$

$= 0.00256 + 0.000833 + 0.000556 + 0.000455$

$= 0.004407 = 4.407$ mS

(b) $R_{eq} = \dfrac{1}{G_T} = \dfrac{1}{4.407 \times 10^{-3}} = 0.227 \times 10^3$

$= 227\ \Omega$

9-5. EQUIVALENT RESISTANCE: TWO-BRANCHED CIRCUIT

In electronics, the two-branch parallel circuit is one we work with very often. We can analyze this circuit using the formulas relating to the analysis of any parallel circuit. However, there is one special formula that we all use for two-branch parallel circuits. It is known as the product-over-sum formula and derives from Eq. (9-3).

$$R_{eq} = \dfrac{1}{1/R_1 + 1/R_2}$$

If we use $R_1 R_2$ as a common denominator and then take the required reciprocal, we have

$$R_{eq} = \dfrac{R_1 R_2}{R_1 + R_2} \tag{9-6}$$

This formula is readily used with the scientific calculator. We must be careful in working with the denominator if we are to avoid making a mathematics mistake. One cannot simply divide the product by R_1 and then by R_2. Remember that the denominator is the sum of R_1 and R_2. We have two ways to work the problem correctly.

1. For calculators with parenthesis keys, we use a direct approach to working with the formula. The sequence is as follows: Enter R_1, press $\boxed{\times}$, enter R_2, press $\boxed{\div}$, press $\boxed{(}$, enter R_1, press $\boxed{+}$, enter R_2, press $\boxed{)}$, press $\boxed{=}$.
2. For calculators without parenthesis keys, the most direct approach is to make use of the reciprocal key. Remember that division by a number may be done by multiplication with the reciprocal of the number. The sequence is as follows: Enter R_1, press $\boxed{+}$, enter R_2, press $\boxed{=}$, press $\boxed{1/x}$, press $\boxed{\times}$, enter R_1, press $\boxed{\times}$, enter R_2,

press $\boxed{=}$. What we have done in this case is to take the sum of R_1 and R_2, and then the reciprocal of this sum. We multiplied the reciprocal of the sum by R_1 and then this product by R_2. The result is R_{eq}.

Some examples should make both processes clear.

EXAMPLE 9-7:
Given that $R_1 = 760\ \Omega$ and $R_2 = 1500\ \Omega$, solve for R_{eq} of the two-branch parallel circuit.

Solution:
(a) Using parenthesis keys:

Enter 760	760
Press $\boxed{\times}$	760
Enter 1500	1500
Press $\boxed{\div}$	1140000
Press $\boxed{(}$	1140000
Enter 760	760
Press $\boxed{+}$	760
Enter 1500	1500
Press $\boxed{)}$	2260
Press $\boxed{=}$	504.4247788

$$R_{eq} = 504\ \Omega$$

(b) Using the $\boxed{1/x}$ key:

Enter 760	760
Press $\boxed{+}$	760
Enter 1500	1500
Press $\boxed{=}$	2260
Press $\boxed{1/x}$	.0004424779
Press $\boxed{\times}$	.0004424779
Enter 760	760
Press $\boxed{\times}$	.3362831858
Enter 1500	1500
Press $\boxed{=}$	504.4247788

$$R_{eq} = 504\ \Omega$$

It is common practice to use the symbol $\|$ to mean "in parallel with." Thus, if we read $R_a \| R_b$, it means R_a in parallel with R_b. The symbol may be used with any number of parallel branches. For example, $R_1 \| R_2 \| R_3 \| R_4$ means that resistors R_1, R_2, R_3, and R_4 are in parallel with each other.

EXAMPLE 9-8:
Solve for R_{eq}, given 12 kΩ ∥ 15 kΩ.

Solution:

$$R_{eq} = \frac{12 \text{ k}\Omega \times 15 \text{ k}\Omega}{12 \text{ k}\Omega + 15 \text{ k}\Omega} = 6.67 \text{ k}\Omega$$

Note that there was no need to enter any powers of 10 because all quantities were in kilohms; therefore, the result is in kilohms.

Calculator solution:

$$12 \times 15 \div (12 + 15) = 6.6666$$

9-6. EQUIVALENT RESISTANCE: EQUAL-BRANCH RESISTANCE

Another special formula is used when branch resistances are equal. We begin with Eq. (9-13) and n equal branch resistances.

$$R_{eq} = \frac{1}{1/R + 1/R + \cdots + 1/R}$$

Because all the resistances are equal, the common denominator is simply branch resistance, R. The sum of the fractions is n/R. This gives us

$$R_{eq} = \frac{1}{n/R}$$

Solving,

$$R_{eq} = \frac{R}{n} \qquad (9\text{-}7)$$

where n = number of equal branches
R = resistance of one branch

EXAMPLE 9-9:
A parallel circuit consists of three branches, each with $R = 150$ Ω. (a) What is the equivalent resistance of the circuit? (b) What current will be drawn from a 10-V source?

Solution:

(a) $R_{eq} = \dfrac{R}{n} = \dfrac{150}{3} = 50$ Ω

(b) $I_T = \dfrac{E}{R_{eq}} = \dfrac{10}{50} = 0.2$ A

EXAMPLE 9-10:
A fixture has eight lamp bulbs that are parallel connected to a 120-V source. Each lamp has a rated resistance of 240 Ω. (a) What is the total load current? (b) What is the current per lamp? (c) What is the equivalent resistance of the fixture?

Solution: We begin by solving for R_{eq}.

$$R_{eq} = \frac{R}{n} = \frac{240}{8} = 30 \: \Omega$$

(a) $I_T = \dfrac{E}{R_{eq}} = \dfrac{120}{30} = 4 \text{ A}$

(b) Current per lamp can be obtained by dividing total current by 8 or by dividing the line voltage by branch resistance. Both give the same result.

$$I = \frac{I_T}{n} = \frac{4}{8} = 0.5 \text{ A}$$

or

$$I = \frac{E}{R} = \frac{120}{240} = 0.5 \text{ A}$$

(c) R_{eq} has been solved for and is 30 Ω.

We should recognize that there are other orders of steps to get these solutions. But, the results cannot differ. For example, we can find branch current. This current, multiplied by 8, gives us total current. Line voltage divided by total current gives us the value of R_{eq}.

9-7. MORE USEFUL FORMULAS

We can set up many special cases of two-branch circuits. We shall examine only one special case, where the resistance of one branch is twice the resistance of the other. We find that the equivalent resistance is two-thirds of the smaller resistance.

Given $R_1 \parallel R_2$ and that $R_1 = 2(R_2)$, then

$$R_{eq} = \frac{2R_2}{3} = 0.667(R_2) \qquad (9\text{-}8)$$

This is readily derived by use of Eq. (9-6).

$$R_{eq} = \frac{R_1 R_2}{R_1 + R_2}$$

But $R_1 = 2R_2$.
Therefore,

$$R_{eq} = \frac{2R_2 \times R_2}{2R_2 + R_2} = \frac{2R_2 \times R_2}{3R_2} = 2/3 R_2$$

It should be clear that other formulas for other ratios can be derived. These are not as useful as the 2 : 1 ratio formula and will not be covered.

EXAMPLE 9-11:
What is the equivalent resistance for each of the following two-branch parallel circuits?

(a) 150 Ω ∥ 300 Ω
(b) 300 Ω ∥ 600 Ω
(c) 45 Ω ∥ 90 Ω
(d) 500 Ω ∥ 250 Ω
(e) 120 Ω ∥ 60 Ω
(f) 200 Ω ∥ 100 Ω

Solution: In each pair or resistances, R_{eq} is equal to two-thirds of the smaller resistance.
(a) $R_{eq} = 2/3(150) = 100\ \Omega$
(b) $R_{eq} = 2/3(300) = 200\ \Omega$
(c) $R_{eq} = 2/3(45) = 30\ \Omega$
(d) $R_{eq} = 2/3(250) = 167\ \Omega$
(e) $R_{eq} = 2/3(60) = 40\ \Omega$
(f) $R_{eq} = 2/3(100) = 66.7\ \Omega$

Suppose that we need to know how much resistance must be placed in parallel with another resistance in order to have some specified equivalent resistance. In other words, if we know R_{eq} and R_1, how do we compute R_2? A little algebraic manipulation of Eq. (9-6) gives us the answer.

$$R_2 = \frac{R_1 \times R_{eq}}{R_1 - R_{eq}} \tag{9-9}$$

EXAMPLE 9-12:
Given the circuit of Fig. 9-6, what resistance must be connected in parallel with 200 Ω in order for R_{eq} to be 132 Ω?

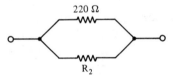

Figure 9-6

Solution:

$$R_2 = \frac{220 \times 132}{220 - 132}$$

$$= \frac{29{,}040}{88} = 330\ \Omega$$

We may use any of the formulas in the solution of parallel circuits. We can use the product-over-sum formula in circuits with more than two branches. Simply reduce the circuit two branches at a time until the final reduction brings us to two remaining branches, which are solved for equivalent resistance. We can use the formula that applies to equal branch resistance, which allows us to reduce many equal branches to a single branch.

We may solve for an unknown branch resistance by means of Eq. (9-9), provided that we reduce the remainder of the parallel circuit to a single branch.

EXAMPLE 9-13:

In the circuit of Fig. 9-7, solve for R_{eq}.

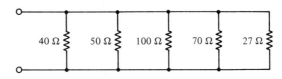

Figure 9-7

Solution:

1. Solve for 40 || 50:

$$40 \parallel 50 = \frac{40 \times 50}{40 + 50} = 22.2 \ \Omega$$

2. Solve for 100 || 70:

$$100 \parallel 70 = \frac{100 \times 70}{100 + 70} = 41.2 \ \Omega$$

At this point we have reduced the circuit from five branches to three branches (Fig. 9-8).

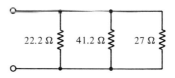

Figure 9-8

3. Solve for 41.2 || 27:

$$41.2 \parallel 27 = \frac{41.2 \times 27}{41.2 + 27} = 16.3 \ \Omega$$

The circuit has been reduced to two branches and R_{eq} can be found (Fig. 9-9).

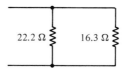

Figure 9-9

4. Solve for 22.2 || 16.3:

$$22.2 \parallel 16.3 = \frac{22.2 \times 16.3}{22.3 + 16.3}$$
$$= 9.4 \ \Omega = R_{eq}$$

Check:

$$R_{eq} = \frac{1}{1/40 + 1/50 + 1/100 + 1/70 + 1/27}$$
$$= 9.4 \, \Omega$$

It should be clear from Example 9-13 that making repeated two-branch reductions for R_{eq} is certainly not a time saver! But that was because we really had to make some calculations. Suppose that we have a circuit that allows us to make mental calculations, as in Example 9-14.

EXAMPLE 9-14:
Given the circuit of Fig. 9-10, solve for R_{eq}.

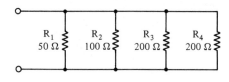

Figure 9-10

Solution:

1. $R_3 \parallel R_4 = \dfrac{200}{2} = 100 \, \Omega$ (Fig. 9-11)

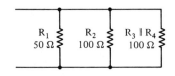

Figure 9-11

2. $R_2 \parallel (R_3 \parallel R_4) = 100 \parallel 100 = 50 \, \Omega$ (Fig. 9-12)

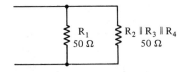

Figure 9-12

3. Finally, R_1 in parallel with $50 \, \Omega$:

$$50 \parallel 50 = 25 \, \Omega = R_{eq}$$

EXAMPLE 9-15:
Solve for R_{eq}, given the following four-branch parallel circuit:

$$R_1 = 50 \, \Omega, \, R_2 = R_3 = R_4 = 150 \, \Omega.$$

Solution:

$$R_2 \parallel R_3 \parallel R_4 = \frac{150}{3} = 50 \ \Omega$$
$$R_{eq} = 50 \parallel 50 = 25 \ \Omega$$

9-8. CIRCUIT ANALYSIS

We make use of Ohm's law in the analysis of parallel circuits. We may solve for current, voltage, and resistance as in the series circuit. We must be careful to use the total circuit values in dealing with R_{eq} and branch values when working with branch currents.

Because we may choose to solve for conductance in parallel circuits, it can be useful to write the Ohm's law formulas in terms of conductance.

Total current:

$$I_T = E \times G_T \tag{9-10}$$

where $G_T = 1/R_{eq}$.

Branch current:

$$I_x = E \times G_x \tag{9-11}$$

where x represents any branch and $G_x = 1/R_x$.

Current division. In series circuits we have the *voltage-division rule*; current in a series circuit is constant.

In parallel circuits we have the *current-division rule*; voltage in a parallel circuit is constant.

$$I_x = I_T \frac{G_x}{G_T} \tag{9-12}$$

In the special case of a two-branch parallel circuit, it is convenient to write the current-division rule using R_1 and R_2 rather than conductances.

$$I_1 = I_T \frac{R_2}{R_1 + R_2} \tag{9-13}$$

$$I_2 = I_T \frac{R_1}{R_1 + R_2} \tag{9-14}$$

We have covered a wealth of material on parallel resistance circuits. Let us now go through many examples of how we may use Ohm's law with these circuits. We begin with an example of solutions for branch currents and total current.

EXAMPLE 9-16:
In the circuit of Fig. 9-13, solve for each branch current and the total current. Applied voltage is 200 V. $R_1 = 40 \ \Omega$, $R_2 = 50 \ \Omega$, and $R_3 = 100 \ \Omega$.

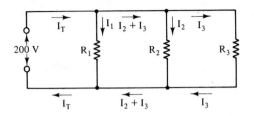

Figure 9-13

Solution:

$$I_1 = \frac{E}{R_1} = \frac{200}{40} = 5 \text{ A}$$

$$I_2 = \frac{E}{R_2} = \frac{200}{50} = 4 \text{ A}$$

$$I_3 = \frac{E}{R_3} = \frac{200}{100} = 2 \text{ A}$$

$$I_T = I_1 + I_2 + I_3 = 5 + 4 + 2 = 11 \text{ A}$$

Check:
(a) Solve for R_{eq}:

$$R_{eq} = \frac{1}{1/40 + 1/50 + 1/100}$$
$$= 18.18 \, \Omega$$

(b) Solve for I_T:

$$I_T = \frac{E}{R_{eq}} = \frac{200}{18.18} = 11 \text{ A}$$

EXAMPLE 9-17:
Given the circuit of Fig. 9-14, use Ohm's law to solve for R_{eq}.

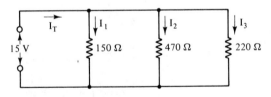

Figure 9-14

Solution:

$$R_{eq} = \frac{E}{I_T} = \frac{E}{I_1 + I_2 + I_3}$$

$$I_1 = \frac{E}{R_1} = \frac{15}{150} = 0.1 \text{ A}$$

$$I_2 = \frac{E}{R_2} = \frac{15}{470} = 0.0319 \text{ A}$$

$$I_3 = \frac{E}{R_3} = \frac{15}{220} = 0.0682 \text{ A}$$

$$I_T = 0.1 + 0.0319 + 0.0682 = 0.2 \text{ A}$$

$$R_{eq} = \frac{15}{0.2} = 75 \text{ }\Omega$$

EXAMPLE 9-18:

In the circuit of Fig. 9-15, current in the 39-kΩ branch is 1.5 mA. (a) What is the voltage across the parallel circuit? (b) What is the current in the 47-kΩ resistor?

Solution: The voltage across *each* branch of a parallel circuit must be the same. Therefore, a solution for voltage across one branch is a solution for voltage across any other branch. Once voltage is known, current in any branch is readily found by Ohm's law.

(a) $V = IR$
    $= 1.5 \text{ mA} \times 39 \text{ k}\Omega$
    $= 58.5 \text{ V}$

(b) $I = \frac{V}{R}$
    $= \frac{58.5}{47} \times 10^3 = 1.24 \text{ mA}$

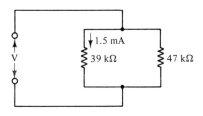

Figure 9-15

EXAMPLE 9-19:

In the circuit of Fig. 9-16, $I_T = 0.3$ A. Solve for R_2 using two methods.

Solution: One method is to solve for R_{eq} by Ohm's law; then use Eq. (9-9) to solve for R_2. Another method is to use Ohm's law to find the current in R_1. The difference between this current and the total current is the current in R_2. Once we know this current, we have an Ohm's law solution for R_2.

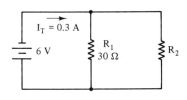

Figure 9-16

(a) $R_{eq} = \dfrac{E}{I_T} = \dfrac{6}{0.3} = 20\ \Omega$

$R_2 = \dfrac{R_1 R_{eq}}{R_1 - R_{eq}} = \dfrac{30 \times 20}{10} = 60\ \Omega$

The calculator sequence for this step is $R_1\ \times\ R_2\ \div\ (\ R_1\ -\ R_2\ )\ =$

(b) *Alternate Method:*

$$I_1 = \dfrac{E}{R_1} = \dfrac{6}{30} = 0.2\ \text{A}$$

$$I_2 = I_T - I_1 = 0.3 - 0.2 = 0.1\ \text{A}$$

$$R_2 = \dfrac{E}{I_2} = \dfrac{6}{0.1} = 60\ \Omega$$

In Example 9-20 we shall use two methods for finding an unknown resistance. In the first, we use Ohm's law to find the resistance, once its current has been calculated. In the second method, we reduce the circuit to a two-branch parallel circuit, and then use Eq. (9-9) to find the unknown resistance.

EXAMPLE 9-20:
Given the circuit of Fig. 9-17, solve for R_4. Make use of two solutions in solving for the resistor.

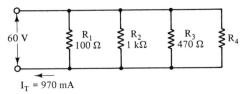

Figure 9-17

Solution:

(a) $R_4 = \dfrac{E}{I_4}$

$I_4 = I_T - (I_1 + I_2 + I_3)$

$I_1 = \dfrac{E}{R_1} = \dfrac{60}{100} = 0.6\ \text{A} = 600\ \text{mA}$

$I_2 = \dfrac{E}{R_2} = \dfrac{60}{1000} = 60\ \text{mA}$

$I_3 = \dfrac{E}{R_3} = \dfrac{60}{470} = 0.128\ \text{A} = 128\ \text{mA}$

$I_4 = 970 - (600 + 60 + 128)$
$ = 970\ \text{mA} - 788\ \text{mA}$
$ = 182\ \text{mA}$

$$R_4 = \frac{60}{0.182} = 330 \, \Omega$$

(b) Solve for R_{eq} of the entire circuit:

$$R_{eq} = \frac{E}{I_T} = \frac{60}{0.97} = 61.86 \, \Omega$$

Reduce $R_1 \parallel R_2 \parallel R_3$ to an equivalent resistance.

$$R_1 \parallel R_2 \parallel R_3 = \frac{1}{1/R_1 + 1/R_2 + 1/R_3}$$
$$= \frac{1}{1/100 + 1/1000 + 1/470}$$
$$= 76.18 \, \Omega$$

We now have a two-branch parallel circuit. One branch is 76.18 Ω, the other branch (R_4) is unknown, and $R_{eq} = 61.86 \, \Omega$. We find R_4 by use of Eq. (9-9).

$$R_4 = \frac{76.18 \times 61.86}{76.18 - 61.86} = 329 \simeq 330 \, \Omega$$

9-9. TROUBLESHOOTING

When connected to a proper voltage source, any circuit will draw some specific load current. This current is determined by the components of the circuit and the operating conditions.

There are only a few things that can go wrong with a branch of a parallel circuit. The branch may open, it may change in resistance value, it may be shorted, or it may not be connected into the circuit properly. This last problem will give symptoms similar to an open-circuited branch until voltage and resistance measurements are made.

A measurement of total current is usually one of the first steps in troubleshooting a parallel circuit. Let us consider the possible problems that variations from the normal value of total current might indicate.

1. *Total current lower than normal.* This can be caused by the following:
   a. Lower than normal source voltage. A voltage measurement will quickly check if the source voltage is the problem.
   b. An open or disconnected branch resistor. We can test the branch for voltage. If voltage is present across the resistor, it obviously is not disconnected. If poor wiring is not the problem, we must remove power and measure the branch resistor with an ohmmeter.
   c. A branch resistance that has increased in value. Remove power and check all branches with an ohmmeter.
2. *Total current is greater than normal.* This can be caused by the following:

a. Lower than normal branch resistance. Remove power and check the resistance of the branches with an ohmmeter.
b. Source voltage that is higher than normal. Clearly, a voltmeter check will get us the needed information.
c. Shorted branches need not be considered at this time. Any short circuits will cause excessive loading on the source. This will cause the circuit breaker or fuse to disconnect the circuit from the source. Also, a shorted branch will cause the voltage across a parallel circuit to be zero.

SUMMARY

1. There is more than one path for current in a parallel circuit.
2. The voltage across any branch of a parallel circuit is the same as the voltage across the other branches.
3. The total current of a parallel circuit is the sum of the branch currents.
4. The total resistance of a parallel circuit is less than the least branch resistance.
5. We refer to the total resistance of a parallel circuit as *equivalent* resistance (R_{eq}).
6. We use the product-over-sum formula for the equivalent resistance of a two-branched circuit.
7. When a parallel circuit consists of equal branch resistances, the equivalent resistance equals branch resistance divided by the number of branches.
8. It is often convenient to reduce a many-branched circuit, two branches at a time, in order to solve for R_{eq}.

PROBLEMS

9-1. Redraw the circuit of Fig. 9-18 so that it conforms to a standard sketch of a parallel circuit.

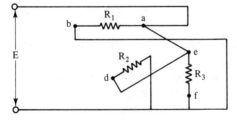

Figure 9-18

9-2. Redraw the circuit of Fig. 9-19 to conform to a standard sketch of a parallel circuit.

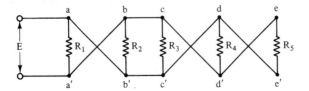

Figure 9-19

9-3. In a three-branch parallel circuit, there are 50 V across R_3. What are the voltages across R_1 and R_2?

9-4. In the circuit of Fig. 9-20, if the voltage across R_a is 5V, what is the voltage across R_b?

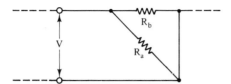

Figure 9-20

9-5. In a two-branch parallel circuit, the current in R_1 is 0.5 A. What is the voltage across R_2 given that $R_1 = 22\,\Omega$ and $R_2 = 47\,\Omega$?

9-6. What is the total current in the circuit of Fig. 9-21?

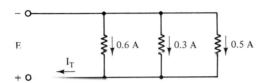

Figure 9-21

9-7. Solve for total current in the circuit of Fig. 9-22.

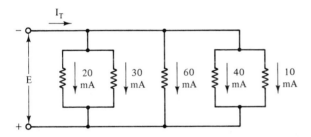

Figure 9-22

9-8. Solve for I_1 in the circuit of Fig. 9-23.

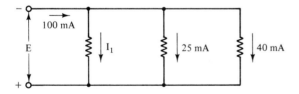

Figure 9-23

9-9. A two-branch parallel circuit has branch currents of 2.4 mA and 5.11 mA when connected to a 24-V source. What is the total resistance of the circuit?

9-10. A three-branch parallel circuit draws a total current of 7.25 A from a 100-V source. Given that $R_1 = 25\,\Omega$, $R_2 = 50\,\Omega$, and R_3 is not specified.
  (a) Solve for total resistance.
  (b) Solve for each branch current.
  (c) Solve for R_3.

9-11. Given the sketch of Fig. 9-24, solve for the currents in R_1 and in R_2.

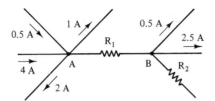

Figure 9-24

9-12. Write the Kirchhoff current equations for nodes A and B in Fig. 9-24.

9-13. Given a two-branch parallel circuit with $R_1 = 100\ \Omega$ and $R_2 = 400\ \Omega$, solve for (a) R_{eq} and (b) total conductance.

9-14. Solve for R_{eq} in each of the circuits of Fig. 9-25.

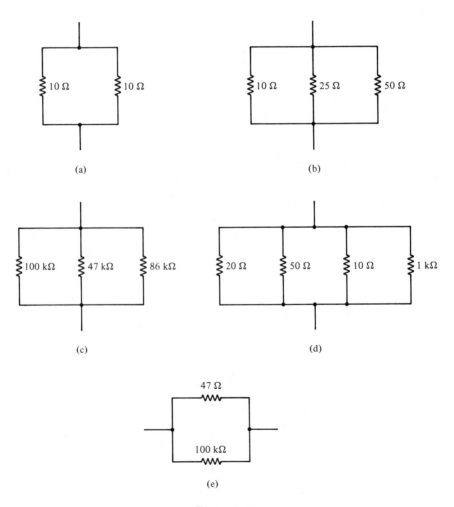

Figure 9-25

9-15. Solve for R_{eq} in each circuit of Fig. 9-26.

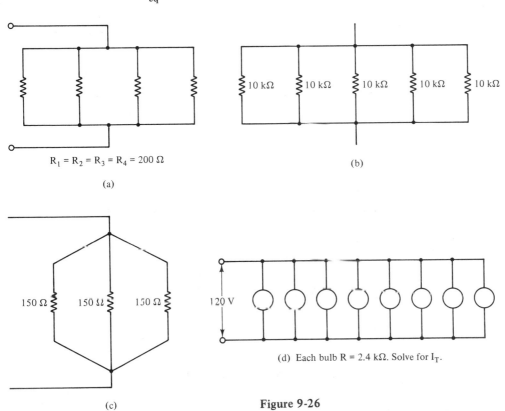

Figure 9-26

9-16. Solve for R_{eq} in each circuit of Fig. 9-27.

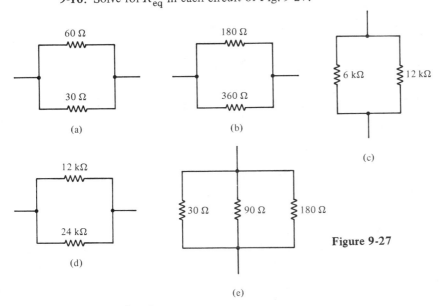

Figure 9-27

Chap. 9   Problems   147

9-17. Solve for the unknown resistance in each circuit of Fig. 9-28.

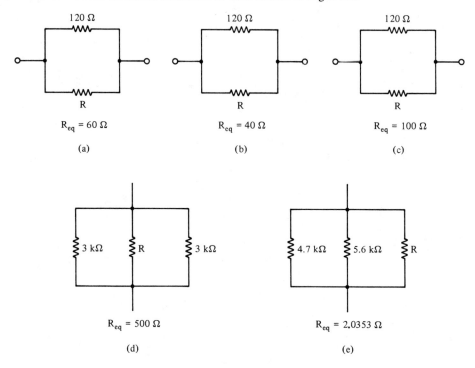

Figure 9-28

9-18. Solve for the total conductance of each circuit of Fig. 9-29.

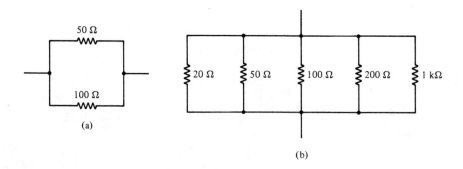

Figure 9-29

9-19. Solve for the total (equivalent) resistance of each network in Fig. 9-30.

9-20. Use the current-division formula for a two-branch parallel circuit and solve for each branch current in the circuits of Fig. 9-31. Verify your results by testing for equal branch voltages for $I_1 R_1$ and $I_2 R_2$.

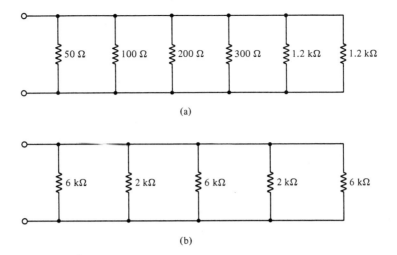

Figure 9-30

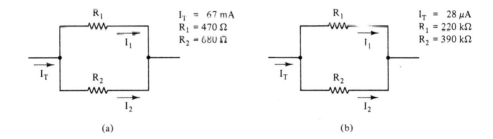

Figure 9-31

9-21. Given the circuit of Fig. 9-32:
  (a) Solve for total conductance in microsiemens.
  (b) Use the current-division rule and solve for each branch current, given that $I_T = 1.6$ mA.

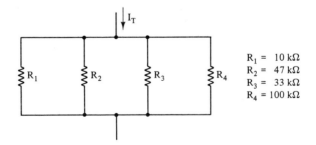

Figure 9-32

9-22. In each circuit of Fig. 9-33, solve for the unknown quantities.

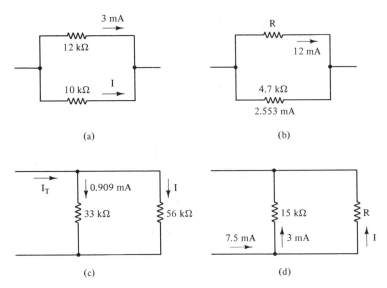

Figure 9-33

9-23. In the circuit of Fig. 9-34, the ammeter reads 13.37 mA. (a) What is the most likely cause of the malfunction? (b) Assuming normal operation, what should the ammeter read?

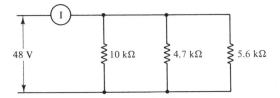

Figure 9-34

9-24. Suppose that the ammeter in the circuit of Fig. 9-34 reads 40 mA. Describe how you would use an ammeter to locate the circuit fault. Assume that source voltage is normal.

9-25. Reduce the following to equivalent resistances:
   (a) 56 Ω || 72 Ω || 100 Ω
   (b) 1 kΩ || 680 Ω || 2.2 kΩ || 6.8 kΩ
   (c) 2.2 MΩ || 1.5 MΩ || 470 kΩ

10

Series–Parallel Circuits

10-1. INTRODUCTION

We have studied series circuits and parallel circuits. We shall now begin the study of the *series–parallel* circuit, the circuit arrangement that is the most widely used and most common in electronics. In reality, all circuits are series–parallel circuits, although we may choose to ignore series or parallel elements. In this chapter we shall consider practical examples of this type of circuit; working with purely resistive circuits.

The series–parallel circuit is basically a series circuit, with some of the elements of the series circuit consisting of parallel circuits, as shown in Fig. 10-1. We find that there is a total current drawn from a voltage source. This current, in following series paths, flows through the series elements of the circuit. However, the current (I_T) divides in a parallel circuit, recombining into a single current wherever there is only a single path for the current.

EXAMPLE 10-1:
Given the circuit of Fig. 10-2, identify the currents and circuit relationships.

Solution:
Resistors R_1 and R_4 are in series with the source. Therefore, each carries the same current.

151

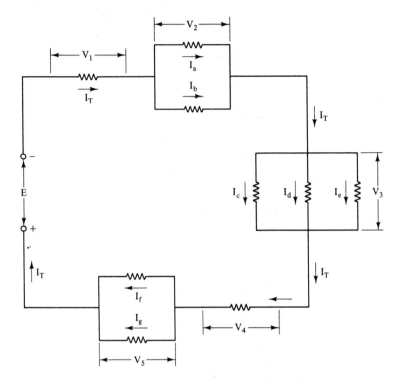

Figure 10-1 Voltage and current in a series-parallel circuit.

Resistors R_2 and R_3 are in parallel. The total current must divide. Using the current-division rule,

$$I_2 = I_T \frac{R_3}{R_2 + R_3}$$

$$I_3 = I_T \frac{R_2}{R_2 + R_3}$$

The sum of the branch currents is equal to the total current.

$$I_T = I_2 + I_3$$

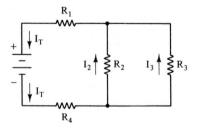

Figure 10-2

152   Chap. 10   Series-Parallel Circuits

We find that $R_2 \parallel R_3$ are in series with R_1 and R_4.
  We may follow the current path:
  (a) Current leaves the negative terminal of the source. This is total current and flows through R_4.
  (b) This total current arrives at the junction of R_4, R_2, and R_3. The current divides, becoming I_2 and I_3.
  (c) The two currents combine at the junction of R_1, R_2, and R_3, becoming total current.

We shall find that circuit solutions are not difficult, once we have learned how to relate the various parts of a circuit to each other. In actual practice, the circuits are not as easy to read as in Example 10-1.

EXAMPLE 10-2:
Determine the current paths and circuit relationships in the circuit of Fig. 10-3.

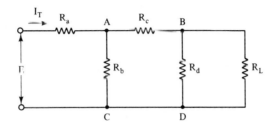

Figure 10-3

Solution:
  1. The most effective way to analyze the circuit is to begin with the load resistor (R_L) and work back toward the source.
     (a) Resistors R_d and R_L are in parallel.
     $$R_d \parallel R_L$$
     (b) This parallel combination is in series with R_c.
     $$R_c + (R_d \parallel R_L)$$
     (c) This series-parallel circuit is in parallel with R_b.
     $$R_b \parallel [R_c + (R_d \parallel R_L)]$$
     (d) The parallel system is in series with R_a.
     $$R_a + R_b \parallel [R_c + (R_d \parallel R_L)]$$
  2. We may now trace currents.
     (a) The total current flows through R_a.
     $$I_a = I_T$$
     (b) Total current divides at node A.
     $$I_T = I_b + I_c$$

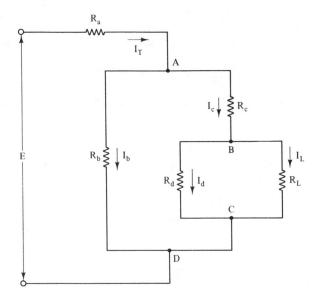

Figure 10-4 Redrawing a circuit is frequently a good way to "see" its operation. This is the circuit of Example 10-2.

(c) The current I_c divides at node B.

$$I_c = I_d + I_L$$

(d) The current at node C is the sum of I_d and I_L. This is the same as I_c and flows between nodes C and D.

(e) The currents I_b, I_d, and I_L combine at node D and become total current.

$$I_T = I_b + I_d + I_L$$

It is sometimes easier to see circuit action when a circuit is redrawn. For the sake of clarity, we identify the same nodes and redraw the circuit of Example 10-2. In this redrawn circuit (Fig. 10-4), we can easily see the series and parallel circuit relationships.

EXAMPLE 10-3:
Redraw the circuit of Fig. 10-5 so that it is more easily analyzed; then explain the current relationships.

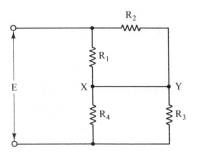

Figure 10-5

Solution:
1. (a) Terminals X and Y are shorted and really represent a single node, not two as drawn.
   (b) Resistors R_1 and R_2 are in parallel.
   (c) Resistors R_3 and R_4 are in parallel.
   (d) The two parallel circuits are in series (Fig. 10-6).

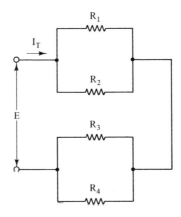

Figure 10-6

2. The current relationships are

$$I_T = I_1 + I_2 = I_3 + I_4$$

10-2. VOLTAGE AND CURRENT RELATIONSHIPS

We have learned that the sum of the voltage drops around a series circuit is equal to the source voltage. This is also true for the series–parallel circuit. However, we must recognize that the voltage drop across a parallel circuit is the same for each branch; therefore, we must consider a parallel circuit as a *single* voltage drop in taking the sum of the voltage drops.

We have seen that current divides into branch currents in parallel circuits. The sum of the branch currents is equal to the total current. It is the total current that flows through each section of a series–parallel circuit.

We shall use some examples to reinforce these ideas.

EXAMPLE 10-4:
Given the circuit of Fig. 10-7, what is the source voltage, E?

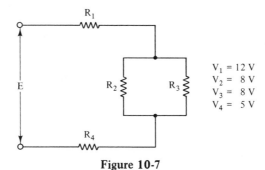

$V_1 = 12$ V
$V_2 = 8$ V
$V_3 = 8$ V
$V_4 = 5$ V

Figure 10-7

Solution:

$$E = V_1 + V_2 + V_4 = V_1 + V_3 + V_4$$
$$= 12 + 8 + 5 = 25 \text{ V}$$

EXAMPLE 10-5:
In the circuit of Fig. 10-8, solve for the voltage drop across the parallel circuit.

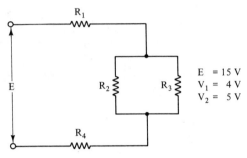

Figure 10-8

Solution:

$$V_2 = V_3 = E - (V_1 + V_4)$$
$$= 15 - (4 + 5) = 6 \text{ V}$$

EXAMPLE 10-6:
Given the circuit of Fig. 10-9, solve for I_2.

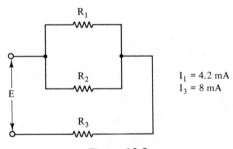

Figure 10-9

Solution:

$$I_3 = I_T = I_1 + I_2$$
$$I_2 = I_3 - I_1$$
$$= 8 - 4.2 = 3.8 \text{ mA}$$

EXAMPLE 10-7:
Given the circuit of Fig. 10-10, solve for (a) source voltage, (b) total current, and (c) the current in R_5.

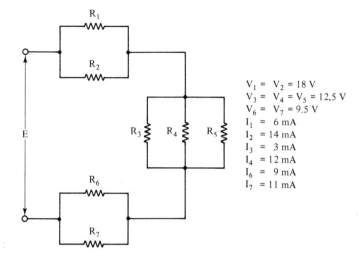

Figure 10-10

Solution:
(a) $E = V_1 + V_3 + V_6$
   $= 18 + 12.5 + 9.5$
   $= 40$ V
(b) $I_T = I_1 + I_2 = 6$ mA $+ 14$ mA $= 20$ mA
   $I_T = I_6 + I_7 = 9$ mA $+ 11$ mA $= 20$ mA
(c) $I_T = I_3 + I_4 + I_5$
   Solving for I_5,

$$I_5 = I_T - (I_3 + I_4)$$
$$= 20 \text{ mA} - (3 + 12) \text{ mA}$$
$$= 5 \text{ mA}$$

10-3. SERIES-EQUIVALENT CIRCUITS

The concept of equivalent circuits is very important in electric circuit analysis. This is especially true in electronics. We have learned about the equivalent resistance of a parallel circuit. Now we take the next step; we replace, for the purposes of analysis, all parallel circuits by their equivalent resistances. This results in a series circuit that may be solved for total resistance, total current, and voltage drops. Once we know the voltage drop across the equivalent resistance of a parallel circuit, we can solve for each branch current.

The following examples show how to solve for total resistance of a series–parallel circuit. Study each carefully, as we proceed from elementary to complex circuits.

EXAMPLE 10-8:
Solve for the total resistance of the circuit of Fig. 10-11.

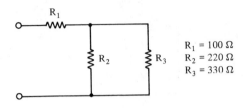

Figure 10-11

Solution:
1. Solve for the equivalent resistance of the parallel circuit.

$$R_{eq} = \frac{220 \times 330}{220 + 330} = 132 \text{ }\Omega$$

2. Redraw the circuit (Fig. 10-12).

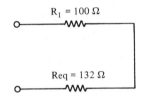

Figure 10-12

3. Solve for total resistance.

$$R_T = R_1 + R_{eq}$$
$$= 100 + 132 = 232 \text{ }\Omega$$

EXAMPLE 10-9:
Solve for total resistance in the circuit of Fig. 10-13.

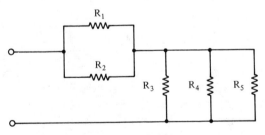

Figure 10-13

Solution:
1. Solve for the equivalent resistance of each parallel circuit.
   (a) $R_{eq1} = R_1 \parallel R_2$
   $$= \frac{R_1 R_2}{R_1 + R_2}$$
   $$= \frac{180 \times 270}{180 + 270}$$
   $$= 108 \text{ }\Omega$$

(b) $R_{eq2} = R_3 \parallel R_4 \parallel R_5$

$$= \frac{1}{1/1000 + 1/1500 + 1/2700}$$

$$= 491 \, \Omega$$

2. The total resistance is the sum of the equivalent resistances.

$$R_T = R_{eq1} + R_{eq2}$$
$$= 108 + 491$$
$$= 599 \, \Omega$$

The next example deals with a much more complex series–parallel circuit. In this instance we have circuits within circuits. In this type of circuit, we solve the circuit by beginning with the smallest portion of the circuit. This is that portion "farthest" from the voltage source. At each step, a parallel circuit is reduced to an equivalent resistance. In most of the steps, the equivalent resistance is added to a series resistor, which then forms another branch of a parallel circuit. The process is repeated until we solve the entire circuit. Follow Example 10-10 very carefully; you have much to gain in understanding circuits. The resistors have been selected so as to keep arithmetic to a minimum; your interest must be in the *analysis* of the circuit.

EXAMPLE 10-10:
Solve for the total resistance of the circuit of Fig. 10-14.

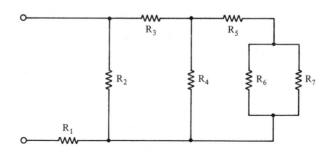

Figure 10-14

Solution:
1. The equivalent resistance of $R_6 \parallel R_7$ is found.

$$100 \parallel 100 = 50 \, \Omega$$

2. This equivalent resistance is in series with R_5.

$$R_5 + R_6 \parallel R_7 = 50 + 50 = 100 \, \Omega$$

3. This series combination is in parallel with R_4.

$$R_4 \parallel (R_5 + R_6 \parallel R_7) = 100 \parallel 100 = 50 \, \Omega$$

4. This resulting resistance is in series with R_3.

$$R_3 + R_4 \parallel (R_5 + R_6 \parallel R_7) = 50 + 50 = 100 \, \Omega$$

5. The resistance of R_2 is in parallel with this combination of resistors.

$$R_2 \| [R_3 + R_4 \| (R_5 + R_6 \| R_7)] = 100 \| 100 = 50 \ \Omega$$

6. We find total resistance by adding R_1.

$$\begin{aligned} R_T &= R_1 + R_2 \| [R_3 + R_4 \| (R_5 + R_6 \| R_7)] \\ &= 50 + 50 \\ &= 100 \ \Omega \end{aligned}$$

10-4. CIRCUIT ANALYSIS

The analysis of series–parallel circuits simply requires the use of the laws and rules that we have applied to series and parallel circuits. Usually, we reduce the series–parallel circuit to a series circuit by replacing the parallel circuits with equivalent resistances. Once we have a series equivalent circuit for a series–parallel circuit, solutions for current and voltage are easily made.

The procedure that we shall use depends upon the type of problem to be solved. Follow the examples very carefully. The method of solution in each example depends upon the requirements of the problem.

EXAMPLE 10-11:

Solve for the voltage drop across each resistor in the circuit of Fig. 10-15 (a) by Ohm's law: (b) by the voltage-division rule (VDR).

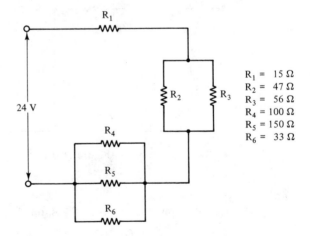

Figure 10-15

Solution:

1. Solve for the equivalent resistance of each parallel circuit.

$$\begin{aligned} R_{eq1} &= R_2 \| R_3 \\ &= \frac{47 \times 56}{47 + 56} \\ &= 25.6 \ \Omega \end{aligned}$$

$$R_{eq2} = R_4 \| R_5 \| R_6$$
$$= \frac{1}{1/100 + 1/150 + 1/33}$$
$$= 21.3 \,\Omega$$

2. Ohm's law solution:
   (a) Solve for total resistance.
   $$R_T = R_1 + R_{eq1} + R_{eq2}$$
   $$= 15 + 25.6 + 21.3$$
   $$= 61.9 \,\Omega$$

   (b) Solve for total current.
   $$I_T = \frac{E}{R_T} = \frac{24}{61.9} = 0.388 \text{ A}$$

   (c) Solve for the voltage drops.
   $$V_1 = I_T R_1 = 0.388 \times 15 = 5.82 \text{ V}$$
   $$V_2 = V_3 = I_T R_{eq1} = 0.388 \times 25.6$$
   $$= 9.92 \text{ V}$$
   $$V_4 = V_5 = V_6 = I_T R_{eq2}$$
   $$= 0.388 \times 21.3 = 8.26 \text{ V}$$

   The sum of the voltage drops should equal the applied voltage.
   $$E = V_1 + V_2 + V_4$$
   $$= 5.82 + 9.92 + 8.26 = 24 \text{ V}$$

3. Solution by the use of VDR:
   (a) Solve for total resistance.
   $$R_T = 61.9 \,\Omega$$

   (b) Solve for each drop by use of the formula
   $$V = E\frac{R}{R_T}$$
   $$V_1 = E\frac{R_1}{R_T} = 24 \times \frac{15}{61.9}$$
   $$= 5.82 \text{ V}$$
   $$V_2 = V_3 = E\frac{R_{eq1}}{R_T}$$
   $$= 24 \times \frac{25.6}{61.9}$$
   $$= 9.92 \text{ V}$$
   $$V_4 = V_5 = V_6 = E\frac{R_{eq2}}{R_T}$$
   $$= 24 \times \frac{21.3}{61.9}$$
   $$= 8.26 \text{ V}$$

EXAMPLE 10-12:
Demonstrate that the sum of the branch currents is equal to the total current in the circuit of Example 10-11.

Solution:
1. The total current is known.
$$I_T = 0.388 \text{ A}$$

2. $I_2 = \dfrac{V_2}{R_2} = \dfrac{9.92}{47} = 0.211 \text{ A}$

   $I_3 = \dfrac{V_2}{R_3} = \dfrac{9.92}{56} = 0.177 \text{ A}$

   $I_T = I_2 + I_3 = 0.211 + 0.177 = 0.388 \text{ A}$

3. $I_4 = \dfrac{V_4}{R_4} = \dfrac{8.26}{100} = 0.0826 \text{ A}$

   $I_5 = \dfrac{V_4}{R_5} = \dfrac{8.26}{150} = 0.0551 \text{ A}$

   $I_6 = \dfrac{V_6}{R_6} = \dfrac{8.26}{33} = 0.250 \text{ A}$

   $I_T = I_4 + I_5 + I_6$
   $= 0.0826 + 0.0551 + 0.250$
   $= 0.3877 \text{ A} = 0.388 \text{ A}$

In the next example, we are given the voltage across a branch of a parallel circuit. We are required to solve for source voltage. We begin by finding the currents in the other branches of the parallel circuit. Remember that the voltage across each branch of a parallel circuit is the same. The sum of the branch currents is equal to the total current.

We then find the equivalent resistance of the other parallel circuit. We can calculate the voltage drop across the parallel circuit and also the remaining series resistor. The sum of these drops is the applied voltage.

Another way to find the applied voltage, once total current is known, is to find the total resistance of the series–parallel circuit. The product of total current and total resistance is the source voltage. We shall work the example both ways so that you can increase your experience with circuit analysis.

EXAMPLE 10-13:
In the circuit of Fig. 10-16, $V_4 = 28.2$ V. Solve for the source voltage.

Solution:
1. Solve for total current. We have that $V_4 = 28.2$ V. This voltage is also across R_3 and R_5 ($V_3 = V_4 = V_5$). We find total current by solving for the current in each resistor. The sum of these currents is total current.

$$I_T = I_3 + I_4 + I_5$$

$$I_3 = \dfrac{28.2}{5.6 \text{ k}\Omega} = 5.04 \text{ mA}$$

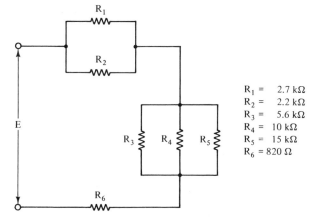

Figure 10-16

$$I_4 = \frac{28.2}{10 \text{ k}\Omega} = 2.82 \text{ mA}$$

$$I_5 = \frac{28.2}{15 \text{ k}\Omega} = 1.88 \text{ mA}$$

$$I_T = 5.04 + 2.82 + 1.88$$
$$= 9.74 \text{ mA}$$

We may also find total current by first finding the equivalent resistance of $R_3 \parallel R_4 \parallel R_5$. Total current is the ratio of V_4/R_{eq}.

(a) $R_{eq} = \dfrac{1}{1/R_3 + 1/R_4 + 1/R_5}$

$= \dfrac{1}{\dfrac{1}{5.6 \times 10^3} + \dfrac{1}{10 \times 10^3} + \dfrac{1}{15 \times 10^3}}$

$= 2.896 \text{ k}\Omega$

(b) $I_T = \dfrac{V_4}{R_{eq}}$

$= \dfrac{28.2}{2.896 \times 10^3}$

$= 9.74 \text{ mA}$

2. Solve for the voltage drop across $R_1 \parallel R_2$.

$$V_1 = V_2 = I_T(R_1 \parallel R_2)$$
$$= I_T \frac{R_1 \times R_2}{R_1 + R_2}$$
$$= 9.74 \times 10^{-3} \frac{2.7 \times 10^3 \times 2.2 \times 10^3}{2.7 \times 10^3 + 2.2 \times 10^3}$$
$$= 9.74 \times 10^{-3} \times 1.21 \times 10^3$$
$$= 11.8 \text{ V}$$

3. Solve for the voltage drop across R_6.

$$V_6 = I_T R_6$$
$$= 9.74 \times 10^{-3} \times 0.82 \times 10^3$$
$$= 7.99 \text{ V}$$

4. The supply voltage is found by taking the sum of the voltage drops.

$$E = V_1 + V_4 + V_6$$
$$= 11.8 + 28.2 + 7.99$$
$$= 47.99 = 48 \text{ V}$$

We should note that we could have found the supply voltage by first finding total current, then total resistance. The product of these quantities is the supply voltage, E.

$$E = I_T R_T$$

$$I_T = 9.74 \text{ mA}$$

$$R_T = R_6 + R_{eq1} + R_{eq2}$$
$$= 820 + 1210 + 2896$$
$$= 4926 \text{ }\Omega$$

$$E = 9.74 \times 10^{-3} \times 4.926 \times 10^3$$
$$= 47.97 = 48 \text{ V}$$

EXAMPLE 10-14:
Given the circuit of Fig. 10-17, solve for each voltage drop. The measured current, I_7, equals 0.2 A.

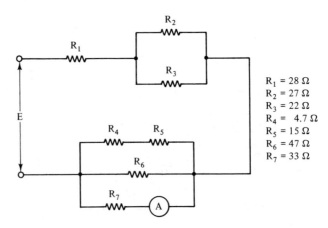

Figure 10-17

Solution:
1. Given the current in R_7, we find the voltage across *that* parallel circuit:

$$V_7 = V_6 = V_4 + V_5$$

$$V_7 = I_7 R_7$$
$$= 0.2 \times 33 = 6.6 \text{ V}$$

2. Solve for the remaining branch currents of the three-branch parallel circuit.

$$I_6 = \frac{V_6}{R_6} = \frac{6.6}{47} = 0.14 \text{ A}$$

$$I_{4-5} = \frac{V_4 + V_5}{R_4 + R_5} = \frac{6.6}{4.7 + 15} = 0.335 \text{ A}$$

3. Solve for total current.

$$\begin{aligned} I_T &= I_{4-5} + I_6 + I_7 \\ &= 0.335 + 0.14 + 0.2 \\ &= 0.675 \text{ A} \end{aligned}$$

4. Voltage drops across R_4 and R_5 are found.
   (a) By Ohm's law:

$$V_4 = I_{4-5} R_4 = 0.335 \times 4.7 = 1.57 \text{ V}$$

$$V_5 = I_{4-5} R_5 = 0.335 \times 15 = 5.03 \text{ V}$$

   (b) By use of the VDR:

$$V_4 = 6.6 \times \frac{4.7}{19.7} = 1.57 \text{ V}$$

$$V_5 = 6.6 \times \frac{15}{19.7} = 5.03 \text{ V}$$

5. Solve for the equivalent resistance of $R_2 \| R_3$.

$$R_{eq} = \frac{R_2 R_3}{R_2 + R_3} = \frac{27 \times 22}{27 + 22} = 12.12 \text{ }\Omega$$

6. The voltage drop across each branch of a parallel circuit is the same.

$$\begin{aligned} V_2 = V_3 &= I_T R_{eq} \\ &= 0.675 \times 12.12 = 8.18 \text{ V} \end{aligned}$$

7. The voltage drop across R_1 is then calculated.

$$V_1 = I_T R_1 = 0.675 \times 18 = 12.15 \text{ V}$$

One way to check the accuracy of our work is to look for other combinations of currents or voltages that may be checked against our calculations. In this case we can check the currents through R_2 and R_3. If the example was correctly solved, the sum of these currents must be equal to the total current.

$$I_T = I_2 + I_3$$

$$I_2 = \frac{V_2}{R_2} = \frac{8.18}{27} = 0.303 \text{ A}$$

$$I_3 = \frac{V_3}{R_3} = \frac{8.18}{22} = 0.372 \text{ A}$$

$$I_T = 0.303 + 0.372 = 0.675 \text{ A} \quad \text{(checks)}$$

10-5. VOLTAGE AND CURRENT MEASUREMENTS

Current measurements are made for branch current and for total current. In *all* cases, *the meter must be connected in series* at the desired point in a circuit.

When voltages are measured, we must consider polarities and reference points. Voltage is a difference of potential and exists *between* points in a circuit, *never at one point only*. We would measure zero volts if both leads of a voltmeter were connected to the same point in a circuit. So far in our studies we have been concerned with voltage drops; we now consider voltages with respect to a circuit reference point.

Suppose that we have a series–parallel circuit with the negative side of the voltage source as the reference. We connect the negative lead of the voltmeter to the reference point. If we connect the positive lead of the meter to the positive terminal of the source, we read the supply voltage. Moving around the circuit from this terminal, we find that the readings decrease with each voltage drop. Example 10-15 illustrates this situation.

EXAMPLE 10-15:
Given the circuit of Fig. 10-18 and the specified voltage drops, what is the voltage at each point in the circuit with respect to the negative terminal of the supply?

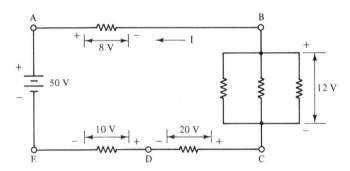

Figure 10-18

Solution: The direction of electron flow determines the polarity of the voltage drops. Where electrons enter a resistor, the assigned polarity is negative. Where the electrons leave the resistor, the polarity is positive. Placing these signs on the diagram so that we know the polarity of the voltage drops makes it easy to calculate voltages with respect to the reference.

$V_A = 50$ V We are simply measuring supply voltage.

$V_B = 42$ V We subtract 8 V from supply voltage.

$V_C = 30$ V In this case we subtract the 8-V and 12-V drops.

$V_D = 10$ V This is $50 - (8 + 12 + 12)$.

$V_E = 0$ Voltage cannot exist at a point.

Note that we could have started from the reference terminal and proceeded to make measurements around the circuit from point E.

$$V_E = 0$$
$$V_D = 10 \text{ V}$$
$$V_C = 30 \text{ V} \quad (10 + 20)$$
$$V_B = 42 \text{ V} \quad (10 + 20 + 12)$$
$$V_A = 50 \text{ V}$$

Suppose that the reference point is changed in Example 10-15. This does not change the circuit, only the point to which we refer for differences of potential. We would have some very different results. In this instance we shall use point C as the reference.

$$V_A = 20 \text{ V} \quad (8 + 12)$$
$$V_B = 12 \text{ V}$$
$$V_C = 0 \qquad \text{No voltage at a point.}$$
$$V_D = -20 \text{ V}$$
$$V_E = -30 \text{ V} \quad (-10 + -20)$$

Many electronic circuits require a *common* reference point. Thus we find a common line or area on a printed circuit board is used to provide this reference point. It is usual practice to refer to this common point as *ground*. The name comes from the safety practice in electrical wiring of using an earth ground for one side of the ac power line in order to prevent a voltage from existing between that side of the power line and ground. This grounding is done by connecting the line to a water pipe or to a metal rod that is driven into the earth. To eliminate shock hazard, typical practice is to cause the circuit ground to be connected to the earth ground if the electronic equipment is operated from the ac power lines. If the electronic equipment is battery operated, its ground is not related to earth ground. Automobile wiring is an example. The auto chassis is ground for all its various circuits (lighting, radio, tape player, etc.); however, this ground has no relationship to earth ground.

The most widely used symbol for ground is ⏚ . A chassis or earth ground may be shown as ⏚ . This is the usual practice if earth ground is not the same as circuit ground, and both connections are used in the system.

In Fig. 10-19 we make voltage measurements with respect to ground. Be sure that you understand the voltage readings given before you continue on in this chapter.

Sec. 10-5.   Voltage and Current Measurements

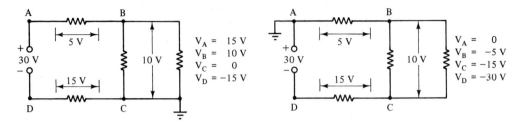

Figure 10-19 Voltage measurements in a series-parallel circuit. All measurements made at the indicated points with respect to the ground.

10-6. VOLTAGE DIVIDERS

Now that we have an understanding of series–parallel circuits, we can turn to a practical example. Many electronic circuits require different amounts of operating voltage for different loads. When the currents drawn by these loads are relatively small, on the order of a few milliamperes or less, it is convenient to use a resistive network called a *voltage divider* to develop the various operating voltages.

An example of such an application may be found in an amplifier. We can have one section, the power amplifier, requiring a high voltage and current. The preamplifier stage operates at a lower voltage and current. For example, we may need 24 V at 200 mA for the power amplifier, while only 8 V at 2 mA is needed for the preamplifier. The circuit problem is sketched in Fig. 10-20. R_{L_1} represents the load resistance equal to the preamplifier requirements, which are 8 V at 2 mA and a load resistance of 4 kΩ.

Note that there is a current in R_1 and R_2 whether or not R_{L_1} is connected. This current is known as the *bleeder current*. The current in R_2 is the sum of the bleeder current and the preamplifier load current.

Note also that R_{L_2} is connected between the 24-V source and ground and is independent of the voltage divider. The values of R_1 and R_2 for a given load voltage and current depend upon the *selected* value of bleeder current. If it is likely that load current will vary, but we wish to hold the

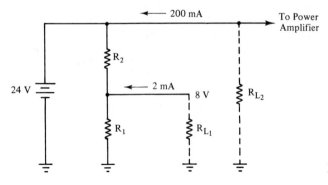

Figure 10-20

168   Chap. 10   Series-Parallel Circuits

voltage fairly constant at or near the required value at the tap, the bleeder current must be chosen as a value that is at *least 10 times load current*.

If we select a small value of bleeder current, we will find that the divider voltages will vary widely with changes in load current. The following examples illustrate these concepts, using a three-voltage system.

EXAMPLE 10-16:
A 36-V source must supply 24 V at 10 mA to one load and 8 V at 5 mA to another load. (a) Sketch the circuit. (b) Calculate the currents through the voltage divider when the bleeder current is 150 mA. (c) Calculate the effective load resistances.

Solution:
(a) Voltage-divider circuit and loads (Fig. 10-21).

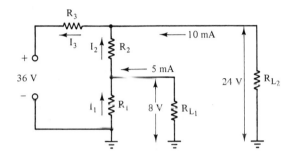

Figure 10-21

(b) I_1 = bleeder current = 150 mA
$I_2 = I_1 + I_{L1}$ = 150 mA + 5 mA = 155 mA
$I_3 = I_2 + I_{L2}$ = 155 mA + 10 mA = 165 mA

(c) The effective load resistances are found by Ohm's law:

$$R_{L1} = \frac{V_{L1}}{I_{L1}} = \frac{8}{5 \times 10^{-3}} = 1.6 \text{ k}\Omega$$

$$R_{L2} = \frac{V_{L2}}{I_{L2}} = \frac{8}{10 \times 10^{-3}} = 2.4 \text{ k}\Omega$$

EXAMPLE 10-17:
Given the circuit of Example 10-16, (a) solve for the voltage-divider resistors; (b) use the equivalent circuit to demonstrate that the voltages across R_{L1} and R_{L2} are correct.

Solution:
(a) $R_1 = \dfrac{V_1}{I_1} = \dfrac{8}{0.15} = 53.33 \ \Omega$

$R_2 = \dfrac{V_2}{I_2} = \dfrac{24 - 8}{0.155} = 103.23 \ \Omega$

$R_3 = \dfrac{V_3}{I_3} = \dfrac{36 - 24}{0.165} = 72.73 \ \Omega$

(b)
1. $R_{eq1} = R_{L1} \| R_1$

$$= \frac{53.33 \times 1600}{53.33 + 1600} = 51.61 \, \Omega$$

$R_{eq2} = (R_{eq1} + R_2) \| R_{L2}$

$$= \frac{(51.61 + 103.23) \times 2400}{51.61 + 103.23 + 2400}$$

$$= 145.46 \, \Omega$$

2. Draw the equivalent circuit and solve for V_{L2}. (Fig. 10-22).

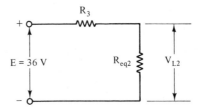

Figure 10-22

$$V_{L2} = E \frac{R_{eq2}}{R_{eq2} + R_3}$$

$$= 36 \times \frac{145.46}{145.46 + 72.73}$$

$$= 24 \, V$$

3. In the equivalent circuit, V_{L1} is derived from V_{L2}.

$$V_{L1} = V_{L2} \frac{R_{eq1}}{R_{eq1} + R_2}$$

$$= 24 \times \frac{51.61}{51.61 + 103.23}$$

$$= 7.999+ = 8 \, V$$

EXAMPLE 10-18:

Assume that the load resistances in Example 10-16 change. Let $R_{L1} = 2 \, k\Omega$ and $R_{L2} = 4 \, k\Omega$. Use the voltage-divider circuit of Example 10-17 and calculate the load voltages.

Solution:
1. Calculate the equivalent resistances.

$$R_{eq1} = \frac{53.33 \times 2000}{53.33 + 2000} = 51.94 \, \Omega$$

$$R_{eq2} = \frac{(51.94 + 103.23) \times 4000}{51.94 + 103.23 + 4000}$$

$$= 149.38 \, \Omega$$

2. Use the VDR and calculate V_{L2}.

$$V_{L2} = E \frac{R_{eq2}}{R_{eq2} + R_3}$$
$$= 36 \times \frac{149.38}{149.38 + 72.73}$$
$$= 24.2 \text{ V}$$

This represents a voltage change of only 0.83%.

3. Calculate V_{L1}.

$$V_{L1} = V_{L2} \frac{R_{eq1}}{R_{eq1} + R_2}$$
$$= 24.2 \times \frac{51.94}{51.94 + 103.23}$$
$$= 8.1 \text{ V}$$

This represents a voltage change of only 1.25%.

Note that the output voltages remained constant (for all practical purposes) despite a very large change in load resistance.

The following examples will demonstrate the effect on output voltage stability of a greatly reduced bleeder current.

EXAMPLE 10-19:

Given the same load currents and voltages as in the previous examples, calculate the voltage-divider currents if the bleeder current is 1.5 mA. Source voltage remains at 36 V.

Solution:

$$I_1 = \text{bleeder current} = 1.5 \text{ mA}$$
$$I_2 = I_1 + I_{L1} = 1.5 \text{ mA} + 5 \text{ mA}$$
$$= 6.5 \text{ mA}$$
$$I_3 = I_2 + I_{L2} = 6.5 \text{ mA} + 10 \text{ mA}$$
$$= 16.5 \text{ mA}$$

EXAMPLE 10-20:

Given the load voltages and currents of Example 10-19, (a) calculate the voltage-divider resistors; (b) calculate the series–parallel equivalent circuit; (c) sketch the equivalent circuit (Fig. 10-23).

Solution:

(a) $R_1 = \dfrac{V_1}{I_1} = \dfrac{8}{1.5 \text{ mA}} = 5.33 \text{ k}\Omega$

$R_2 = \dfrac{V_2}{I_2} = \dfrac{24 - 8}{6.5 \text{ mA}} = 2.46 \text{ k}\Omega$

$R_3 = \dfrac{V_3}{I_3} = \dfrac{36 - 24}{16.5 \text{ mA}} = 727.3 \text{ }\Omega$

(b) $R_{eq1} = R_1 \| R_{L1}$
$= \dfrac{5.333 \text{ k}\Omega \times 1.6 \text{ k}\Omega}{5.333 \text{ k}\Omega + 1.6 \text{ k}\Omega}$
$= 1.23 \text{ k}\Omega$

$R_{eq2} = (R_{eq1} + R_2) \| R_{L2}$
$= \dfrac{(1.23 \text{ k}\Omega + 2.46 \text{ k}\Omega) \times 2.4 \text{ k}\Omega}{1.23 \text{ k}\Omega + 2.46 \text{ k}\Omega + 2.4 \text{ k}\Omega}$
$= 1.45 \text{ k}\Omega$

(c)

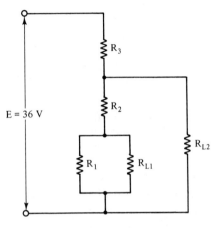

Figure 10-23

EXAMPLE 10-21:

Use the voltage-division rule to calculate V_{L1} and V_{L2} of the preceding examples.

Solution:

1. $V_{L2} = E \dfrac{R_{eq2}}{R_{eq2} + R_3}$
$= 36 \times \dfrac{1450}{1450 + 727.3}$
$= 24 \text{ V}$

2. V_{L1} is derived from V_{L2}.

$V_{L1} = V_{L2} \dfrac{R_{eq1}}{R_{eq1} + R_2}$
$= 24 \times \dfrac{1230}{1230 + 2460}$
$= 8 \text{ V}$

We shall now determine the output voltages when the load resistances change.

EXAMPLE 10-22:
Assume that the load resistances change as in Example 10-18. Let $R_{L1} = 2$ kΩ and $R_{L2} = 4$ kΩ. Use the voltage-divider circuit of Example 10-20 and calculate the load voltages.

Solution:
1. Calculate the equivalent resistances.

$$R_{eq1} = 5.333 \text{ k}\Omega \parallel 2 \text{ k}\Omega$$
$$= 1.45 \text{ k}\Omega$$

$$R_{eq2} = (R_{eq1} + R_2) \parallel R_{L2}$$
$$= (1.45 \text{ k}\Omega + 2.46 \text{ k}\Omega) \parallel 4 \text{ k}\Omega$$
$$= 1.98 \text{ k}\Omega$$

2. $V_{L2} = E \dfrac{R_{eq2}}{R_{eq2} + R_3}$
$$= 36 \times \dfrac{1980}{1980 + 727.3}$$
$$= 26.3 \text{ V}$$

The voltage changed by nearly 10%.

3. $V_{L1} = V_{L2} \dfrac{R_{eq1}}{R_{eq1} + R_2}$
$$= 26.3 \times \dfrac{1450}{1450 + 2460}$$
$$= 9.75 \text{ V}$$

This is a voltage change of 22%!

It should be clear after all these examples that a reasonably constant voltage is obtained from the voltage divider only when (1) there is a bleeder current and (2) the bleeder current is much larger than the sum of the load currents. This is practical only with small load currents. For larger load currents, we must use electronic regulating circuits and separate supplies.

SUMMARY

1. A series–parallel circuit may be analyzed as a series circuit by replacing each parallel circuit by its equivalent resistance. The resulting circuit is the series equivalent of a series–parallel circuit.
2. The voltage drop across each parallel circuit is found by using total current and equivalent resistance.
3. Problems in series–parallel circuits are solved following the rules for series and parallel circuits.
4. The voltage-division rule (VDR) is used to find the voltage drop across a part of a series circuit. $V_x = E(R_x/R_T)$.

5. Ground is the term used for the common reference point in a circuit. Normally, one side of the source is connected to ground.
6. A voltage divider is a series–parallel circuit. Currents consist of load currents plus a bleeder current.
7. The greater the amount of bleeder current, the more stable are the voltages in the divider, despite changes in load current. Bleeder current should be much greater than any load currents in any of the divider resistors.

PROBLEMS

10-1. Redraw the circuit of Fig. 10-24 so that it appears as a series system of parallel circuits.

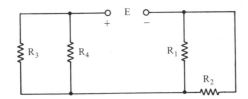

Figure 10-24

10-2. Redraw the network of Fig. 10-25 so that the series–parallel relationships are clearer.

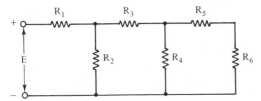

Figure 10-25

10-3. Redraw the circuit of Fig. 10-26 so that the series–parallel relationships are clearer.

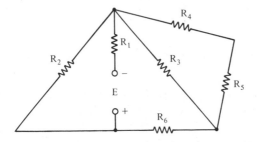

Figure 10-26

10-4. Write the equations for total current in terms of branch currents for each of the circuits of Fig. 10-27.
10-5. In the circuit of Fig. 10-28, $I_1 = 670$ mA and $I_2 = 320$ mA. What is I_3?
10-6. In the circuit of Fig. 10-29, $I_4 = 200$ mA. What is I_1?
10-7. Given the data of Prob. 10-6, what is the sum of I_2 and I_3? the sum of I_5, I_6, and I_7?
10-8. In the circuit of Fig. 10-29, what is V_4?

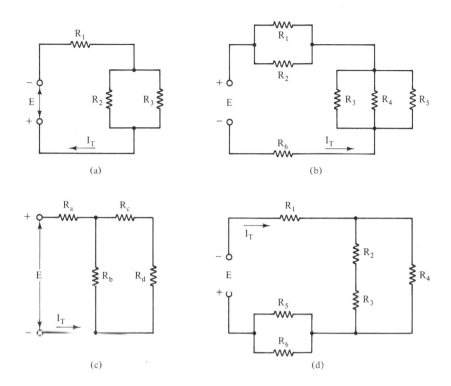

Figure 10-27

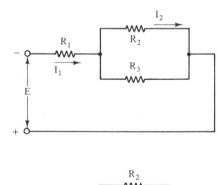

Figure 10-28

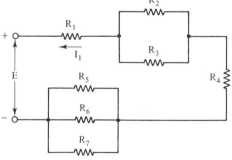

Figure 10-29

10-9. Given some of the voltage drops in the circuit of Fig. 10-30, what is the applied voltage?

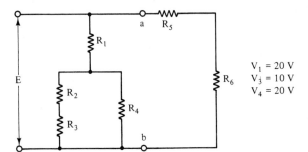

Figure 10-30

10-10. Given the data of Prob. 10-9, what is the voltage drop across R_2?

10-11. What is the voltage between points a and b, given the data of Prob. 10-9?

10-12. Given the circuit of Fig. 10-31, what is the series equivalent resistance of parallel circuit B-B'?

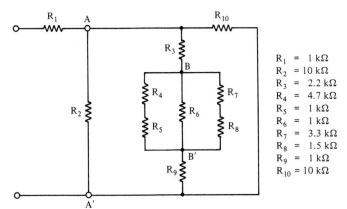

Figure 10-31

10-13. Solve the circuit of Fig. 10-31 for the series equivalent resistance of parallel circuit A-A'.

10-14. Solve for the total resistance of the circuit of Fig. 10-32.

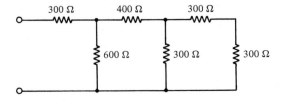

Figure 10-32

10-15. Given the equation for R_T, sketch the circuit.
    (a) $R_1 \parallel R_2 + R_3 = R_T$
    (b) $R_1 \parallel R_2 \parallel R_3 + (R_4 + R_5) \parallel R_6 = R_T$
    (c) $R_1 + R_2 + R_3 \parallel R_4 \parallel (R_5 + R_6) = R_T$

10-16. Solve for voltage drops and applied voltage in the circuit of Fig. 10-33, given that $I_2 = 1.5$ A.

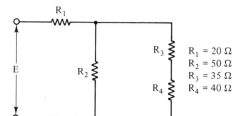

Figure 10-33

10-17. Suppose that 300 V is applied to the circuit of Fig. 10-34. Solve for the new value of I_2.

10-18. Using the data of Prob. 10-16, what is the voltage drop across R_4?

10-19. In the circuit of Fig. 10-34, solve for R_2.

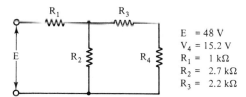

Figure 10-34

10-20. The amplifier of Fig. 10-35 is operated from a single 24-V source. Solve for R_1, R_2, R_L, and R_S, given the operating data.

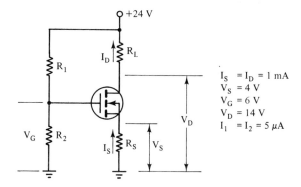

Figure 10-35

10-21. The switching circuit of Fig. 10-36 operates from two voltage sources. When the input is low, $V_B = -2$ V, $I_C = 0$, and $I_1 = I_2 = 0.24$ mA. When the input is high, $I_C = 10$ mA and $V_C \simeq 0$. Solve for R_1, R_2, and R_L.

Chap. 10   Problems

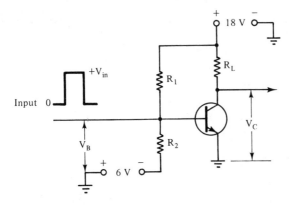

Figure 10-36

10-22. In the circuit of Fig. 10-37, $E = 36$ V. Solve for the voltage with respect to ground at points A and B.

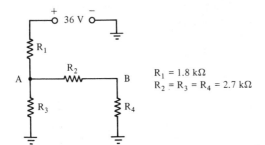

$R_1 = 1.8$ kΩ
$R_2 = R_3 = R_4 = 2.7$ kΩ

Figure 10-37

10-23. In the circuit of Fig. 10-38, solve for the indicated voltages with respect to ground.

For the circuit of Figure 10-39, solve all the following unrelated statements.

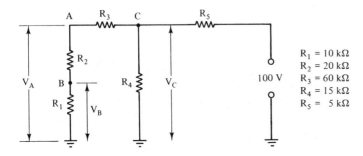

$R_1 = 10$ kΩ
$R_2 = 20$ kΩ
$R_3 = 60$ kΩ
$R_4 = 15$ kΩ
$R_5 = 5$ kΩ

Figure 10-38

178   Chap. 10   Series-Parallel Circuits

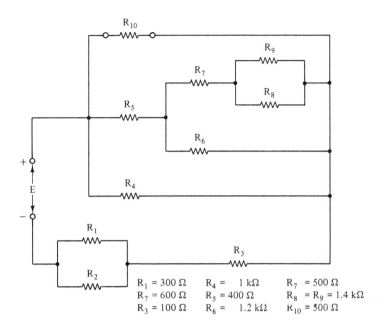

Figure 10-39

10-24. If $I_9 = 2$ mA, solve for E.
10-25. Solve for V_{10} if $V_2 = 48$ V.
10-26. If the supply voltage is 400 V, solve for V_5.
10-27. Solve for I_9 if $V_3 = 10$ V.
10-28. Solve for V_4 if $I_8 = 10$ mA.

11

Network Analysis

11-1. KIRCHHOFF'S CURRENT LAW

You will find that several of the topics we study in this chapter are already familiar to you. They are important in network analysis and their review is worthwhile. Kirchhoff's current law (KCL) is also not new to you; only the name is!

We know that the sum of the currents entering a point must equal the sum of the currents leaving that point. Suppose that we assign polarity signs to the currents so that entering currents are considered *negative* and leaving currents are *positive*. Then at any instant in time the total current in this node must be zero.

Kirchhoff's law of currents simply states that the algebraic sum of the currents in any node is zero.

$$I_1 + I_2 + \cdots + I_n = 0 \qquad (11\text{-}1)$$

Figure 11-1 illustrates KCL with a four-current node. We shall make use of the KCL in a form of circuit analysis called nodal or node-voltage analysis.

11-2. KIRCHHOFF'S VOLTAGE LAW

We have learned that the sum of the voltage drops around a closed circuit is equal to applied voltage. Kirchhoff's law of voltages (KVL) carries this concept further.

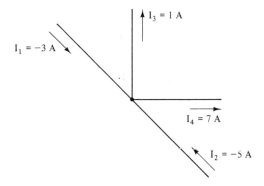

Figure 11-1 Kirchhoff's current law.

The algebraic sum of all the voltages in any circuit loop is zero.

A loop is simply any closed path around a circuit that proceeds from a given node, around the loop, and back to the start node. The direction of the loop is not important. We may move with current direction or against it. It helps to assign polarity to voltage drops. We assign negative polarity to the side where electron current enters a resistor and positive polarity to the other side of the resistor. We then write loop equations based upon the polarity signs first reached at all voltages, whether drops or sources.

If the loop direction is the same as the current direction, the voltage drops have *negative* signs. If loop direction is opposite to that of the current, voltage drops have *positive* signs. In both cases, voltage sources are given the same sign as the polarity of the terminal found in going around the loop.

$$E + V_1 + V_2 + \cdots + V_n = 0 \qquad (11\text{-}2)$$

Some examples will clarify these concepts.

EXAMPLE 11-1:
Given the circuit of Fig. 11-2, begin at the negative terminal of the source and write the KVL equations for (a) loop direction the same as current direction, and (b) loop direction opposite to current direction.

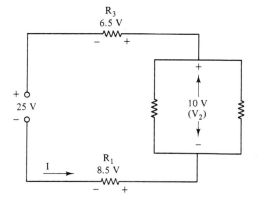

Figure 11-2

Sec. 11-2. Kirchhoff's Voltage Law

Solution:

$$V_1 + V_2 + V_3 + E = 0$$

(a) Loop direction with the current:

$$-8.5 + (-10) + (-6.5) + 25 = 0$$
$$-25 + 25 = 0$$

(b) Loop direction against the current:

$$-25 + 6.5 + 10 + 8.5 = 0$$
$$-25 + 25 = 0$$

EXAMPLE 11-2:
In the series-parallel circuit of Fig. 11-3, demonstrate Kirchhoff's voltage law for (a) loop *a-b-f-g-a*, (b) loop *b-c-d-e-f-b*, (c) loop *e-d-e*, and (d) loop *a-g-f-b-a*.

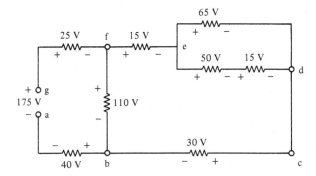

Figure 11-3

Solution:
(a) $-40 - 110 - 25 + 175 = 0$
$-175 + 175 = 0$
(b) $-30 - 65 - 15 + 110 = 0$
$-110 + 110 = 0$
Another equation is possible:

$$-30 - 15 - 50 - 15 + 110 = 0$$
$$-110 + 110 = 0$$

(c) Clockwise loop:

$$65 - 15 - 50 = 0$$
$$65 - 65 = 0$$

Counterclockwise loop:

$$50 + 15 - 65 = 0$$
$$65 - 65 = 0$$

(d) $-175 + 25 + 110 + 40 = 0$
$-175 + 175 = 0$

Note: The loops in parts (a) and (b) were with the current direction. The loop in (d) is against the current.

We shall find the KVL very useful for circuit analysis. The method used is called *mesh current analysis*. It can be used to solve circuits containing several sources of voltage.

11-3. VOLTAGE DIVISION RULE (VDR)

The VDR is not new to us. We review it in this chapter because it is so important in circuit analysis.

> The voltage drop across any portion of a series circuit is equal to the applied voltage times the ratio of the resistance of that part of the circuit to the resistance of the entire circuit.

$$V_x = E \frac{R_x}{R_T} \qquad (11\text{-}3)$$

Keep in mind that the original circuit need not be a series circuit. We may reduce a series-parallel circuit to a series equivalent circuit and then apply the VDR.

EXAMPLE 11-3.
Solve for the voltage drop across R_1 in Fig. 11-4.

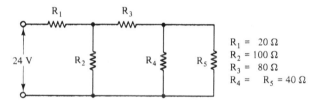

Figure 11-4

Solution:
1. $R_4 \parallel R_5 = 40 \parallel 40 = 20 \ \Omega$
2. $R_3 + R_4 \parallel R_5 = 80 + 20 = 100 \ \Omega$
3. $R_2 \parallel (R_3 + R_4 \parallel R_5) = 100 \parallel 100 = 50 \ \Omega$
4. The series equivalent circuit is shown in Fig. 11-5.

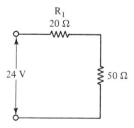

Figure 11-5

5. Use the VDR to find V_1:

$$V_1 = E\frac{R_1}{R_T}$$

$$= 24\frac{20}{70} = 6.857 \text{ V}$$

11-4. CURRENT DIVISION RULE (CDR)

Current division in any parallel circuit depends upon the ratio of branch conductance to total conductance. Actual branch current is the product of total current and the conductance ratio.

$$I_x = I_T \frac{G_x}{G_T} \tag{11-4}$$

While the CDR is not as widely used as the voltage division rule, it has value in circuit analysis. Certainly, it helps in the understanding of circuit behavior.

EXAMPLE 11-4:
Write the CDR equation for I_3, given the circuit of Fig. 11-6.

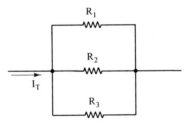

Figure 11-6

Solution:
By use of Eq. (11-4):

$$I_3 = I_T \frac{G_3}{G_T}$$

$$= I_T \frac{1/R_3}{1/R_1 + 1/R_2 + 1/R_3}$$

11-5. APPLICATIONS OF KIRCHHOFF'S LAWS

We can have several sources of voltage in a complex circuit. A complete solution for this kind of circuit may be made by use of Kirchhoff's laws. The methods of analysis make use of algebraic techniques, regardless of our approach. However, we shall make use of the scientific calculator to help us solve some of the equations.

There are three methods of analysis that use Kirchhoff's laws.

1. *Branch current analysis.* This is based upon Kirchhoff's current law and the use of Kirchhoff's voltage law.
2. *Mesh current analysis.* This is very similar to branch current analysis, except that the circuit equations are based upon loop currents, rather than the actual branch currents.
3. *Node voltage analysis.* In this method, we write equations based upon Kirchoff's current law and Ohm's law.

Branch Current Analysis

1. Identify the branch currents and assign current directions. The choice of current direction is *not* important. When we complete our circuit solution, any "negative" currents simply tell us that we chose a wrong direction for that current. The actual current direction is opposite to the one that we chose.
2. Write the current equations using the KCL.
3. Assign polarity to the voltage drops around the circuit according to the assumed currents.
4. Write the voltage equations for each loop using Kirchhoff's voltage law.
5. Solve the equations for currents. Many algebraic methods are possible. We shall use different ones in the examples in order to help you.

Mesh Current Analysis

1. Sketch loop currents in the circuit. Once again, direction is not important.
2. Assign polarity to voltage drops around the circuit as in branch current analysis.
3. Write the KVL equations.
4. Solve for the loop currents. This is an algebraic solution.
5. Determine the branch currents by use of KCL.

Node Voltage Analysis

1. Select a node to be used as the reference node. This is usually the node with the greatest number of connections. Frequently, it is circuit ground. We are seeking a point of reference for as many voltages as possible.
2. Identify voltages at other nodes with respect to the reference node.
3. Write equations for voltage drops. These are written as sums or differences of node voltages.

Sec. 11-5.   Applications of Kirchhoff's Laws

4. Write equations for branch currents with Ohm's law. Use the voltage drops determined in step 3.
5. Make an algebraic solution for the currents in the equations of step 4.

In the following examples we shall use the same circuit. In this way we shall demonstrate how each method is used. You may find that one method is more attractive to you than some of the others.

EXAMPLE 11-5:
Given the circuit of Fig. 11-7, solve for each branch current. Use branch current analysis. Use the KVL to demonstrate the accuracy of the solutions.

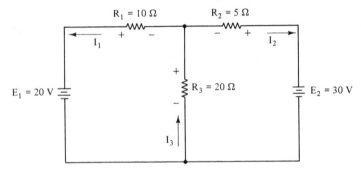

Figure 11-7

Solution:
1. Branch currents are sketched on the circuit program.
2. Write KCL equations:
   (a) $I_1 + I_2 - I_3 = 0$
   (b) $I_1 = I_3 - I_2$
   (c) $I_2 = I_3 - I_1$
3. Sketch polarity of drops on the circuit.
4. Write KVL equations:
   (a) $20 - 20I_3 - 10I_1 = 0$
   (b) $30 - 20I_3 - 5I_2 = 0$
   Rewriting,
   (a') $20I_3 + 10I_1 = 20$
   (b') $20I_3 + 5I_2 = 30$
5. Algebraic solution: In this example we shall use the method known as substitution.
   Step 1: Rewrite equation (a') by using the KCL equivalent of I_1.
   (a') $20I_3 + 10I_1 = 20$
   Therefore, $20I_3 + 10(I_3 - I_2) = 20$
   (a'') $30I_3 - 10I_2 = 20$
   Solving for I_2,

$$30I_3 - 20 = 10I_2$$
$$I_2 = 3I_3 - 2$$

We then take this value for I_2 and substitute into (b') of step 4. We solve for I_3.

(b') $\quad 20I_3 + 5I_2 = 30$
$20I_3 + 5(3I_3 - 2) = 30$
$20I_3 + 15I_3 - 10 = 30$
$35I_3 = 40$
$I_3 = 1.143 \text{ A}$

We use this value for I_3 in equation (a') of step 4 and solve for I_1.

(a') $\quad 20I_3 + 10I_1 = 20$
$20(1.143) + 10I_1 = 20$
$22.86 + 10I_1 = 20$
$10I_1 = 20 - 22.86$
$10I_1 = -2.86$
$I_1 = -0.286 \text{ A}$

The negative current tells us our selected current direction is incorrect. However, we do not make any change until we finish solving for I_2.

From step 2 we have

$$I_2 = I_3 - I_1$$
$$= 1.143 - (-0.286)$$
$$= 1.429 \text{ A}$$

Note that I_1 is a charging current for E_1. The drop across R_1 is in series aiding with E_1.

KVL loops:
(1) $-20 - (10 \times 0.286) + 20 \times 1.143 - 0$
$-20 - 2.86 + 22.86 = 0$ checks
(2) $-(20 \times 1.143) - (5 \times 1.429) + 30 = 0$
$-22.86 - 7.14 + 30 = 0$ checks

EXAMPLE 11-6:

Use mesh current analysis to find currents and voltages in the circuit of Example 11-5.

Solution:
1. Loop currents and polarities are sketched on the circuit.
2. Write the KVL equations:
    (a) $-20I_1 - 20I_2 - 10I_1 + 20 = 0$
    (b) $-20I_2 - 20I_1 - 5I_2 + 30 = 0$
    (Note that because I_1 and I_2 are assumed to flow in the 20-Ω resistor, each had to be part of that voltage drop. The equations show this as separate terms.
    Rewriting and collecting terms:
    (a') $-30I_1 - 20I_2 + 20 = 0$
    (b') $-20I_1 - 25I_2 + 30 = 0$
    (a'') $20 = 30I_1 + 20I_2$
    (b'') $30 = 20I_1 + 25I_2$
3. We shall make an algebraic solution by the method known as elimination. If we multiply equation (b'') by 1.5, we have
    (b'') $\quad 45 = 30I_1 + 37.5I_2$
    $-$(a'') $\quad \underline{20 = 30I_1 + 20I_2}$
    $\quad\quad\quad 25 = 17.5I_2$
    $\quad\quad\quad I_2 = 1.429 \text{ A}$

Sec. 11-5. Applications of Kirchhoff's Laws

Substitute this value of I_2 into equation a″.

$$20 = 30I_1 + 20(1.429)$$
$$20 = 30I_1 = 28.58$$
$$20 - 28.58 = 30I_1$$
$$I_1 = -0.286$$

We now know the magnitude of the currents I_1 and I_2. To find I_3, we take the algebraic sum of these currents.

$$I_3 = I_1 + I_2$$
$$= -0.286 + 1.429$$
$$= 1.143 \text{ A}$$

The results are identical to those found using branch current analysis.

EXAMPLE 11-7:
Use nodal analysis on the circuit of Examples 11-5 and 11-6. Solve for the branch currents.

Solution:
1. Draw the circuit and identify nodes (Fig. 11-8). We have marked one node as Ⓐ and the other, the principal node, as ground.

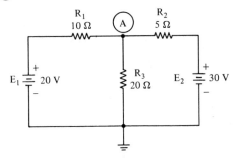

Figure 11-8

2. The voltage drops across the resistors are identified.

$$V_1 = E_1 - V_A$$
$$V_2 = E_2 - V_A$$
$$V_3 = V_A$$

3. $I_1 = \dfrac{V_1}{R_1} = \dfrac{E_1 - V_A}{R_1}$

$= \dfrac{20 - V_A}{10}$

$I_2 = \dfrac{V_2}{R_2} = \dfrac{E_2 - V_A}{R_2}$

$= \dfrac{30 - V_A}{5}$

$$I_3 = \frac{V_A}{R_3} = \frac{V_A}{20}$$

4. From Kirchhoff's current law,

$$I_3 = I_1 + I_2$$

$$\frac{V_A}{20} = \frac{20 - V_A}{10} + \frac{30 - V_A}{5}$$

Multiplying both sides by 20,

$$V_A = 40 - 2V_A + 120 - 4V_A$$

$$7V_A = 160$$

$$V_A = 22.68 \text{ V}$$

Once V_A is known, we find the currents:

$$I_1 = \frac{20 - 22.86}{10}$$
$$= -0.286 \text{ A}$$

$$I_2 = \frac{30 - 22.86}{5}$$
$$= 1.428 \text{ A}$$

$$I_3 = \frac{22.86}{20}$$
$$= 1.143 \text{ A}$$

When we solve multisource circuits by use of Kirchhoff's laws, we eventually solve a set of equations involving more than one unknown. In the examples so far we have used *substitution* (Example 11-5) and *elimination* (Example 11-6). Both methods require algebraic skills. There is another way to solve for the unknowns that is routine, easily done with a calculator, and requires little knowledge of algebra. This process is the solution of simultaneous equations by *determinants*.

Given the KVL equations:

$$aI_1 + bI_2 = K$$
$$cI_1 + dI_2 = L$$

where the coefficients of I_1 are a and c, the coefficients of I_2 are b and d, and the constant quantities are K and L.

The solution of two equations with two unknowns by the use of determinants requires that we work with the coefficients and constants.

1. List the constants and unknowns in columns:

$$\begin{matrix} a & b & K \\ c & d & L \end{matrix}$$

2. To solve for an unknown, we set up a fraction. The denominator is the array of the coefficients. The numerator is also a two-column array. *We substitute the constants for the coefficients of the unknown in this array.*
3. The diagonal pairs of each array are multiplied. The product of the lower left and upper right pair is subtracted from the other product.
4. The multiplication and subtraction result in the numerator and denominator of a fraction that represent the unknown quantity, a fraction easily solved with the calculator.

Solution by Determinants

1. To solve for I_1, set up an array of coefficients and constants in the form of the fraction:

$$I_1 = \frac{\begin{array}{cc} K & b \\ L & d \end{array}}{\begin{array}{cc} a & b \\ c & d \end{array}}$$

Note that the coefficients of I_1 are replaced by the constants for the numerator.

2. We multiply and subtract:

$$I_1 = \frac{Kd - Lb}{ad - bc}$$

3. I_2 is found by replacing its coefficients with constants for the numerator of the fraction. We repeat the multiplication and subtraction to obtain the numerator and denominator of the fraction that is equal to I_2.

$$I_2 = \frac{\begin{array}{cc} a & K \\ c & L \end{array}}{\begin{array}{cc} a & b \\ c & d \end{array}}$$

$$= \frac{aL - cK}{ad - bc}$$

Note that the denominator is the same in both fractions. You really need solve for the denominator only once.

Let us try the procedure on some algebraic problems.

1. $2x + 3y = 15$
   $x + y = 4$

Solution for x:

$$\frac{\begin{matrix}15 & 3\\ 4 & 1\end{matrix}}{\begin{matrix}2 & 3\\ 1 & 1\end{matrix}} = \frac{15-12}{2-3} = \frac{3}{-1} = -3$$

$$x = -3$$

Solution for y:

$$\frac{\begin{matrix}2 & 15\\ 1 & 4\end{matrix}}{\begin{matrix}2 & 3\\ 1 & 1\end{matrix}} = \frac{8-15}{2-3} = \frac{-7}{-1} = 7$$

$$y = 7$$

Check:

$$2x = -6; \quad 3y = 21$$
$$2x + 3y = 15; \quad -6 + 21 = 15 \quad \text{checks}$$
$$x + y = 4; \quad -3 + 7 = 4 \quad \text{checks}$$

2. $3x + 2y = 19$
   $5x - 6y = -15$

Solution for x:

$$\frac{\begin{matrix}19 & 2\\ -15 & -6\end{matrix}}{\begin{matrix}3 & 2\\ 5 & -6\end{matrix}} = \frac{-114-(-30)}{-18-10} = \frac{-114+30}{-28}$$

$$= \frac{-84}{-28}$$

$$x = 3$$

Solution for y:

$$\frac{\begin{matrix}3 & 19\\ 5 & -15\end{matrix}}{-28} = \frac{-45-95}{-28} = \frac{-140}{-28}$$

$$y = 5$$

Check:

$$3x + 2y = 9 + 10 = 19 \quad \text{checks}$$
$$5x - 6y = 15 - 30 = -15 \quad \text{checks}$$

EXAMPLE 11-8:

Solve for branch currents in the circuit of Fig. 11-9. Use determinants in making the solution. Check the results by use of the loop equations and by Kirchhoff's voltage law.

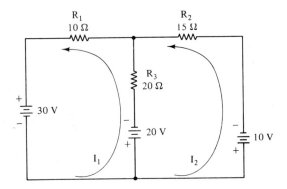

Figure 11-9

Solution:
1. Assume that the mesh currents flow in counterclockwise loops, as shown in Fig. 11-9. Assign polarity to the voltage drops around each mesh.
2. Write the mesh equations. For loop 1,

$$30 + 20 - 20I_1 + 20I_2 - 10I_1 = 0$$
$$50 - 30I_1 + 20I_2 = 0$$
$$30I_1 - 20I_2 = 50$$

For loop 2,

$$10 - 15I_2 - 20I_2 + 20I_1 - 20 = 0$$
$$-10 + 20I_1 - 35I_2 = 0$$
$$-20I_1 + 35I_2 = -10$$

3. Arrange the equations for solution by determinants:

$$30I_1 - 20I_2 = 50$$
$$-20I_1 + 35I_2 = -10$$

Solution for I_1:

$$\frac{\begin{vmatrix} 50 & -20 \\ -10 & 35 \end{vmatrix}}{\begin{vmatrix} 30 & -20 \\ -20 & 35 \end{vmatrix}} = \frac{1750 - 200}{1050 - 400} = \frac{1550}{650}$$

$$I_1 = 2.385 \text{ A}$$

Solution for I_2:

$$\frac{\begin{vmatrix} 30 & 50 \\ -20 & -10 \end{vmatrix}}{650} = \frac{-300 - (-1000)}{650} = \frac{700}{650}$$

$$I_2 = 1.077 \text{ A}$$

4. Redraw the circuit and calculate the branch currents.

$$I_1 = 2.385 \text{ A}$$
$$I_2 = 1.077 \text{ A}$$
$$\begin{aligned} I_3 &= I_1 - I_2 \\ &= 2.385 - 1.077 \\ &= 1.308 \text{ A} \end{aligned}$$

We can check the results. First, by the loop equtions, for loop 1,

$$50 - 30I_1 + 20I_2 = 0$$
$$50 - 30(2.385) + 20(1.077) = 0 \quad \text{checks}$$

For loop 2,

$$-10 + 20I_1 - 35I_2 = 0$$
$$-10 + 20(2.385) - 35(1.077) = 0 \quad \text{checks}$$

By KVL,

$$30 + 20 - 20(1.308) - 10(2.385) = 0$$
$$0 = 0$$
$$10 - 15(1.077) + 20(1.308) - 20 = 0$$
$$0 = 0$$

11-6. SUPERPOSITION THEOREM

We have used Kirchhoff's laws in the analysis of multisource circuits. This has required that we write sets of equations in order to solve for voltages and currents. The superposition theorem enables us to solve multisource networks by use of Ohm's law and our knowledge of series–parallel circuits.

Given a multisource circuit, the currents and therefore the voltage drops are found by treating each source separately, with all other sources replaced by their internal resistances.

Superposition Theorem. The branch currents in a multisource circuit are found by determining the response of the circuit to each source, one at a time, with all other sources replaced by their internal resistances. The resultant branch current is the algebraic sum of the currents found when using one source at a time.

The replacement of sources with internal resistances usually means that voltage sources will be replaced by short circuits. Current sources have very high internal resistance; our usual practice will be to replace them with open circuits.

Some examples will help make this theorem clear. Let us return to Example 11-8 and use the superposition theorem.

EXAMPLE 11-9:
Given the circuit of Fig. 11-10, apply the principle of superposition and solve for $I_1, I_2,$ and I_3.

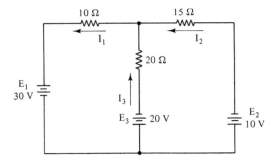

Figure 11-10

Solution:
1. Let E_1 be the source; replace E_2 and E_3 by short circuits. (Fig. 11-11).

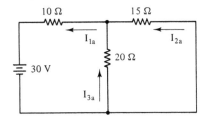

Figure 11-11

(a) Solve for I_{1a}:

$$I_{1a} = \frac{30}{10 + 20 \parallel 15} = 1.615 \text{ A}$$

(b) Use the current-division rule to solve for I_{3a}:

$$I_{3a} = I_{1a} \frac{R_2}{R_2 + R_3} = 1.615 \frac{15}{35}$$
$$= 0.692 \text{ A}$$

(c) Solve for I_{2a}:

$$I_{2a} = I_{1a} - I_{3a} = 1.615 - 0.692$$
$$= 0.923 \text{ A}$$

2. Let E_2 be the source, and short E_1 and E_3 (Fig. 11-12).

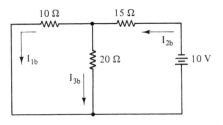

Figure 11-12

(a) Solve for I_{2b}:

$$I_{2b} = \frac{10}{15 + 10 \parallel 20} = 0.461 \text{ A}$$

(b) Use the CDR to solve for I_{1b}:

$$I_{1b} = I_{2b} \frac{R_3}{R_1 + R_3} = 0.461 \frac{20}{30}$$
$$= 0.307 \text{ A}$$

(c) $I_{3b} = I_{2b} - I_{1b} = 0.461 - 0.307$
$$= 0.154 \text{ A}$$

3. Let E_3 be the source, and short E_1 and E_2. This causes R_1 and R_2 to be in parallel (Fig. 11-13).

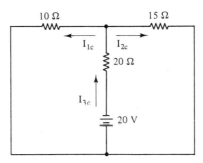

Figure 11-13

(a) Solve for I_{3c}:

$$I_{3c} = \frac{20}{20 + 10 \parallel 15} = 0.769 \text{ A}$$

(b) Solve for I_{1c} by the CDR:

$$I_{1c} = I_{3c} \frac{R_3}{R_1 + R_3} = 0.769 \frac{15}{35}$$
$$= 0.461 \text{ A}$$

(c) $I_{2c} = I_{3c} - I_{1c} = 0.769 - 0.461$
$$= 0.308 \text{ A}$$

4. Take the algebraic sum of the currents in each branch.
  (a) With each source, the current in R_1 had the same direction.

$$I_1 = I_{1a} + I_{1b} + I_{1c}$$
$$= 1.615 + 0.307 + 0.461$$
$$= 2.383 \text{ A}$$

(b) The current direction in R_3 was not the same for each supply. Therefore, the actual current is the net (algebraic sum) amount.

$$I_3 = I_{3a} - I_{3b} + I_{3c}$$
$$= 0.692 - 0.154 + 0.769$$
$$= 1.307 \text{ A}$$

Sec. 11-6. Superposition Theorem

(c) I_2 is the algebraic sum of the currents in R_2.

$$I_2 = I_{2a} + I_{2b} + I_{2c}$$
$$= 0.923 + 0.461 - 0.308$$
$$= 1.076 \text{ A}$$

The currents are almost exactly the same as those calculated in Example 11-8. The only reason for any differences is rounding in the computations.

We shall go into detail on current sources and how to convert voltage sources to current sources in Chapter 12. For now, simply consider that a current source supplies a constant amount of current regardless of the loads connected to it. An ideal current source has an infinite amount of internal resistance, which is in *parallel* with the source. We make use of current sources mixed with a voltage source in our next example.

EXAMPLE 11-10:
Given the circuit of Fig. 11-14, solve for I_1, I_2, and I_3. Use the superposition theorem.

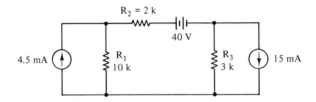

Figure 11-14

Solution:
1. Apply superposition beginning with the 4.5-mA source. Short out the 40-V source and open the 15-mA source (Fig. 11-15).

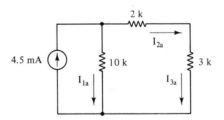

Figure 11-15

(a) Use the CDR to find I_{1a}:

$$I_{1a} = 4.5 \text{ mA} \frac{2 \text{ k}\Omega + 3 \text{ k}\Omega}{15 \text{ k}\Omega} = 1.5 \text{ mA}$$

(b) $I_{2a} = I_{3a} = 4.5 \text{ mA} - 1.5 \text{ mA} = 3 \text{ mA}$

2. Use the 15-mA source and solve for the currents. Remove the other sources (Fig. 11-16).

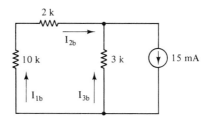

Figure 11-16

(a) Use the CDR to find I_{3b}:

$$I_{3b} = 15 \text{ mA} \frac{2 \text{ k}\Omega + 10 \text{ k}\Omega}{15 \text{ k}\Omega} = 12 \text{ mA}$$

(b) $I_{1b} = I_{2b} = 15 \text{ mA} - 12 \text{ mA} = 3 \text{ mA}$

3. Use the 40-V source, removing the other sources as in Fig. 11-17. Solve for the currents. Note that we have a simple series circuit.

$$I_{1c} = I_{2c} = I_{3c} = \frac{40}{15 \text{ k}\Omega} = 2.67 \text{ mA}$$

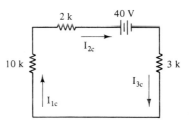

Figure 11-17

4. $I_1 = I_{1a} - I_{1b} - I_{1c}$
   $= 1.5 \text{ mA} - 3 \text{ mA} - 2.67 \text{ mA} = -4.17 \text{ mA}$

   The negative sign simply indicates that the actual current direction is "up" rather than "down."

   $I_2 = I_{2a} + I_{2b} + I_{2c}$
   $= 3 \text{ mA} + 3 \text{ mA} + 2.67 \text{ mA} = 8.67 \text{ mA}$

   $I_3 = I_{3a} - I_{3b} + I_{3c}$
   $= 3 \text{ mA} - 12 \text{ mA} + 2.67 \text{ mA} = -6.33 \text{ mA}$

5. A check of the current nodes demonstrates the validity of the solution.

   node A: $4.5 \text{ mA} + 4.17 \text{ mA} - 8.67 \text{ mA} = 0$ checks

   node B: $8.67 \text{ mA} + 6.33 \text{ mA} - 15 \text{ mA} = 0$ checks

   node C: $15 \text{ mA} - 8.67 \text{ mA} - 6.33 \text{ mA} = 0$ checks

   node D: $8.67 \text{ mA} - 4.5 \text{ mA} - 4.17 \text{ mA} = 0$ checks

Frequently, we need to know about current and voltage in part of a circuit, rather than the entire circuit. The superposition theorem may be used for this purpose. The following example illustrates this kind of application.

EXAMPLE 11-11:
Solve for the voltage drop across R_L in the circuit of Fig. 11-18.

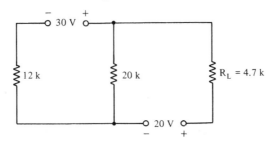

Figure 11-18

Solution:
1. Begin the circuit analysis with the 30-V source. Replace the 20-V source with a short circuit.
   (a) Solve for total current.

   $$I_T = \frac{30}{12 \text{ k}\Omega + 20 \text{ k}\Omega \parallel 4.7 \text{ k}\Omega} = \frac{30}{15.81 \text{ k}\Omega}$$
   $$= 1.898 \text{ mA}$$

   (b) Use the current-division rule to find the current in R_L for this source.

   $$I_{L1} = 1.898 \text{ mA} \frac{20 \text{ k}\Omega}{20 \text{ k}\Omega + 4.7 \text{ k}\Omega}$$
   $$= 1.537 \text{ mA}$$

2. Use the 20-V source and replace the 30-V source by a short circuit. Solve for the total current. This is the current in R_L.

   $$I_{L2} = \frac{20}{4.7 \text{ k}\Omega + 12 \text{ k}\Omega \parallel 20 \text{ k}\Omega}$$
   $$= 1.6393 \text{ mA}$$

3. The current in R_L is the algebraic sum of I_{L1} and I_{L2}. They are in opposite directions, so we take the difference. The current direction through the load is that of I_{L2} because it is larger than I_{L1}.

   $$I_L = I_{L2} - I_{L1} = 1.6393 \text{ mA} - 1.537 \text{ mA}$$
   $$= 0.1023 \text{ mA}$$

4. The voltage across the load is calculated.

   $$V_L = I_L R_L = 0.1023 \text{ mA} \times 4.7 \text{ k}\Omega$$
   $$= 0.481 \text{ V}$$

11-7. THEVENIN'S THEOREM

Thevenin's theorem is a method of circuit analysis that is accurate *only* for a specific part of a circuit. We shall call this portion of a circuit the load and use R_L to identify it. Thevenin's theorem is widely used in electronic circuit analysis. It permits us to replace a complex network with a very simple series circuit consisting of a source, an internal resistance, and R_L. The portion of a complex circuit, R_L, for which we wish to know current and voltage is connected to the Thevenin source, an *equivalent* source. The current in R_L and the voltage across it are the same as when R_L is a part of the complex circuit.

Thevenin's Theorem. Any resistive network may be replaced by an equivalent emf, V_{Th}, and an equivalent internal resistance, R_{Th}. The equivalent source will provide the same voltage and current to a load as the original network, provided that V_{Th} and R_{Th} are determined according to the following rules:
1. V_{Th} is the voltage appearing at the *open-circuited* load terminals of the original network (R_L is "removed").
2. R_{Th} is the resistance computed between the open circuited terminals of the original network, with all sources replaced by their internal resistances. This usually means that we replace voltage sources with short circuits. We simply open current sources.

It is important that we recognize that Thevenin's theorem is an *equivalent* circuit theorem. *This means equivalency only for the load in the circuit of Fig. 11-9. Load current and load voltage, and therefore load power, are the same as with the actual circuit.* There is no equivalency anywhere else in the analysis. Any other calculations of total power, total current, voltage drops, or efficiency are meaningless if we make them on the equivalent circuit. If one could replace complex circuits totally by simple series circuits, would there ever be a need for complex circuits? We shall demonstrate these ideas with some examples in Chapter 12 when we study power in greater detail.

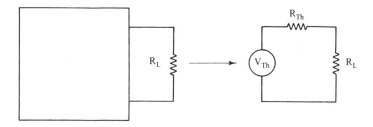

Figure 11-19 Thevenin's theorem enables us to replace a complex network by a simple voltage source with its internal resistance.

Let us begin with some examples of how to use Thevenin's theorem. To make your studies more meaningful, we shall first learn how to find V_{Th}. Then we shall work examples in solving for R_{Th}. Finally, we shall put it all together and solve circuit problems.

EXAMPLE 11-12:
In each of the single-source resistive circuits (Figs. 11-20 through 11-24), solve for V_{Th}. In each circuit the load is identified as R_L.

(a)

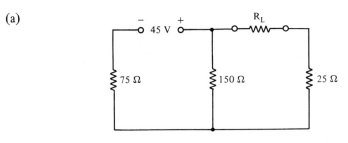

Figure 11-20

Solution:
With R_L removed, there is zero current in the 25-Ω resistor; therefore, there is no voltage drop across it. V_{Th} is the same as the voltage drop across the 150-Ω resistor.

$$V_{Th} = 45 \frac{150}{75 + 150} = 30 \text{ V}$$

(b)

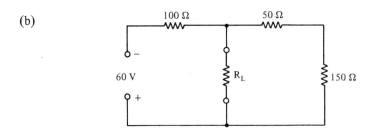

Figure 11-21

Solution:
When we remove R_L, we have a series circuit. V_{Th} is equal to the sum of the voltage drops across the 50-Ω and 150-Ω resistors. It is also equal to the difference between the 60-V source and the drop across the 100-Ω resistor.

$$V_{Th} = 60 \frac{50 + 150}{100 + 50 + 150} = 40 \text{ V}$$

$$I = \frac{60}{300} = 0.2 \text{ A}$$

$$V_{Th} = 60 - 0.2 \times 100 = 60 - 20 = 40 \text{ V}$$

(c)

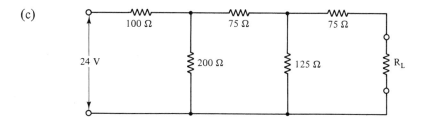

Figure 11-22

Solution:
When R_L is removed, V_{Th} is the same as the voltage drop across the 125-Ω resistor (Fig. 11-23).

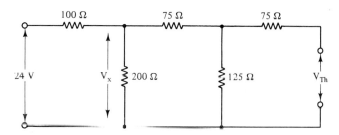

Figure 11-23

$$V_x = 24 \frac{200 \parallel (75 + 125)}{100 + 200 \parallel (75 + 125)}$$

$$= 24 \frac{200 \parallel 200}{200}$$

$$= 24 \frac{100}{200} = 12 \text{ V}$$

$$V_{Th} = V_x \frac{125}{125 + 75}$$

$$= 12 \frac{125}{200} = 7.5 \text{ V}$$

(d)

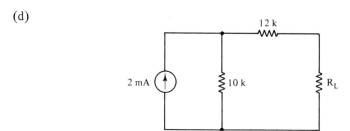

Figure 11-24

Sec. 11-7. Thevenin's Theorem

Solution:
With R_L removed, V_{Th} is equal to the voltage drop across the internal resistance of the current source.

$$V_{Th} = 2 \text{ mA} \times 10 \text{ k}\Omega = 20 \text{ V}$$

EXAMPLE 11-13:
Solve for V_{Th} in the circuit of Fig. 11-25.

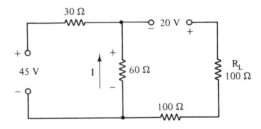

Figure 11-25

Solution:
1. Remove R_L and redraw the circuit (Fig. 11-26).

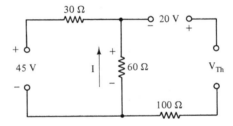

Figure 11-26

2. With the load removed, there is no voltage drop across the 100-Ω resistor. V_{Th} is the algebraic sum of the voltage drop across the 60-Ω resistor and the 20-V source.
   (a) The voltage drop across the 60-Ω resistor is found by the VDR:

$$V = 45 \frac{60}{90} = 30 \text{ V}$$

   (b) The polarity of the drop is in series aiding with the 20-V source.

$$V_{Th} = 30 + 20 = 50 \text{ V}$$

EXAMPLE 11-14:
Given the circuit Fig. 11-27, solve for V_{Th}.

Solution:
1. Redraw the circuit with R_L removed (Fig. 11-28).
2. The resulting circuit is a series circuit with two sources. The sources are series opposing.

$$I = \frac{30 - 20}{60 + 40} = \frac{10}{100} = 0.1 \text{ A}$$

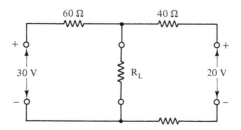

Figure 11-27

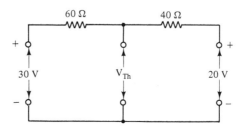

Figure 11-28

V_{Th} is equal to the 30-V source less the voltage drop across the 60-Ω resistor.

$$V_{Th} = 30 - 0.1 \times 60 = 24 \text{ V}$$

We can also find V_{Th} as the sum of the 20-V source and drop across the 40-Ω resistor.

$$V_{Th} = 20 + 0.1 \times 40 = 24 \text{ V}$$

Solving for R_{Th} can be more difficult than solving for V_{Th}. Remember, this is the resistance that we calculate at the load terminals, looking back into the circuit. Problems come up on how to remove sources, and especially whether or not certain parts of the circuit are in series or in parallel. Some guidelines will help:

1. *Voltage sources:* The ideal source has zero internal resistance. In actual practice, we can ignore the internal resistance of voltage sources because they are so small in comparison to the resistance of the external circuit. In calculations for R_{Th}, we replace voltage sources by short circuits as in Fig. 11-29.
2. *Current sources:* The ideal source has an infinite internal resistance. In practice, the actual internal resistance of a current source is shown in parallel with the symbol for the source. To solve for R_{Th}, we simply open the current source, keeping its internal resistance as part of the circuit as in Fig. 11-30.
3. A good way to determine the series and parallel relations that exist when we attempt to find R_{Th} is to begin by labeling the load termi-

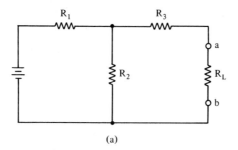

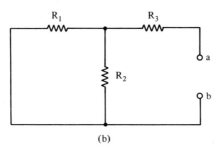

Figure 11-29 Solving for the Thevenin equivalent resistance. (a) To find R_{Th} we replace the voltage source by a short circuit. This places R_1 in parallel with R_2. (b) $R_{Th} = R_3 + R_1 \parallel R_2$.

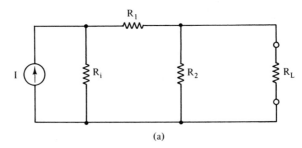

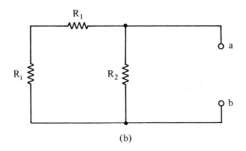

Figure 11-30 Solving for the Thevenin equivalent resistance when the circuit has a current source. (a) To find R_{Th} we remove the current source, keeping its internal resistance. The internal resistance of a current source is always shown as a resistor across the source. (b) $R_{Th} = R_2 \parallel (R_i + R_1)$

nals. Trace complete paths from one terminal to the other. If we find that there are separate paths, clearly these paths are in parallel.

EXAMPLE 11-15:
Given the circuit of Fig. 11-31, solve for R_{Th}.

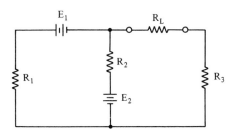

Figure 11-31

Solution:
1. (a) Remove R_L and label the load terminals.
   (b) Replace E_1 and E_2 by short circuits.
2. R_1 is in parallel with R_2; this combination is in series with R_3.

$$R_{Th} = R_3 + R_1 \parallel R_2$$

EXAMPLE 11-16:
Solve for R_{Th} in the circuit of Fig. 11-32.

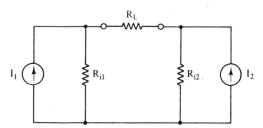

Figure 11-32

Solution:
1. Redraw the circuit.
   (a) Remove R_L and label the load terminals.
   (b) Remove I_1 and I_2, keeping R_{i1} and R_{i2}.
2. R_{Th} is the sum of the internal resistances.

$$R_{Th} = R_{i1} + R_{i2}$$

A category of circuits called *bridge* circuits can be difficult to solve for R_{Th}. The following examples will help you to understand these circuit problems.

Sec. 11-7.   Thevenin's Theorem

EXAMPLE 11-17:
Given the bridge circuit of Fig. 11-33, solve for R_{Th}.

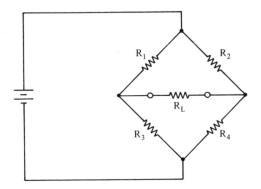

Figure 11-33

Solution:
1. Replace the source with a short circuit, remove R_L, and redraw the circuit (Fig. 11-34).

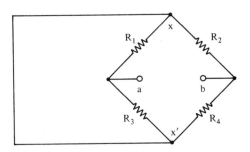

Figure 11-34

2. The short circuit connects the points labeled x and x'. This places R_1 in parallel with R_3 and R_4 in parallel with R_2.

$$R_{Th} = R_1 \| R_3 + R_2 \| R_4$$

EXAMPLE 11-18:
Given the circuit of Fig. 11-35, solve for R_{Th}.

Solution:
1. The internal resistance for I_1 is $R_3 + R_1$. The source I_2 has an internal resistance of $R_4 + R_2$.
   (a) Remove R_L and the current sources.
   (b) Redraw the circuit (Fig. 11-36).
2. $R_{Th} = (R_1 + R_2) \| (R_3 + R_4)$

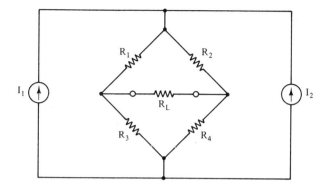

Figure 11-35

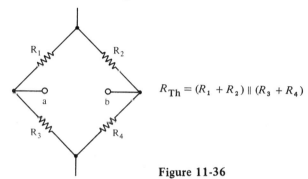

$R_{Th} = (R_1 + R_2) \parallel (R_3 + R_4)$

Figure 11-36

Let us proceed to some complete problems. To be able to compare Thevenin's theorem to other methods for working with multiple-source problems, we shall work with the same circuits we used for the examples of Kirchhoff's laws and the superposition theorem. Following these examples, we shall use Thevenin's theorem with a bridge circuit.

EXAMPLE 11-19:
Given the circuit of Example 11-5, use R_2 as R_L and solve for I_2 (Fig. 11-37).

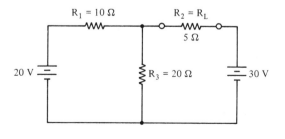

Figure 11-37

Sec. 11-7. Thevenin's Theorem

Solution:
1. Solve for V_{Th}.
   (a) Remove R_L.
   (b) The only current flow is in the 10- and 20-Ω resistors. Solve for the voltage drop across the 20-Ω resistor.
   $$V = 20 \times \frac{20}{30} = 13.33 \text{ V}$$
   (c) The polarity of the voltage drop causes it to be in series opposition to the 30-V source.
   $$V_{Th} = 30 - 13.33 = 16.67 \text{ V}$$
2. Solve for R_{Th}. Replace the sources with short circuits as in Fig. 11-38.
   $$R_{Th} = 10 \parallel 20 = 6.67 \text{ Ω}$$

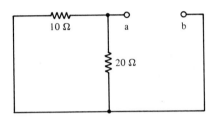

Figure 11-38

3. Draw the Thevenin equivalent circuit connected to R_L (Fig. 11-39).

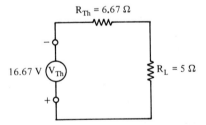

Figure 11-39

4. Solve for current.
   $$I = \frac{16.67}{6.67 + 5} = 1.428 \text{ A}$$

EXAMPLE 11-20:
Solve for the current in R_3 in the circuit of Example 11-8 (Fig. 11-40).

Solution:
1. Solve for V_{Th}:
   (a) The 30- and 10-V sources are in series aiding in the circuit.
   $$I = \frac{30 + 10}{10 + 15} = \frac{40}{25} = 1.6 \text{ A}$$

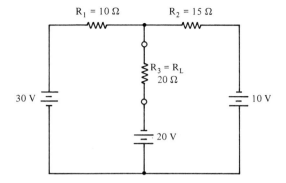

Figure 11-40

(b) The voltage drop across the 10-Ω resistor is calculated.

$$V = 1.6 \times 10 = 16 \text{ V}$$

(c) V_{Th} is the algebraic sum of the voltages around the mesh.

$$V_{Th} = -16 + 30 + 20 = 34 \text{ V}$$

2. Solve for R_{Th} (Fig. 11-41).

$$R_{Th} = 10 \parallel 15 = \frac{150}{25} = 6 \text{ Ω}$$

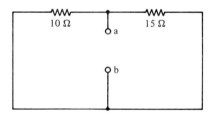

Figure 11-41

3. Solve for I_L:

$$I_L = \frac{34}{20 + 6} = 1.308 \text{ A}$$

EXAMPLE 11-21:
Solve for the load voltage in the circuit of Fig. 11-42.

Figure 11-42

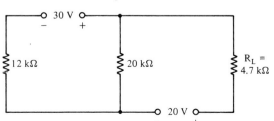

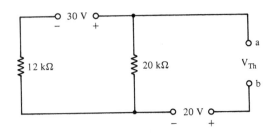

Figure 11-43

Solution:
1. Redraw the circuit (Fig. 11-43).
2. Solve the mesh for the voltage drop across the 20-kΩ resistor.

$$V_{Th} = 30 \frac{20 \text{ k}\Omega}{12 \text{ k}\Omega + 20 \text{ k}\Omega}$$
$$= 18.75 \text{ V}$$

3. V_{Th} is the difference between the 20-V source and the 18.75-V drop because they are in series opposition across the load terminals. Terminal a is negative with respect to b.

$$V_{Th} = 20 - 18.75 = 1.25 \text{ V}$$

4. Solve for R_{Th}.
   (a) Replace the sources by short circuits.
   (b) $R_{Th} = 12 \text{ k}\Omega \parallel 20 \text{ k}\Omega = 7.5 \text{ k}\Omega$
5. Solve for V_L.

$$V_L = V_{Th} \frac{R_L}{R_T} = 1.25 \frac{4.7 \text{ k}\Omega}{7.5 \text{ k}\Omega + 4.7 \text{ k}\Omega}$$
$$= 0.481 \text{ V}$$

EXAMPLE 11-22:
Solve for load current in the bridge circuit of Fig. 11-44.

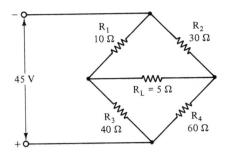

Figure 11-44

Solution:
1. Solve for V_{Th}:
   (a) Redraw the circuit, removing R_L (Fig. 11-45).

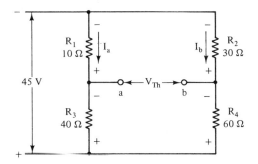

Figure 11-45

(b) Solve for the branch currents:

$$I_a = \frac{45}{10+40} = 0.9 \text{ A}$$

$$I_b = \frac{45}{30+60} = 0.5 \text{ A}$$

(c) Solve for the voltage drops:

$$V_1 = I_a R_1 = 0.9 \times 10 = 9 \text{ V}$$
$$V_3 = I_a R_3 = 0.9 \times 40 = 36 \text{ V}$$
$$V_2 = I_b R_2 = 0.5 \times 30 = 15 \text{ V}$$
$$V_4 = I_b R_4 = 0.5 \times 60 = 30 \text{ V}$$

(d) V_{Th} is the algebraic sum of V_1 and V_2. It is also the algebraic sum of V_3 and V_4. Starting at terminal a,

$$V_{Th} = V_1 + V_2 = 9 + -15 = -6 \text{ V}$$
$$V_{Th} = V_3 + V_4 = -36 + 30 = -6 \text{ V}$$

Terminal a is the negative side of V_{Th}.

2. Solve for R_{Th}.
   (a) Replace the voltage source with a short circuit.
   (b) $R_{Th} = 10 \parallel 40 + 30 \parallel 60$
   $= 8 + 20 = 28 \text{ }\Omega$
3. Draw the Thevenin equivalent circuit (Fig. 11-46).

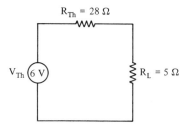

Figure 11-46

Sec. 11-7. Thevenin's Theorem

4. Solve for I_L.

$$I_L = \frac{V_{Th}}{R_{Th} + R_L} = \frac{6}{28 + 5}$$
$$= 0.182 \text{ A}$$

In the actual circuit, load current is from left to right. We know this because we found terminal a to be the negative terminal of V_{Th}.

11-8. NORTON'S THEOREM

Once again, we have an *equivalent* circuit theorem that is accurate *only* for a portion of a circuit. With this theorem we replace a complex network by a current source and its internal resistance. When the source is properly determined, a load connected to such a source will have the same current as when it is part of a complex network. Norton's theorem is the current source duplicate of Thevenin's theorem and is shown in Fig. 11-47.

Norton's Theorem. Any resistive network may be replaced by an equivalent current source, I_N, and an equivalent internal resistance, R_N. The equivalent current source will provide the same current and voltage to a load as the original network, provided that I_N and R_N are determined according to the following rules:

1. I_N is the current flow between the load terminals when those terminals are *shorted*.
2. R_N is the resistance between the load terminals of the original network, with the load removed and all sources replaced by their internal resistances. We calculate this equivalent resistance in exactly the same way as the Thevenin equivalent resistance.

$$R_N = R_{Th} \tag{11-5}$$

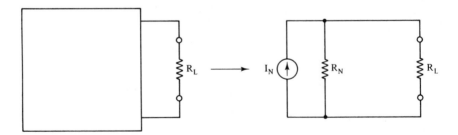

Figure 11-47 Norton's theorem allows us to replace a complex network by a simple current source with its internal resistance.

In the Norton equivalent circuit, R_N is connected in parallel with the current source, I_N. The load current is calculated readily by the current-division rule.

$$I_L = I_N \frac{R_N}{R_N + R_L} \qquad (11\text{-}6)$$

Norton's theorem is very useful in the understanding of semiconductor circuits, particularly transistor amplifiers. We shall use the theorem with some of the same examples we have used for the other circuit analysis methods so that we may compare processes and results.

EXAMPLE 11-23:
Solve for the load current in the circuit of Fig. 11-48 by use of Norton's theorem.

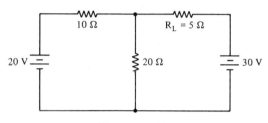

Figure 11-48

Solution:
1. Solve for I_N. Replace R_L by a short circuit and solve for the current in the shorted terminals.
   (a) Redraw the circuit (Fig. 11-49).

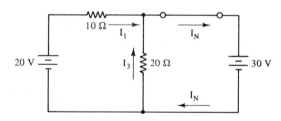

Figure 11-49

(b) Examination of the circuit indicates

$$I_N = I_1 + I_3$$

In solving for I_1, note that the voltage sources are opposing across the 10-Ω resistor.

$$I_1 = \frac{30 - 20}{10} = 1 \text{ A}$$

$$I_3 = \frac{30}{20} = 1.5 \text{ A}$$

$$I_N = 1 + 1.5 = 2.5 \text{ A}$$

2. Solve for R_N. The Norton equivalent resistance is the same as the Thevenin equivalent resistance. From the example, $R_{Th} = 6.67\ \Omega$.

$$R_N = 6.67\ \Omega$$

3. Sketch the Norton equivalent circuit and solve for I_L (Fig. 11-50).

$$I_L = I_N \frac{R_N}{R_N + R_L} = 2.5 \frac{6.67}{11.67}$$
$$= 1.429\ \text{A}$$

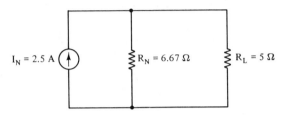

Figure 11-50

EXAMPLE 11-24:

Given the circuit of Example 11-20, solve for the load current by use of Norton's theorem.

Solution:
1. Solve for I_N.
   (a) Redraw the circuit (Fig. 11-51).

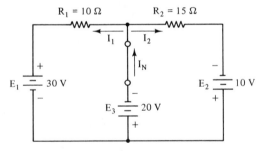

Figure 11-51

(b) The current directions for I_1 and I_2 are selected because E_1 and E_2 are series aiding, while E_2 and E_3 are in series opposition, with E_2 greater than E_3.

(c) Solve for the currents. The voltage across R_1 is the sum of E_1 and E_2.

$$I_1 = \frac{E_1 + E_2}{R_1} = \frac{50}{10} = 5\ \text{A}$$

The voltage across R_2 is $E_2 - E_3$.

$$I_2 = \frac{E_2 - E_3}{R_2} = \frac{10}{15} = 0.667 \text{ A}$$

$$I_N = I_1 + I_2 = 5.667 \text{ A}$$

2. Solve for R_N. From Example 11-20, $R_{Th} = 6 \, \Omega$.

$$R_N = R_{Th} = 6 \, \Omega$$

3. Solve for I_L.

$$I_L = 5.667 \frac{6}{26} = 1.308 \text{ A}$$

EXAMPLE 11-25:

Use Norton's theorem on the circuit of Example 11-21 and solve for the load voltage.

Solution:
1. Solve for I_N (Fig. 11-52).

(a)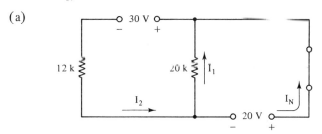

Figure 11-52

(b) $I_1 = \dfrac{20}{20 \text{ k}\Omega} = 1 \text{ mA}$

$$I_2 = \frac{30 - 20}{12 \text{ k}\Omega} = 0.833 \text{ mA}$$

$$I_N = I_1 - I_2 = 0.167 \text{ mA}$$

2. Solve for R_N.

$$R_N = R_{Th} = 7.5 \text{ k}\Omega$$

3. Solve for I_L (Fig. 11-53).

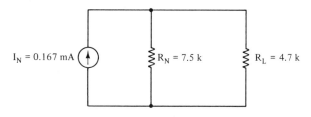

Figure 11-53

$$I_L = 0.167 \text{ mA } \frac{7.5 \text{ k}\Omega}{12.2 \text{ k}\Omega}$$
$$= 0.1026 \text{ mA}$$

4. $V_L = I_L R_L$
   $= 0.1026 \text{ mA} \times 4.7 \text{ k}\Omega$
   $= 0.482 \text{ V}$

There are a great many additional topics in network analysis. This chapter covered the basic tools that you need for work with complex resistive networks. We shall add more network analysis and conversion theorems in our studies of alternating-current circuits. We shall also review the Thevenin and Norton theorems, using networks that are not entirely resistive.

SUMMARY

1. Kirchhoff's current law (KCL) confirms that the algebraic sum of the currents at a node is zero.
2. The algebraic sum of the voltages around any closed loop is zero. This is known as Kirchhoff's voltage law (KVL).
3. Branch current analysis requires that we identify branch currents before we write KVL equations.
4. Mesh current analysis requires that we write a KVL equation for the voltages around each closed loop in a circuit. Branch currents are resolved from the loop currents.
5. Node voltage analysis requires that we write KCL equations based on Ohm's law statements of currents.
6. Determinants are a very useful tool for solving simultaneous equations.
7. Superposition allows us to analyze a complex circuit one source at a time, with all other sources replaced by internal resistances.
8. Actual current in any part of a circuit is the algebraic sum of the currents found by the use of the superposition theorem.
9. Thevenin's theorem is an equivalent circuit theorem. It allows us to replace part of a circuit by a simple voltage source and a series resistance. The resulting solution, in terms of output voltage and current, is accurate *only* for the output.
10. Norton's theorem is the current source equivalent of Thevenin's theorem.

$$I_N = V_{Th}/R_{Th}, \quad R_N = R_{Th}$$

PROBLEMS

11-1. Use Kirchhoff's current law to solve for the unknown current at a node.
   (a) $4 - 1.5 + 2 - I_4 = 0$, $I_4 =$
   (b) $I_T - (5 + 7.2 + 3) \text{ mA} = 0$, $I_T =$
   (c) $40 \text{ mA} - 920 \text{ μA} - I_2 = 0$, $I_2 =$

11-2. Given the circuit of Fig. 11-54, write the Kirchhoff voltage law equations.

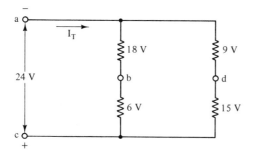

Figure 11-54

11-3. Given the circuit of Fig. 11-55, write the Kirchhoff voltage law equations for the indicated loops.
 (a) loop *a-e-f-d-c-a*
 (b) loop *e-a-c-d-e*
 (c) loop *e-f-d-e*
 (d) loop *a-b-c-a*

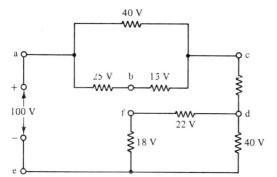

Figure 11-55

11-4. Use the voltage-divider rule (VDR) and write the equation for V_2 in the circuit of Fig. 11-56.

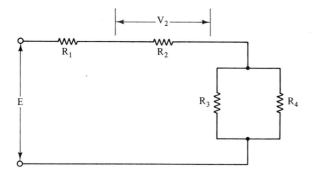

Figure 11-56

11-5. In the circuit of Fig. 11-56, solve for V_2, given that $E = 15$ V, $R_1 = 22$ kΩ, $R_2 = 12$ kΩ, $R_3 = 15$ kΩ, and $R_4 = 30$ kΩ.

11-6. Given the data of Prob. 11.5, solve for the voltage drop across the parallel circuit.

11-7. Use the current-division rule (CDR) and write the equation for I_1 in the circuit of Fig. 11-57.

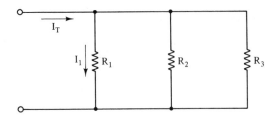

Figure 11-57

11-8. In the circuit of Fig. 11-57, $I_T = 32$ mA, $R_1 = 10$ kΩ, $R_2 = 4.7$ kΩ, and $R_3 = 18$ kΩ. Use the CDR to solve for each branch current.

11-9. In the circuit of Fig. 11-58, use branch current analysis and write the KCL equations.

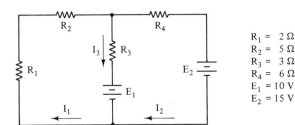

$R_1 = 2\ \Omega$
$R_2 = 5\ \Omega$
$R_3 = 3\ \Omega$
$R_4 = 6\ \Omega$
$E_1 = 10$ V
$E_2 = 15$ V

Figure 11-58

11-10. Given the data, write the KVL equations for the circuit of Figure 11-58.

11-11. Solve for the branch currents in the circuit of Figure 11-58.

11-12. Use determinants to solve for currents in each of the following pairs of equations:
(a) $10I_1 + 15I_2 = 20$
$3I_1 + 7I_2 = -4$
(b) $2.7I_a + 1.2I_b = 1.87$
$3.9I_a + 4.7I_b = 4.5$
(c) $330I_1 - 220I_2 = 1.43$
$860I_1 + 150I_2 = 47.9$

11-13. Write the mesh current equations for the circuit of Fig. 11-59.

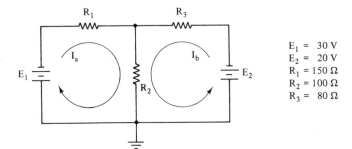

$E_1 = 30$ V
$E_2 = 20$ V
$R_1 = 150\ \Omega$
$R_2 = 100\ \Omega$
$R_3 = 80\ \Omega$

Figure 11-59

11-14. Given the data, solve for the current in each resistor in the circuit of Fig. 11-59.

11-15. Use node voltage analysis on the circuit of Fig. 11-59 and solve for the branch currents. Assume ground is the principal node. Compare the results with those found in Prob. 11.14.

11-16. Use the principle of superposition to solve for currents in the circuit of Fig. 11-60.

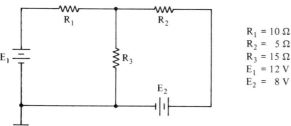

$R_1 = 10\ \Omega$
$R_2 = 5\ \Omega$
$R_3 = 15\ \Omega$
$E_1 = 12\ V$
$E_2 = 8\ V$

Figure 11-60

11-17. Use Thevenin's theorem to solve the circuit of Fig. 11-61 for V_L and I_L.

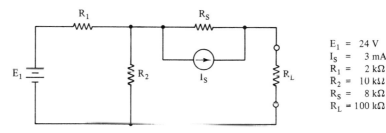

$E_1 = 24\ V$
$I_S = 3\ mA$
$R_1 = 2\ k\Omega$
$R_2 = 10\ k\Omega$
$R_S = 8\ k\Omega$
$R_L = 100\ k\Omega$

Figure 11-61

11-18. Use Thevenin's theorem to solve the circuit of Fig. 11-62 for V_L and I_L.

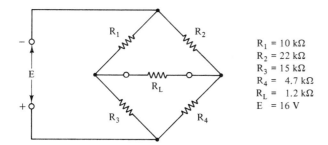

$R_1 = 10\ k\Omega$
$R_2 = 22\ k\Omega$
$R_3 = 15\ k\Omega$
$R_4 = 4.7\ k\Omega$
$R_L = 1.2\ k\Omega$
$E = 16\ V$

Figure 11-62

11-19. Solve for the Norton equivalent current and resistance in each of the circuits of Fig. 11-63.

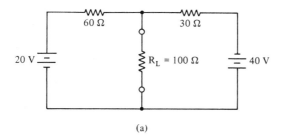

(a)

Figure 11-63

Chap. 11   Problems

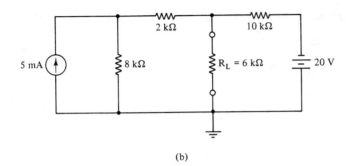

(b)

Figure 11-63 *(continued)*

11-20. Use the data found in Prob. 11-19 and solve for V_L and I_L in each circuit of Fig. 11-63.

11-21. Solve for V_L and I_L in the circuits of Fig. 11-64. In each case, use both Thevenin's and Norton's theorems. The reader will note that each is more advantageous to use, depending upon the circuit and the location of R_L.

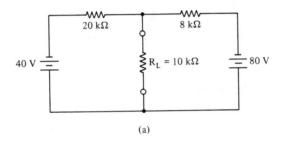

(a)

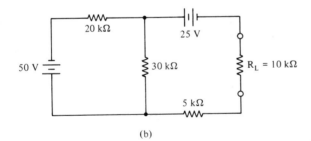

(b)

Figure 11-64

12

Power in DC Circuits

12-1. BASIC POWER FORMULAS

Recall that power is the *rate* of doing work, the rate of transforming energy. In electricity, work in joules is the product of voltage and charge displacement in coulombs.

$$W = EQ$$

The rate of work, power in watts (W), is the product of current and voltage.

$$P = \frac{W}{t} = E\frac{Q}{t} = EI$$

We have the three basic power formulas, each resulting in the same answer. The choice of formula depends upon the information contained in a given problem.

1. Given voltage and current:

$$P_T = EI \qquad (12\text{-}1)$$

$$P = VI \qquad (12\text{-}1\text{a})$$

2. Given current and resistance:

$$P = I^2 R \qquad (12\text{-}2)$$

This formula is derived from Eq. (12-1a) by substituting IR for V.

3. Given voltage and resistance, we derive another formula from Eq. (12-1a) by substituting V/R for I.

$$P = VI = V\frac{V}{R} = \frac{V^2}{R}$$

$$P = \frac{V^2}{R} \tag{12-3}$$

$$P = \frac{E^2}{R_T} \tag{12-3a}$$

EXAMPLE 12-1:
The voltage drop across a resistor is 15 V. The current in the resistor is 100 mA. How much power is dissipated in the resistor?

Solution:

$$P = VI$$
$$= 15 \times 0.1 = 1.5 \text{ W}$$

EXAMPLE 12-2:
Given that the voltage drop across a 220-Ω resistor is 20 V, what is the power dissipated in the resistor?

Solution:

$$P = \frac{V^2}{R}$$
$$= \frac{20^2}{220} = \frac{400}{220} = 1.82 \text{ W}$$

EXAMPLE 12-3:
The current in an 82-Ω resistor is 0.5 A. What is the power dissipated in the resistor?

Solution:

$$P = I^2 R$$
$$= 0.5^2 \times 82 = 0.25 \times 82 = 20.5 \text{ W}$$

Input power can be calculated by two methods.

1. We can use any of the power formulas, provided that we work with quantities that relate to the total circuit.
2. We can calculate the power dissipation in each resistor. The sum of these dissipations is equal to the total power, the input power. It does not matter whether we are working with series or parallel circuits.

$$P_T = P_1 + P_2 + \cdots + P_n \tag{12-4}$$

EXAMPLE 12-4:
Given the circuit of Fig. 12-1, (a) calculate total resistance and total current; (b) calculate the power dissipated in each resistor; and (c) calculate total power by Eqs. (12-1) and (12-4).

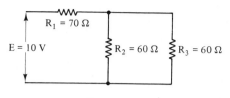

Figure 12-1

Solution:
(a) $R_T = 70 + 60 \parallel 60$
$ 70 + 30 = 100 \, \Omega$
$I_T = \dfrac{E}{R_T} = \dfrac{10}{100} = 0.1 \, A$

(b) $P_1 = I_T^2 R_1$
$ = 0.1^2 \times 70 = 0.01 \times 70 = 0.7 \, W$
$P_2 = I_2 R_2 \quad (I_2 = 1/2 I_T = 0.05 \, A)$
$ = 0.05^2 \times 60 = 0.0025 \times 60 = 0.15 \, W$
$P_3 = P_2 = 0.15 \, W$

(c) By Eq. (12-4),

$$P_T = P_1 + P_2 + P_3$$
$$= 0.7 + 0.15 + 0.15 = 1 \, W$$

By Eq. (12-1),

$$P_T = E \times I_T$$
$$= 10 \times 0.1 = 1 \, W$$

12-2. EFFICIENCY

When power is dissipated in electrical circuits, it is in the form of heat. In most electronic circuits, power requirements and power dissipations are relatively small. However, efficiency is an important consideration because of the effect of heat upon electronic devices.

Efficiency is also extremely important in larger circuits, as in industrial applications, commercial broadcasting, and power distribution systems.

Efficiency is the ratio of output power to input power. The quantity is usually expressed as a percentage. However, we use the decimal equivalent of percentage when solving problems. The symbol for efficiency is the Greek letter η (eta).

$$\eta = \dfrac{P_{out}}{P_{in}} \qquad (12\text{-}5)$$

In this form, η is a decimal fraction. Simply multiply by 100 to change it to a percentage. *Efficiency is always less than 100%.*

We can arrange Eq. (12-5) to get two more formulas. Given input power and efficiency, we can calculate the output power.

$$P_{out} = \eta P_{in} \qquad (12\text{-}6)$$

Given output power and efficiency, we solve for input power.

$$P_{in} = \frac{P_{out}}{\eta} \qquad (12\text{-}7)$$

EXAMPLE 12-5:
The power developed in a load is 25 W. The input power to the system is 75 W. What is the efficiency of the system?

Solution:
By Eq. (12-5),

$$\eta = \frac{P_{out}}{P_{in}}$$
$$= \frac{25}{75} = 0.333 = 33.3\%$$

EXAMPLE 12-6:
Suppose that the system of Example 12-5 is redesigned so that it is 45% efficient. Assuming that the input power is 75 W, what is the output power?

Solution:

$$P_{out} = \eta P_{in} \qquad (12\text{-}6)$$
$$= 0.45 \times 75 = 33.75 \text{ W}$$

EXAMPLE 12-7:
Suppose that the system of Example 12-6 must develop an output power of 25 W. What is the required input power?

Solution:

$$P_{in} = \frac{P_{out}}{\eta} \qquad (12\text{-}7)$$
$$= \frac{25}{0.45} = 55.56 \text{ W}$$

We may use circuit relationships to find efficiency. In a series circuit,

$$P_{out} = I^2 R_L$$
$$P_{in} = I^2 R_1 + I^2 R_2 + I^2 R_L + \cdots + I^2 R_n$$
$$\eta = \frac{I^2 R_L}{I^2 (R_1 + R_2 + R_L + \cdots + R_n)}$$
$$\eta = \frac{R_L}{R_1 + R_2 + R_L + \cdots + R_n} \qquad (12\text{-}8)$$
$$\eta = \frac{R_L}{R_T} \qquad (12\text{-}8a)$$

In a parallel circuit,

$$P_{out} = \frac{E^2}{R_L}$$

$$P_{in} = \frac{E^2}{R_L} + \frac{E^2}{R_1} + \frac{E^2}{R_2} + \cdots + \frac{E^2}{R_n}$$

$$= E^2 \left(\frac{1}{R_L} + \frac{1}{R_1} + \frac{1}{R_2} + \cdots + \frac{1}{R_n} \right)$$

$$\eta = \frac{E^2/R_L}{E^2(1/R_L + 1/R_1 + 1/R_2 + \cdots + 1/R_n)}$$

$$= \frac{1/R_L}{1/R_L + 1/R_1 + 1/R_2 + \cdots + 1/R_n}$$

$$= \frac{G_L}{G_T} \qquad (12\text{-}9)$$

We shall find these formulas useful when working problems involving voltage and current sources, particularly in terms of equivalent circuits.

EXAMPLE 12-8:
Given the circuit of Fig. 12-2, solve for power output (I'_L) and efficiency.

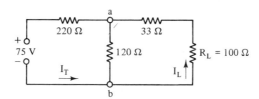

Figure 12-2

Solution:
    1. Solve for the circuit for I_T and I_L.
       (a) Reduce the parallel circuit to its series equivalent.

$$R_{a-b} = 120 \parallel (33 + 100)$$
$$= \frac{120 \times 133}{120 + 133} = 63.1 \; \Omega$$

       (b) $R_T = 220 + 63.1$
$$= 283.1 \; \Omega$$

       (c) $I_T = \frac{E}{R_T} = \frac{75}{283.1} = 0.265 \; A$

       (d) $V_{a-b} = I_T(R_{a-b})$
$$= 0.265 \times 63.1 = 16.72 \; V$$

       (e) The load current is found by dividing V_{a-b} by the branch resistance.

$$I_L = \frac{16.72}{133} = 0.1257 \; A$$

2. Solve for power output.
$$P_L = I_L^2 R_L$$
$$= 0.1257^2 \times 100$$
$$= 1.58 \text{ W}$$

3. Solve for P_{in}.
$$P_{in} = EI_T$$
$$= 75 \times 0.265$$
$$= 19.9 \text{ W}$$

4. Solve for efficiency.
$$\eta = \frac{P_L}{P_{in}} = \frac{1.58}{19.9} = 0.0794$$
$$= 7.94\%$$

EXAMPLE 12-9:
Apply Thevenin's theorem to the circuit of Example 12-8. (a) Demonstrate that the power output is the same as previously calculated. (b) Demonstrate that efficiency calculation based upon the Thevenin equivalent circuit is meaningless.

Solution:
1. Solve for V_{Th}.
$$V_{Th} = 75 \frac{120}{220 + 120}$$
$$= 26.47 \text{ V}$$

2. Solve for R_{Th}
$$R_{Th} = 33 + 120 \parallel 220$$
$$= 110.6 \text{ } \Omega$$

3. Sketch the equivalent circuit and solve for I_L. (Fig. 12-3).
$$I_L = \frac{26.47}{110.6 + 100} = 0.126 \text{ A}$$

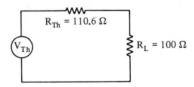

Figure 12-3

4. $P_L = I_L^2 R_L$
$$= 0.126^2 \times 100 = 1.59 \text{ W}$$
5. $P_{in} = V_{Th} I_L$
$$= 26.47 \times 0.126 = 3.34 \text{ W}$$
6. $\eta = \frac{1.59}{3.34} = 0.477 = 47.7\%$

Clearly, the solutions for P_{in} and η are meaningless and wrong! *Equivalent circuit theorems are correct only for the load.*

The power developed in a load is a function of power input and efficiency, as shown by Eq. (12-6). We find that the value of the load resistance affects both of these quantities. In a *series* circuit, increasing R_L results in *reduced* input power and *higher* efficiency. If we use a current source to supply a parallel circuit, we find that increasing R_L results in *increased* input and *lower* efficiency.

1. Given that for a series circuit

$$P_{in} = \frac{E^2}{R_T} \quad \text{and} \quad \eta = \frac{R_L}{R_T}$$

2. For a parallel circuit and a current source,

$$P_{in} = I_T^2 R_{eq} \quad \text{and} \quad \eta = \frac{G_L}{G_T} = \frac{R_{eq}}{R_L}$$

3. Once that we have P_{in} and η, we solve for P_L.

$$P_L = \eta P_{in}$$

It is important to understand that increased efficiency need not result in greater output power. In fact, we shall demonstrate that at maximum possible power output we find efficiency to be 50%.

In the circuit of Fig. 12-4, we shall change R_L in 10-Ω steps from 0 to 100 Ω. We calculate efficiency, input power, and output power using the equations in the previous list. We shall demonstrate some of the calculations, leaving the remainder to the student. Table 12-1 summarizes the results.

1. When $R_L = 0$,

$$\eta = \frac{0}{30} = 0, \quad P_{in} = \frac{50^2}{30} = 83.33 \text{ W}$$

$$P_L = (0)(83.33) = 0$$

2. When $R_L = 10 \text{ Ω}$,

$$\eta = \frac{10}{40} = 25\%, \quad P_{in} = \frac{50^2}{40} = 62.5 \text{ W}$$

$$P_L = 0.25 \times 62.5 = 15.63 \text{ W}$$

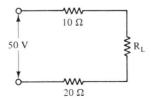

Figure 12-4

3. When $R_L = 30\ \Omega$,

$$\eta = \frac{30}{60} = 50\%, \quad P_{in} = \frac{50^2}{60} = 41.67\ W$$

$$P_L = 0.5(41.67) = 20.83\ W$$

4. When $R_L = 50\ \Omega$,

$$\eta = \frac{30}{80} = 62.5\%, \quad P_{in} = \frac{50^2}{80} = 31.25\ W$$

$$P_L = 0.625 \times 31.25 = 19.53\ W$$

5. When $R_L = 100\ \Omega$,

$$\eta = \frac{100}{130} = 76.92\%, \quad P_{in} = \frac{50^2}{130} = 19.23\ W$$

$$P_L = (0.7692)(19.23) = 14.79\ W$$

We have demonstrated that in a series circuit power input decreases and efficiency increases as the load resistance increases. Table 12-1 lists all the results. Note that for the load resistance of 30 Ω we had a 50% efficiency and maximum power output.

TABLE 12-1

R_L (Ω)	η (%)	P_{in} (W)	P_L (W)
0	0	83.33	0 W
10	25	62.5	15.63
20	40	50	20
30	50	41.67	20.83
40	57.14	35.71	20.4
50	62.5	31.25	19.53
60	66.67	27.78	18.52
70	70	25	17.5
80	72.73	22.73	16.53
90	75	20.83	15.63
100	76.92	19.23	14.79

We shall now demonstrate how changes in R_L affect efficiency and power in a parallel circuit driven by a current source as in Fig. 12-5. The results are listed in Table 12-2.

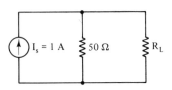

Figure 12-5

TABLE 12-2

R_L (Ω)	η (%)	P_{in} (W)	P_L (W)
10	83.3	8.33	6.94
20	71.45	14.29	10.21
30	62.5	18.75	11.72
40	55.56	22.22	12.34
50	50	25	12.5
60	45.45	27.27	12.39
70	41.67	29.17	12.15
80	38.46	30.77	11.83
90	35.71	32.14	11.48
100	33.33	33.33	11.11

1. When $R_L = 10\ \Omega$,

   $R_{eq} = 10 \parallel 50 = 8.33\ \Omega$

   $P_{in} = I^2 R_{eq} = 8.33\ \text{W}, \quad \eta = \dfrac{8.33}{10} = 83.3\%$

   $P_L = 0.833 \times 8.33 = 6.94\ \text{W}$

2. When $R_L = 30\ \Omega$,

   $R_{eq} = 30 \parallel 50 = 18.75\ \Omega$

   $P_{in} = 18.75\ \text{W}, \quad \eta = \dfrac{18.75}{30} = 62.5\%$

   $P_L = 0.625\,(18.75) = 11.72\ \text{W}$

3. When $R_L = 50\ \Omega$,

   $R_{eq} = 50 \parallel 50 = 25\ \Omega$

   $P_{in} = 25\ \text{W}, \quad \eta = \dfrac{25}{50} = 50\%$

   $P_L = 12.5\ \text{W}$

4. When $R_L = 70\ \Omega$,

   $R_{eq} = 70 \parallel 50 = 29.17\ \Omega$

   $P_{in} = 29.17\ \text{W}, \quad \eta = \dfrac{29.17}{70} = 41.67\%$

   $P_L = 12.16\ \text{W}$

Once again, the results show that maximum power is developed in the load at a particular value of load resistance. Our next topic deals with this subject.

We have compared efficiency and output power with input power for various values of load resistance. It is important for us to realize that we shall have situations where efficiency is much more important than maximum possible power. A classic example would be an electric utility.

12-3. MAXIMUM POWER TRANSFER THEOREM

All sources have internal resistance. As a result, there is only *one* value of load resistance for the source that allows the source to supply *maximum* power. We have seen this in the previous section; in this section we shall go into greater detail.

> *Maximum power is supplied by a source when the resistance connected to the source is equal to the internal resistance of the source.*

Practical dc sources are batteries, power supplies, and generators. It is certainly not practical to load these sources by resistance equal to internal resistance. A source is not usually designed to be operated in this manner. However, the theorem is important to us for use with complex networks. In a complex network, sources are usually isolated from the load by various resistors. We can find the value of load resistance that will develop maximum power in the load by using Thevenin's theorem.

> *A resistive load connected to any network receives maximum power when the load resistance is equal to the Thevenin equivalent resistance of the network as seen from the load terminals.*

Because the Norton equivalent resistance of a network is the same as R_{Th}, we need not modify the maximum power transfer theorem to deal with current sources.

It is important to remember equivalent circuit theorems are accurate only for the load. Any calculation of efficiency must be made using the actual network. We shall find that at maximum power in the load the efficiency is far from 50%, although the power in the load is accurate when using Thevenin or Norton equivalents.

EXAMPLE 12-10:
In the circuit of Fig. 12-6, what value of load resistance is needed for maximum power output?

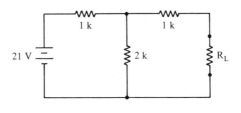

Figure 12-6

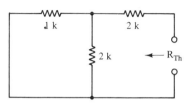

Figure 12-7

Solution:
1. Solve for the Thevenin equivalent resistance (Fig. 12-7).

$$R_{Th} = 1 \text{ k}\Omega + 1 \text{ k}\Omega \parallel 2 \text{ k}\Omega = 1.67 \text{ k}\Omega$$

2. $R_L = R_{Th} = 1.67 \text{ k}\Omega$

EXAMPLE 12-11:
Given the circuit of Example 12-10, with $R_L = 1.67 \text{ k}\Omega$, solve for P_{in}, P_{out}, and η (Fig. 12-6).

Solution:
1. Solve for R_T.

$$R_T = 1 \text{ k}\Omega + 2 \text{ k}\Omega \parallel (1 \text{ k}\Omega + 1.67 \text{ k}\Omega)$$
$$= \frac{(2 \text{ k}\Omega)(2.67 \text{ k}\Omega)}{2 \text{ k}\Omega + 2.67 \text{ k}\Omega} + 1 \text{ k}\Omega$$
$$= 1.143 \text{ k}\Omega + 1 \text{ k}\Omega = 2.143 \text{ k}\Omega$$

2. $P_{in} = \dfrac{E^2}{R_T} = \dfrac{21^2}{2.143 \text{ k}\Omega} = 206 \text{ mW}$

3. Solution for P_L:
   (a) Solve for V_{a-b}.

   $$V_{a-b} = 21 \times \frac{1.143 \text{ k}\Omega}{2.143 \text{ k}\Omega} = 11.2 \text{ V}$$

   (b) Solve for I_L.

   $$I_L = \frac{11.2}{2.67 \text{ k}\Omega} = 4.19 \text{ mA}$$

   (c) $P_L = I_L^2 R_L = (4.19 \times 10^{-3})^2 (1.67 \times 10^3)$
   $= 29.3 \text{ mW}$

4. Solution for efficiency:

$$\eta = \frac{P_L}{P_{in}} = \frac{29.3 \text{ mW}}{206 \text{ mW}} = 0.1422$$
$$= 14.29\%$$

12-4. VOLTAGE AND CURRENT SOURCES

All loaded sources supply current and voltage. We define a source as voltage or current depending upon whether it is load voltage or current that remains relatively constant over a wide range of load resistances. If load volt-

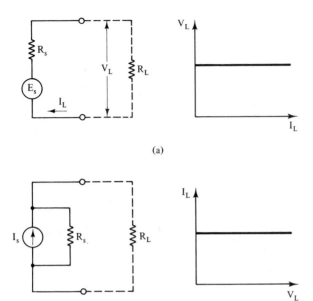

Figure 12-8 (a) A voltage source supplies constant voltage over a wide range of load currents. (b) A current source supplies constant current over a wide range of load resistance.

age remains constant, while load current varies, we refer to the source as a voltage source. When the load current is constant with changes in load resistance, while load voltage varies, we have a current source. These concepts are illustrated in Fig. 12-8.

We may design electronic circuits so that a power supply functions as a current source or a voltage source. We are also capable of operating elementary sources of power, like batteries, so that they function as voltage or current sources. Electronic devices, particularly the bipolar junction transistor, can be made to have constant current characteristics over a wide range of operating voltages.

Voltage Sources

We have said that a voltage source supplies constant voltage and a varying load current as load resistance changes. The ideal voltage source has zero internal resistance. A practical voltage source has an internal resistance that is very much less than the lowest value of load resistance to be used with the source.

EXAMPLE 12-12:
A 30-V source has an internal resistance (R_s) of 0.5 Ω. Calculate load voltage for load resistances of 23, 50, 500, and 1000 Ω.

Solution:

$$V_L = E_s \frac{R_L}{R_s + R_L} \qquad (12\text{-}10)$$

$R_L = 20\ \Omega$:
$$V_L = 30\,\frac{20}{20.5} = 29.27\ \text{V}$$

$R_L = 50\ \Omega$:
$$V_L = 30\,\frac{50}{50.5} = 29.7\ \text{V}$$

$R_L = 500\ \Omega$:
$$V_L = 30\,\frac{500}{500.5} = 29.97\ \text{V}$$

$R_L = 1\ \text{k}\Omega$:
$$V_L = 30\,\frac{1000}{1000.5} = 29.99\ \text{V}$$

Over the range of loads, V_L varied by 0.72 V, a negligible amount. Clearly, the smaller that R_s is in comparison to R_L, the more constant the output voltage. Also, V_L more closely approximates the open-circuit voltage of 30 V.

Current Sources

We defined a current source as one that supplies constant current and varying load voltage with changes in load resistance. The ideal current source is one with an internal resistance that is infinite. A practical current source is one with an internal resistance that is very much greater than any load resistance to be used with the source.

EXAMPLE 12-13:

A current source supplying 200 μA is to be used with a variety of load resistances (Fig. 12-9). The largest amount of load resistance is 1200 Ω. Demonstrate that the load current is approximately 200 μA for load resistances of 100, 500, 750, and 1200 Ω.

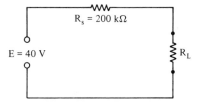

Figure 12-9

Solution:

$$I_L = \frac{E_s}{R_s + R_L} \qquad (12\text{-}11)$$

$R_L = 100\ \Omega$:
$$I_L = \frac{40}{200.1\ \text{k}\Omega} = 0.1999\ \text{mA} = 200\ \mu\text{A}$$

$R_L = 500\ \Omega$:

$$I_L = \frac{40}{200.5\ \text{k}\Omega} = 0.1995\ \text{mA} = 199.5\ \mu\text{A}$$

$R_L = 750\ \Omega$:

$$I_L = \frac{40}{200.75\ \text{k}\Omega} = 0.1993\ \text{mA} = 199.3\ \mu\text{A}$$

$R_L = 1.2\ \text{k}\Omega$:

$$I_L = \frac{40}{201.2\ \text{k}\Omega} = 0.1988\ \text{mA} = 198.8\ \mu\text{A}$$

In this example we demonstrated that the larger that R_s is in comparison to R_L, the more closely I_L equals the short-circuit current of 200 μA.

Note that in both examples, using voltage and current sources, we were concerned with the relationship *between* R_s and R_L. Internal resistance need not be "large" or "small." It is simply a matter of being very much less than load resistance in a voltage source and very much greater than load resistance in a current source.

Suppose that R_L varies from 10 to 100 kΩ. Provided that 10 kΩ is the least amount of load resistance, a source with $R_s = 100\ \Omega$ would funciton as a voltage source.

For load resistance of 1 to 5 Ω, the same source with $R_s = 100\ \Omega$ would work very well as a current source. With these load resistances, it would be a poor voltage source!

Regulation

Regulation is a measure of how well a source maintains its output despite variations in load resistance. The *smaller* the percentage of regulation, the more constant the output is.

Voltage Regulation

$$\%\text{VReg} = \frac{V_{\text{NL}} - V_{\text{FL}}}{V_{\text{FL}}} \times 100 \qquad (12\text{-}12)$$

where V_{NL} = open-circuit voltage (V no-load)
$\quad V_{\text{FL}} = V_L$ with least amount of load resistance (V full-load)
In Example 12-12, the voltage regulation is 2.49%. This is

$$\%\text{VReg} = \frac{30 - 29.27}{29.27} \times 100 = 2.49\%$$

We have electronic circuits for use with voltage sources that result in a voltage regulation of less than 0.1%.

Current Regulation

$$\%\text{IReg} = \frac{I_{sc} - I_{L(\min)}}{I_{L(\min)}} \qquad (12\text{-}13)$$

where I_{sc} = current when $R_L = 0$: $I_{sc} = E_s/R_s$
$I_{L(\min)}$ = load current with the greatest R_L
In Example 12-13, the current regulation is 0.6%.

$$\%\text{IReg} = \frac{(200 - 198.8)\,\mu A}{198.8\,\mu A} \times 100 = 0.604\%$$

12-5. VOLTAGE TO CURRENT SOURCE CONVERSION

It is often convenient to represent a voltage source as a current source. The following process is used:

1. The current of the source is the ratio of E_s to R_s.

$$I_s = \frac{E_s}{R_s} \qquad (12\text{-}14)$$

2. The internal resistance of the current source is R_s, but placed in parallel with I_s.

EXAMPLE 12-14:
Convert the sources of Fig. 12-10 into current sources.

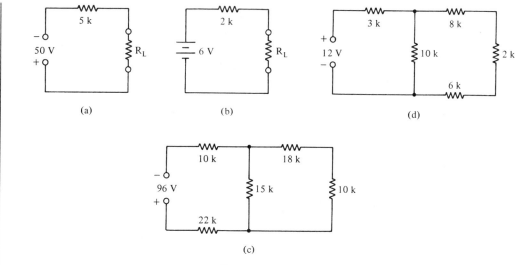

Figure 12-10

Sec. 12-5.   Voltage to Current Source Conversion

Solutions are shown in Fig. 12-11.

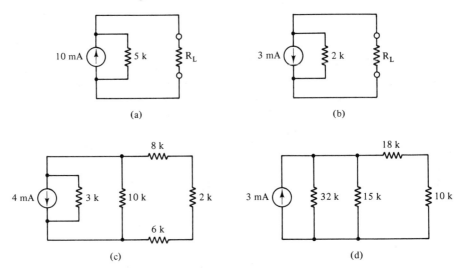

Figure 12-11

The solutions in c and d need further explanation. Whenever you encounter a complex circuit, use only those resistors *directly* in series with the voltage supply as R_s. In c we had one resistor and in d two resistors.

12-6. CURRENT TO VOLTAGE SOURCE CONVERSION

To change a current source to a voltage source, use the following steps:

1. E_s is the product of I_s and R_s. Remember, R_s is the *first* resistor in parallel with I_s.
2. The internal resistance of the voltage source is the same as R_s of the current source. Simply place it in series with E_s.

EXAMPLE 12-15:
Change the current sources of Fig. 12-12 into voltage sources. Solutions are shown in Fig. 12-13. Note that the direction of the current source arrow is consistent with the direction that current will take with the voltage source. This is also followed in the preceding example.

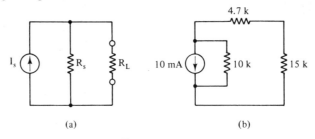

Figure 12-12

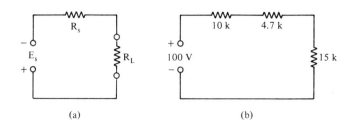

Figure 12-13

SUMMARY

1. Work, in circuits, is the product of voltage and charge displacement.
2. Power is the rate of doing work.
3. Total power is the sum of individual power dissipations.

$$P_T = EI_T = P_1 + P_2 + \cdots + P_n$$

4. Efficiency (η) is the ratio of power output to power input. In a series circuit, $\eta = R_L/R_T$. In a parallel circuit, $\eta = G_L/G_T$.
5. Maximum power is developed in a load when the load resistance is equal to the Thevenin equivalent resistance seen looking back from the load terminals.
6. An ideal voltage source has zero internal resistance.
7. An ideal current source has infinite internal resistance.
8. Good voltage regulation requires that source voltage be constant despite changes in load current.
9. A well-regulated current source maintains constant current despite changes in load voltage.
10. A voltage source can be replaced by a current source. The current is the ratio of source voltage to source resistance: $I = E/R_s$.

PROBLEMS

12-1. Solve for total circuit power, given the following voltages and currents:
  (a) $E = 120 \text{ V}, I_T = 2 \text{ A}$
  (b) $E = 12 \text{ V}, I_T = 30 \text{ mA}$
  (c) $E = 2500 \text{ V}, I_T = 40 \text{ A}$
  (d) $E = 40 \text{ V}, I_T = 250 \text{ mA}$
  (e) $E = 15 \text{ V}, I_T = 12 \text{ mA}$

12-2. Given the following currents and resistances, solve for power dissipation. (Use the appropriate engineering units of power.)
  (a) $R = 10 \text{ k}\Omega, I = 20 \text{ mA}$
  (b) $R = 50 \text{ }\Omega, I = 15 \text{ A}$
  (c) $R = 100 \text{ }\Omega, I = 7.5 \text{ A}$
  (d) $R = 2.2 \text{ k}\Omega, I = 2 \text{ mA}$
  (e) $R = 390 \text{ k}\Omega, I = 50 \text{ }\mu\text{A}$

12-3. A series circuit consists of three resistors: 4.7 kΩ, 2.2 kΩ, and 10 kΩ. Calculate the power dissipated in each resistor if current is 200 mA.

12-4. Solve for the power dissipated in each resistance, given the following voltages:
   (a) $V = 15$ V, $R = 100$ Ω
   (b) $V = 125$ mV, $R = 4.7$ kΩ
   (c) $V = 10$ kV, $R = 200$ Ω
   (d) $V = 120$ V, $R = 240$ Ω
   (e) $V = 15$ mV, $R = 47$ kΩ

12-5. A three-branch parallel circuit has a voltage drop of 30 V. Given that $R_1 = 4.7$ kΩ, $R_2 = 1$ kΩ, and $R_3 = 3.9$ kΩ, calculate the power dissipated in each resistor.

12-6. Calculate the total power dissipation in each of the circuits of Fig. 12-14.

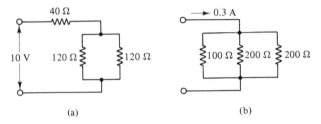

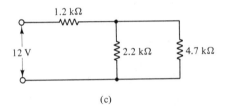

Figure 12-14

12-7. Given the circuit of Fig. 12-15, calculate power input, power output, and efficiency.

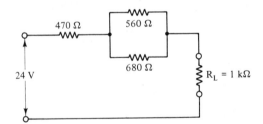

Figure 12-15

12-8. Given each of the circuits of Fig. 12-16, solve for efficiency.

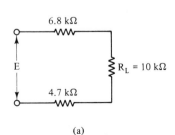

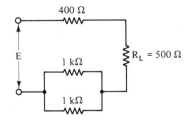

Figure 12-16

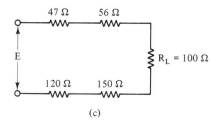

(c)

Figure 12-16 *(continued)*

12-9. Solve for efficiency in each of the circuits of Fig. 12-17.

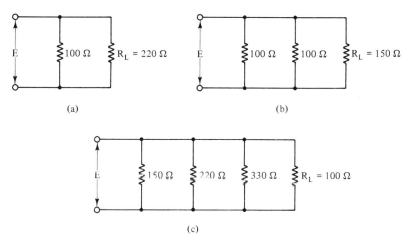

Figure 12-17

12-10. The current in R_5 of Fig. 12-18 is 420 μA. What is the power input to the circuit?

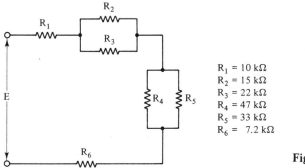

$R_1 = 10\ k\Omega$
$R_2 = 15\ k\Omega$
$R_3 = 22\ k\Omega$
$R_4 = 47\ k\Omega$
$R_5 = 33\ k\Omega$
$R_6 = 7.2\ k\Omega$

Figure 12-18

12-11. In the circuit of Figure 12-18, E has been changed to 12 V. What is the power dissipated in R_6?

12-12. What value of load resistance will result in maximum power output (P_L), given the circuit of Fig. 12-19?

Chap. 12   Problems   239

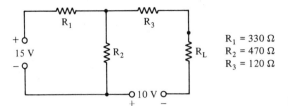

Figure 12-19

12-13. Given the circuit of Fig. 12-20, solve for the value of R_3 that will permit maximum power to be developed in R_L.

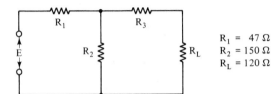

Figure 12-20

12-14. In the circuit of Fig. 12-21, solve for the value of R_2 that permits maximum power dissipation in R_L.

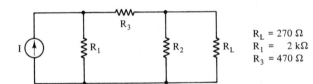

Figure 12-21

12-15. In the circuit of Fig. 12-22, R_2 is a coupling resistor. Solve for the value of R_2 that results in maximum power output in R_L.

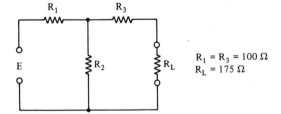

Figure 12-22

12-16. A 45-V source with an internal resistance of 5 Ω is to be used with a variety of load resistances, from 200 to 1200 Ω.
    (a) What is the source output voltage at minimum load current ($R_L = 1.2$ kΩ)?
    (b) What is the source output voltage at maximum load current ($R_L = 200$ Ω)?

12-17. What is the voltage regulation of the source in Prob. 12-16 for the given operating conditions?

12-18. Represent the voltage sources of Fig. 12-23 as current sources.

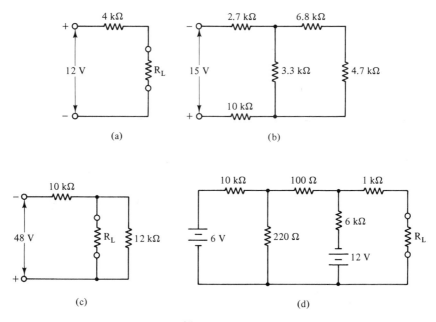

Figure 12-23

12-19. A 100-mA current source with an internal resistance of 2 kΩ is used to supply a variety of load resistances. Maximum load resistance is 47 kΩ. Solve for the current regulation.

12-20. Represent the current sources of Fig. 12-24 as voltage sources.

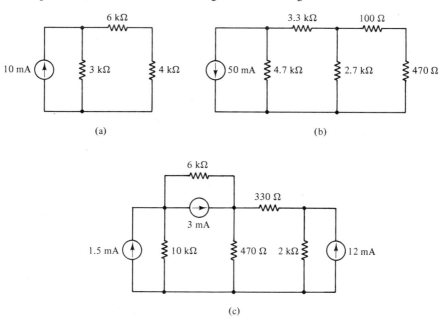

Figure 12-24

Chap. 12   Problems   241

13

Capacitance

13-1. INTRODUCTION

So far we have studied resistive circuits. Resistance is the property of a circuit that opposes current and *dissipates power*. There are two additional properties of passive circuits, *inductance* and *capacitance*. These are energy-storage properties of a circuit. Theoretically, they return as much energy to a circuit as is supplied them. Also, their effects are not observed under steady-state conditions. Rather, the effects of these circuit properties are observed when there is a *change* in voltage or current. We shall study inductance in the next chapter.

Capacitance is usually defined as the property of a circuit that stores energy in the form of an electrostatic field. We shall modify this definition:

> *Capacitance is that circuit property that opposes changes in voltage by means of energy storage in the form of an electrostatic field.* When voltage attempts to rise, capacitance stores energy (charges). When voltage drops, the capacitance returns energy to the circuit (discharges).

13-2. ELECTROSTATIC FIELD

All electric charges have electrostatic fields associated with them. The fields represent the forces of attraction and repulsion by charges upon one another. We know that like charges repel and unlike charges attract each other. By *definition*, we represent the electrostatic field as leaving a positive charge

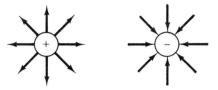

Figure 13-1 Charged particles have electrostatic forces represented by fields. By definition, positive charges radiate force outwards, while negative charges have forces directed toward them.

and ending on a negative charge. Figure 13-1 shows electrostatic flux radiating away from a positive charge and into a negative charge. We are simply agreeing that a positively charged point repels a positive charge, while a negatively charged point attracts the positive charge. This is illustrated in Fig. 13-2.

The force of repulsion or attraction is determined by Coulomb's law.

$$F = \frac{k(Q_1)(Q_2)}{d^2} \text{ N} \qquad (13\text{-}1)$$

where F = force in newtons (N)
$k = 9 \times 10^9$
d = distance between points in meters (m)
Q_1 and Q_2 = charges in coulombs (C)

It is of little practical value to continue discussions of isolated charges. We are interested in the electrostatic field between *charged* conducting plates. For these plates to have charge, we must insulate one from the other and supply energy in the form of voltage.

The field between the plates exerts a force upon charges within the field. This force depends upon the strength of the field. This field strength, sometimes called *field intensity*, depends upon the voltage and the distance between the plates.

$$\mathcal{E} = \frac{V}{d} \qquad (13\text{-}2)$$

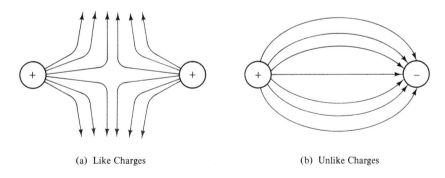

(a) Like Charges (b) Unlike Charges

Figure 13-2 Like charges repel, unlike charges attract. The electrostatic flux drawn is simply an illustration of the forces between charges. The fringing and mutual repulsion between lines of force is much greater than shown.

Sec. 13-2.   Electrostatic Field

where $\mathcal{E}$ = field strength in volts per meter
V = voltage between the plates
d = spacing between the plates in meters

13-3. CAPACITANCE

The amount of capacitance in a circuit depends upon the amount of charge stored for a given voltage.

$$C = \frac{Q}{V} \text{ F} \tag{13-3}$$

where C = capacitance in farads (F)
Q = charge in coulombs (C)
V = voltage applied to plates

A farad is an extremely large unit of capacitance, requiring a charge storage of one coulomb per volt. The common units of capacitance are the microfarad (μF), and the picofarad (pF). A less widely used unit is the nonfarad (nF).

$$1 \ \mu\text{F} = 10^{-6} \text{ F}$$
$$1 \text{ nF} = 10^{-9} \text{ F} = 10^{-3} \ \mu\text{F}$$
$$1 \text{ pF} = 10^{-12} \text{ F} = 10^{-6} \ \mu\text{F}$$

When we wish to insert capacitance into a system, we use *capacitors*. These devices are designed to have a quantity of capacitance. They range in size from a few picofarads to thousands of microfarads.

Just as resistors are used to provide definite amounts of resistance, we use capacitors to cause a circuit to have a specific amount of capacitance. Capacitance also exists anywhere conductors have a difference of potential between them. This is called *stray* capacitance and can be a cause of poor high-frequency performance in electronic circuits.

EXAMPLE 13-1:
A pair of plates are 1 mm apart. When 20 V is applied, the stored charge is 40 mC. What is the capacitance?

Solution:

$$C = \frac{Q}{V}$$
$$= \frac{40 \times 10^{-3}}{20} = 2 \times 10^{-3}$$
$$= 2000 \ \mu\text{F}$$

EXAMPLE 13-2:
What is the amount of charge in coulombs that a 0.05-μF capacitor will store if charged to 100 V?

Solution:
From Eq. (13-3),
$$Q = (C)(V)$$
$$= (0.05 \times 10^{-6})(100) = 5 \times 10^{-6} \text{ C}$$
$$= 5 \text{ } \mu\text{C}$$

EXAMPLE 13-3:

How much voltage is needed to store a charge of 3 mC in a 10-μF capacitor?

Solution:
From Eq. (13-3),
$$V = \frac{Q}{C}$$
$$= \frac{3 \times 10^{-3}}{10 \times 10^{-6}} = 0.3 \times 10^{3}$$
$$= 300 \text{ V}$$

13-4. DETERMINING CAPACITANCE

In this section we examine those factors that influence the amount of capacitance in a capacitor.

In Eq. (13-3), we defined capacitance as the ratio of charge storage to voltage. We can define charge in coulombs in terms of electrostatic flux. In the SI system of units, flux (ψ) is exactly equal to charge in coulombs.

$$\psi \equiv Q \tag{13-4}$$

The flux density of the field is easily found.

$$D = \frac{\psi}{A} = \frac{Q}{A} \tag{13-5}$$

where A is in square meters.

We are able to relate the flux density to field strength in order to consider another quantity. *Permittivity* is a measure of the ease with which an insulating material permits an electrostatic field to be established. It is similar in concept to permeability for magnetic materials.

$$D = \epsilon \frac{V}{d} \tag{13-6}$$

where D = flux density in coulombs per square meter
ϵ = permittivity of the dielectric material
V/d = field strength in volts per meter

We have two expressions for electrostatic flux density. We set them equal to each other and solve for Q/V, capacitance in farads.

$$\frac{Q}{A} = \frac{\epsilon V}{d}$$

$$C = \frac{Q}{V} = \frac{\epsilon A}{d} \tag{13-7}$$

We define the permittivity of a vacuum as ϵ_0.

$$\epsilon_0 = 8.85 \times 10^{-12} \quad \text{F/m}$$

The relative permittivity of a material is the ratio of its permittivity to that of a vacuum.

$$\epsilon_r = \frac{\epsilon}{\epsilon_0} \tag{13-8}$$

Finally, we have the formula for the capacitance of a capacitor.

$$C = \frac{\epsilon_0 \epsilon_r A}{d} \quad \text{F} \tag{13-9}$$

The formula may be written

$$C = \frac{(8.85 \times 10^{-12}) \epsilon_r A}{d} \tag{13-10}$$

The relative permittivity of various dielectric materials is listed in Table 13-1.

TABLE 13-1

Dielectric	Relative Permittivity
Air	1
Bakelite	7
Barium strontium titanate	7500
Glass	6
Mica	5-9
Paper	2.5
Titanium dioxide	90-170
Teflon	2

Capacitance is *proportional* to cross-sectional area and *inversely* proportional to the spacing between the plates. Moving the plates closer results in more capacitance for a given area. Clearly, there are limits on the spacing between the plates. The dielectric, the insulator between the plates, enables us to effectively bring the plates "closer" without the possibility of shorting.

Consider the effect of an electric field upon the molecules of a dielectric material as shown in Fig. 13-3. The dielectric, being an insulator, does not pass measurable current unless the field strength is great enough to cause breakdown. However, the molecules orient themselves in accordance with the field as shown in the figure. The molecules are said to be polarized. The result of the polarized molecules is to increase the effect of the voltage in causing charge storage. The relative permittivity is an indication of the degree of polarization of the material. The greater the proportion of the

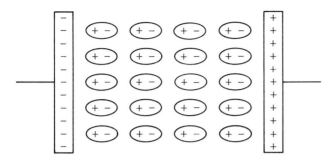

Figure 13-3 Polarization of a dielectric material. Effective dielectrics have large quantities of molecules or atoms that polarize under the influence of an electrostatic field.

molecules that aid in charge storage, the higher the value of the relative permittivity is.

EXAMPLE 13-4:
A capacitor, with paper dielectric, has plates with an area of 0.6 m² and a spacing of 0.1 mm. What is the capacitance?

Solution:
1. ϵ_r for paper = 2.5.
2. Using Eq. (13-10),

$$C = \frac{(8.85 \times 10^{-12})(2.5)(0.6)}{0.1 \times 10^{-3}}$$
$$= 1.33 \times 10^{-7}$$
$$= 0.133 \ \mu F$$

EXAMPLE 13-5:
Suppose that the capacitor of Example 13-4 uses titanium dioxide, with $\epsilon_r = 94$. What is the capacitance?

Solution: We can solve by use of Eq. (13-10), or we can take the ratio of the relative permittivities.

$$\frac{\epsilon_{r2}}{\epsilon_{r1}} = \frac{94}{2.5} = 37.6$$

The capacitance with the ceramic dielectric is 37.6 times greater than the one with the paper dielectric.

$$C_2 = 37.6(C_1)$$
$$= 37.6 \times 1.33 \ \mu F$$
$$= 50 \ \mu F$$

13-5. DIELECTRIC STRENGTH AND LEAKAGE CURRENT

We do not design capacitors, we use them. The previous examples are intended to help you understand principles. We now proceed to an understanding of some of the limitations associated with capacitors.

The *dielectric strength* of a capacitor dielectric has the same meaning as for any insulator. It is a measure of the maximum voltage, measured in volts per mil, that can be applied without breakdown of the insulator. Table 13-2 lists the dielectric strengths of various insulators. These values are approximations and are affected by temperature, humidity, and frequency of applied voltage. Capacitors are rated in terms of capacitance and maximum dc voltage.

TABLE 13-2

Dielectric	Dielectric Strength (V/mil)
Air	75
Barium strontium titanate	75
Porcelain	200
Bakelite	400
Paper	500
Teflon	1500
Glass	3000
Mica	5000

There are no perfect insulators. All insulators have leakage current, however small it may be. We represent the effect of this leakage current, which results in some power dissipation, by a very small series resistance or by a very large parallel resistance. In most types of capacitors, the parallel resistance is on the order of 10^9 Ω. An ohmmeter will simply register infinite resistance when connected across capacitors with low leakage current of this nature. In your studies of alternating-current circuits, you will learn other ways to account for the dissipation that takes place in a capacitor.

13-6. TYPES OF CAPACITORS

Capacitors are manufactured with capacitance values that range from a few picofarads to thousands of microfarads. Some low-voltage types actually exceed 1 F. Maximum operating voltages on commonly used units are as low as 5 V, while others are available with ratings on the order of many kilovolts. A color TV set uses a capacitor with a rating of 50 kV, along with many others whose voltage ratings and capacitances are representative of the range of capacitance values available.

The capacitance and voltage rating are a function of construction. We shall not go into the details of construction, but we should be aware of the various types of capacitors. You will find that we identify most types by the dielectric material.

Mica capacitors are made of sets of parallel plates insulated by thin sheets of mica. These are available in units ranging from a few picofarads to a few nanofarads. They are particularly useful at high radio frequencies. They are characterized by very low leakage current and very high leakage resistance, and they are available with voltage ratings in the kilovolts.

Paper capacitors are made using strips of aluminum foil with treated paper as the dielectric. The foil and dielectric are rolled into a cylindrical form. Leads are brought out each end. The capacitor may be sealed in wax paper or plastic. Usually, plastic encapsulation is used where operating temperatures are high. One side of the capacitor is usually identified, frequently by a small black band. This identifies the lead wire connected to the outer foil of the capacitor. Where possible, this lead is connected to a point of lower signal potential, preferably ground. Like mica, paper capacitors have very low leakage current and very high leakage resistance.

Ceramic capacitors use various types of ceramic dielectrics. Because of the very high permittivity, ceramic capacitors are usually smaller in size than paper or mica for the same amount of capacitance. Capacitors with barium strontium titanate have large amounts of capacitance in a small package. However, they usually have lower breakdown voltages than paper or mica types. Also, they tend to have values of capacitance with large tolerance values, rather than the precision amounts that can be obtained with mica. Leakage resistance is very high, but not as high as for mica or paper capacitors. Some ceramic capacitors are shown in Fig. 13-4.

Electrolytic capacitors are of a special form of construction. One electrode, the *positive* terminal, is made of aluminum foil that has been chemically roughened in order to increase its surface area. For the other plate we use aluminum hydroxide. This is a conductive compound. In the electrolytic capacitor it is usually in paste form, although it can be liquid. The capacitor is packaged in an aluminum cylinder. The aluminum container is in contact with the hydroxide; the negative terminal is connected to the container. In axial lead types, the container is usually covered by a cardboard tube. *When voltage of the correct polarity is applied to the capacitor, a very thin insulating layer of hydrogen atoms forms between the aluminum foil and the aluminum hydroxide.* A reversal of polarity removes the insulating layer, allowing dangerously high currents. If operated with reverse polarity, the capacitor can overheat and explode. In schematic diagrams, one

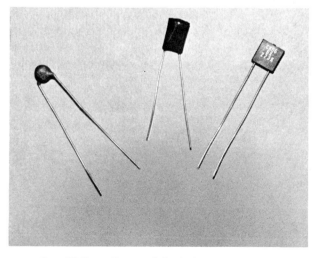

Figure 13-4

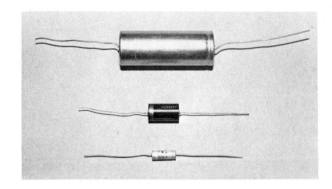

Figure 13-5

must be careful to identify − and + electrodes. So-called nonpolarized electrolytic capacitors are usually two electrolytic capacitors in series in an inverse arrangement so that one of them is always working as a capacitor, and thus prevent destructive currents. Only these types may be used with alternating voltages. We find them in ac motor circuits for start and run operations.

The design of electrolytic capacitors permits great amounts of capacitance in relatively small packages. Typical values of capacitance range from a few to thousands of microfarads. Various electrolytic capacitors are shown in Fig. 13-5.

Figure 13-6

Electrolytic capacitors have lower working voltages than other types. They also have much higher leakage current and much lower leakage resistance. Typical leakage resistance is about 1 MΩ. They also have poor storage life, as they lose the dielectric over long periods of time. It is necessary to "re-form" the dielectric by applying dc of the proper polarity to the capacitor. One must begin with low dc voltages and gradually raise the voltage to the rated voltage of the capacitor.

A variation of the aluminum electrolytic capacitor is the tantalum type, often referred to as the *tantalytic* capacitor. Size and values are similar to standard electrolytic capacitors. They have less leakage current than standard types, but are more expensive. They are made with lower voltage ratings than ordinary electrolytic capacitors because they are used primarily in solid-state circuits.

Variable capacitors permit variation of capacitance by changing the area between the plates or by changing the spacing between the plates. All variable capacitors have low capacitance in the picofarad range.

Capacitors that use a variable area design are used as *tuning capacitors* in variable-frequency circuits. Typical applications are in radios and variable oscillator circuits. Smaller capacitors, usually screwdriver adjustable, are used as *trimmer* capacitors. These are adjusted to provide some specific amount of capacitance, and the setting is not changed until readjustment of the circuit is needed. Air dielectric variable capacitors are shown in Fig. 13-6.

13-7. CAPACITORS IN SERIES

When capacitors are connected in series, the total capacitance is less than that of the smallest capacitor. As in any series circuit, we find that charge movement is the same in all parts of the circuit. Also, the sum of the voltages around the circuit must equal the applied voltage.

If, for example, we have three capacitors in series across a dc source, we find that the total charge is the same as the charge on each capacitor. The sum of the voltages across the capacitors is equal to the applied voltage.

Given that

$$C = \frac{Q}{V}$$

$$V = \frac{Q}{C}$$

Therefore,

$$E = \frac{Q}{C_T}$$

$$V_1 = \frac{Q}{C_1}$$

$$V_2 = \frac{Q}{C_2}$$

$$V_3 = \frac{Q}{C_3}$$

We then have

$$E = V_1 + V_2 + V_3$$

$$\frac{Q}{C_T} = \frac{Q}{C_1} + \frac{Q}{C_2} + \frac{Q}{C_3}$$

Dividing by Q, we have

$$\frac{1}{C_T} = \frac{1}{C_1} + \frac{1}{C_2} + \frac{1}{C_3}$$

The formula for capacitors in series may now be written in general form:

$$C_T = \frac{1}{1/C_1 + 1/C_2 + \cdots + 1/C_n} \qquad (13\text{-}11)$$

Note that the formula for capacitors in series is of the same form as the formula for resistors in parallel. In the same way, we can write a product-over-sum formula for two capacitors in series.

$$C_T = \frac{C_1 C_2}{C_1 + C_2} \qquad (13\text{-}12)$$

Be aware that the voltage across a capacitor is inversely proportional to capacitance. The smaller the capacitor, for a given amount of charge, the greater the voltage across the capacitor. We must be careful when connecting capacitors in series that we do not exceed the voltage rating of the capacitor. For example, connecting two 400-V capacitors in series does *not* provide 800-V capability unless the capacitors have *equal* capacitance.

EXAMPLE 13-6:
Given the circuit of Fig. 13-7, (a) solve for total capacitance, and (b) solve for the voltage across each capacitor.

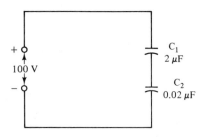

Figure 13-7

Solution:

(a) $C_T = \dfrac{C_1 C_2}{C_1 + C_2}$

$= \dfrac{(2 \times 10^{-6})(2 \times 10^{-8})}{2.02 \times 10^{-6}}$

$= 1.98 \times 10^{-8}$

$= 0.0198 \ \mu F$

(b) $Q = (C_T)(E)$

$= (1.98 \times 10^{-8})(100)$

$= 198 \times 10^{-8}$ C

$V_1 = \dfrac{Q}{C_1}$

$= \dfrac{198 \times 10^{-8}}{2 \times 10^{-6}}$

$= 99 \times 10^{-2}$ V

$= 0.99$ V

$V_2 = \dfrac{Q}{C_2}$

$= \dfrac{198 \times 10^{-8}}{2 \times 10^{-8}}$

$= 99$ V

C_2 is 1/100 of C_1; therefore, it has 100 times as much voltage across it.

EXAMPLE 13-7:

Given the circuit of Fig. 13-8, solve for (a) total capacitance, and (b) the voltage across each capacitor.

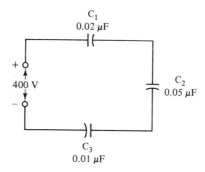

Figure 13-8

Solution:

(a) $C_T = \dfrac{1}{\dfrac{1}{0.02 \times 10^{-6}} + \dfrac{1}{0.05 \times 10^{-6}} + \dfrac{1}{0.01 \times 10^{-6}}}$

$= \dfrac{1}{50 \times 10^6 + 20 \times 10^6 + 100 \times 10^6}$

$$= \frac{1}{170 \times 10^6}$$
$$= 0.00588 \times 10^{-6}$$
$$= 5.88 \text{ nF}$$

(b) $Q = (C_T)(E)$
$$= (5.88 \times 10^{-9})(4 \times 10^2)$$
$$= 23.52 \times 10^{-7} \text{ C}$$

$$V_1 = \frac{Q}{C_1}$$
$$= \frac{23.52 \times 10^{-7}}{2 \times 10^{-8}}$$
$$= 117.6 \text{ V}$$

$$V_2 = \frac{Q}{C_2}$$
$$= \frac{23.52 \times 10^{-7}}{5 \times 10^{-8}}$$
$$= 47.04 \text{ V}$$

$$V_3 = \frac{Q}{C_3}$$
$$= \frac{23.52 \times 10^{-7}}{1 \times 10^{-8}}$$
$$= 235.2 \text{ V}$$

13-8. CAPACITORS IN PARALLEL

The use of capacitors in parallel is a common practice. As in any parallel circuit, the voltage across each capacitor is equal. The total charge is equal to the sum of the charges in each capacitor.

$$Q_T = Q_1 + Q_2 + \cdots + Q_n$$

Given that $Q = CV$, we have

$$Q_T = C_T V$$
$$Q_1 = C_1 V$$
$$Q_2 = C_2 V$$
$$Q_n = C_n V$$
$$C_T V = C_1 V + C_2 V + \cdots + C_n V$$

Dividing by V, we have

$$C_T = C_1 + C_2 + \cdots + C_n \qquad (13\text{-}13)$$

We frequently use the parallel arrangement in order to obtain extremely large amounts of capacitance. Also, we use trimmer capacitors in parallel with tuning capacitors to obtain specific amounts of capacitance over the range of the tuning capacitor.

EXAMPLE 13-8:
Solve for total capacitance in Fig. 13-9.

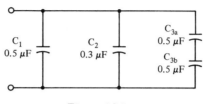

Figure 13-9

Solution:

$$C_T = C_1 + C_2 + 1/2C_3$$
$$= 0.5 + 0.3 + 0.25$$
$$= 1.05 \ \mu F$$

13-9. ENERGY STORAGE

We began our study of capacitance with the understanding that it stores energy in the form of an electrostatic field. The amount of energy stored depends upon capacitance and voltage. The total energy stored is found as

$$W = \frac{CV^2}{2} \ \text{J} \qquad (13\text{-}14)$$

where W = energy stored in joules (J)
C = capacitance in farads (F)
V = voltage across the capacitor

When discharging, the capacitor returns energy to the circuit.

EXAMPLE 13-9:
What is the energy stored in a 200-μF capacitor if the voltage across it is 50 V?

Solution:

$$W = \frac{0.2 \times 10^{-3} \times 50^2}{2}$$
$$= \frac{0.2 \times 10^{-3} \times 2.5 \times 10^3}{2}$$
$$= 0.25 \ \text{J}$$

We may determine energy storage in terms of coulombs of charge stored and capacitance. Using $V = Q/C$, we have

$$W = \frac{Q^2}{2C} \qquad (13\text{-}15)$$

EXAMPLE 13-10:
What is the energy returned to the circuit when a 5-μF capacitor with stored charge of 800 μC is discharged totally?

Solution:

$$W = \frac{Q^2}{2C}$$

$$= \frac{(800 \times 10^{-6})^2}{2 \times 5 \times 10^{-6}}$$

$$= 0.064 \text{ J}$$

SUMMARY

1. Capacitance stores energy in the form of an electrostatic field.
2. The unit of capacitance is the farad, a nonpractical unit. The practical units are microfarads (μF), nanofarads (nF), and picofarads (pF).
3. Electrostatic field strength is measured in volts per meter.
4. The capacitance of a capacitor is greatly affected by the type of dielectric material placed between the plates.
5. Capacitors are rated in capacitance and working voltage.
6. Capacitors in series add like resistors in parallel.
7. Total capacitance of parallel capacitors is the sum of the capacitances: $C_1 \parallel C_2 = C_1 + C_2$.

PROBLEMS

13-1. The charge stored in a capacitor when 24 V is applied is 720 μC. Solve for the capacitance.

13-2. A charge storage of 50 mC occurs when 40 V is applied to a certain capacitor. What is the capacitance?

13-3. A 25-μF capacitor is charged from a 40-V source. Assuming maximum charge, what is the amount of stored charge in microcoulombs?

13-4. A 500-pF capacitor is used in a 25-kV circuit. What is the maximum amount of stored charge?

13-5. What voltage is required to store a charge of 2.4 mC in a capacitance of 4 μF?

13-6. What voltage is needed to store a charge of 60 μC in a 5-nF capacitor?

13-7. An air dielectric capacitor has an area between the plates of 0.018 m² and a spacing between the plates of 2 mm. What is the capacitance of the unit?

13-8. Titanium dioxide, $\epsilon_r = 130$, is used as the dielectric in a capacitor. Total area between the plates is 0.13 m², with a spacing between the plates of 1.5 mm. What is the capacitance?

13-9. Repeat Prob. 13-8, but with paper as the dielectric.

13-10. A certain capacitor has a capacitance of 0.001 μF when mica is the dielectric material. Another capacitor of identical dimensions has a capacitance of 0.02 μF. What is the dielectric material of the second capacitor?

13-11. Given the circuit of Fig. 13-10:
  (a) Solve for total capacitance.
  (b) Assuming 100 V is applied to the circuit, solve for total charge storage.
  (c) Solve for the voltage across each capacitor.

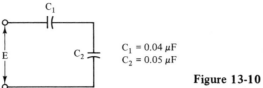

Figure 13-10

13-12. What is the total capacitance of the circuit of Fig. 13-11?

13-13. In the circuit of Fig. 13-11, $E = 150$ V. Solve for the voltage across each capacitor.

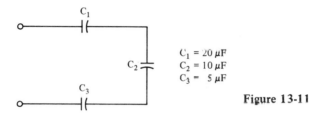

Figure 13-11

13-14. Given the circuit of Fig. 13-12, solve for total capacitance and the voltage across each capacitor. All capacitors are rated at 50 V dc and 100 μF.

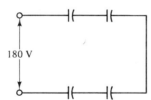

Figure 13-12

13-15. A parallel capacitor combination of 2 μF, 5 μF, and 8 μF is used across a source of 60 V.
  (a) Solve for the total capacitance.
  (b) Solve for the charge stored in each capacitor.
  (c) Solve for total charge storage.

13-16. Solve for total capacitance in the circuit of Fig. 13-13.

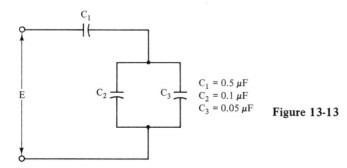

Figure 13-13

13-17. If $E = 80$ V is applied to the circuit of Fig. 13-13, solve for the voltage across each capacitor.

13-18. The source voltage to the circuit of Fig. 13-13 has been changed. The charge stored in C_1 is 4.04 μC. Solve for the applied voltage, E.

13-19. A 500-μF capacitor has charged so that the voltage across it is 90 V. What is the amount of energy stored in the unit?

13-20. What is the amount of energy stored by a 20-μF capacitor if it has a charge storage of 960 μC?

14

Electromagnetism

14-1. INTRODUCTION

The attraction that a magnet has for iron and iron alloys was known as early as the first century. The magnet studied by the ancients was lodestone, an oxide of iron that exhibits magnetic properties. The Chinese were probably the first to use lodestone as a compass. This crude compass helped them sail their ocean-going junks over great distances nearly 2000 years ago. Lodestone is a natural magnet and is of no interest to us except for its historical interest. Our concern is in the relationship between electric current and magnetism or *electromagnetism*.

In the early nineteenth century, Hans Christian Oersted, a Danish physicist, discovered that a compass needle was deflected when brought near a wire carrying a current, as shown in Fig. 14-1. This was the first time that it was learned that electricity and magnetism are related. Modern physics confirms that magnetic forces are always associated with an electric current. In fact, present theory holds that electrons have magnetic properties as well as electrical properties.

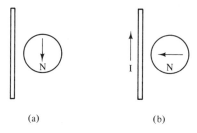

Figure 14-1 The effect of current on the direction of a compass needle. (a) No current in the wire. (b) Current, about 30 A, in the wire. The compass needle is deflected away from its normal direction.

Faraday, the developer of the dynamo, the earliest of the electrical machines, was the first to put electromagnetism to practical use. Other nineteenth-century physicists, notably Ampère, Lenz, Maxwell, and Gauss, developed most of the knowledge we have relating to electromagnetism. The formulas we shall work with are the result of the studies done by these pioneers.

Electromagnetism is used in nearly all applications of electricity and electronics. Stereo systems, automobile electrical systems, ignition systems, relays, telephones, certain computer data storage systems, TV sets, transformers, motors, and generators are just a sample of the many uses of electromagnetism. In fact all electrical generating stations change energy into electrical energy by machines that make use of a basic law of electromagnetism.

14-2. MAGNETIC FIELD

A magnetic field is said to be any region where magnetic force is found. We cannot see magnetic fields, but we can describe such fields by observation of the effects of magnetic fields. For example, if we place stiff paper over a bar magnet and sprinkle iron filings onto the paper, we find that the filings are distributed in a definite pattern. The pattern shows that there appear to be *lines of force* between the ends of the bar magnet. We come to this conclusion because the iron filings have formed groups of lines between the ends of the bar magnet.

We have said that magnetic force in the form of a field exists between the ends of the magnet. We have represented this force by lines. The lines are simply a way of describing the location and amount of magnetic forces. They do not represent the flow of anything.

The ends of a magnet, where the lines of force appear to be most concentrated, are called *poles*. We define the polarity of the poles, and therefore the field, on the basis of an assumed direction for the lines of force. By definition, the lines of force *leave* the *north* pole of a magnet and *enter* the *south* pole of the magnet. Within the magnet the lines of force have a south to north direction. Keep in mind that we use lines of force as a convenient way to describe a magnetic field. This discussion about the direction of the lines of force describes an arbitrary convention that is standard throughout the world; the convention is shown in Fig. 14-2.

We refer to the magnetic lines of force as magnetic flux or simply flux. The symbol for flux is the Greek letter (Φ). The SI unit of flux is the weber (Wb). One weber is equal to 100,000,000 lines of flux.

$$1 \text{ Wb} = 10^8 \text{ lines}$$

Magnetic flux has several important properties:

1. Unlike poles attract each other.
2. Like poles repel each other.

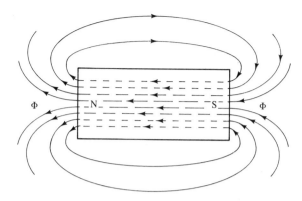

Figure 14-2 The magnetic field of a bar magnet.

3. Lines of force always form complete loops. Therefore, we shall always assume that the amount of flux within a magnet is equal to the amount of flux in its magnetic field.
4. Lines mutually repel each other. This property is observed as a spreading of the flux outside of a magnet. Flux concentrates at the poles of a magnet in order to travel within the magnet from south pole to north pole.
5. The lines of force represent energy along their length. They always tend to shorten so as to establish the greatest amount of flux (for a given situation). The flux follows a path of least opposition, so the shortest path is not one of least distance, but one of least opposition. *We can summarize this last idea by simply realizing that a magnetic field always arranges itself so as to establish maximum possible flux for a given situation.*

14-3. MAGNETIC MATERIALS

We can classify materials in terms of their effect upon a magnetic field. Consider a magnetic field in a vacuum. For a given set of conditions, there would be so many webers of flux. Now consider that the vacuum has been replaced by some material such as wood or paper. We would find that the amount of flux is virtually unchanged from that of a vacuum. Materials that have the same effect upon a magnetic field as does free space (the vacuum) are called *paramagnetic* materials. Air is a classic and widely used example of a paramagnetic substance.

Other materials, such as bismuth, offer slightly more opposition than paramagnetic materials to magnetic flux. This category of materials is called *diamagnetic*.

Certain substances greatly assist in the establishment of magnetic flux. When these types of materials are used in place of paramagnetic substances, the amount of flux increases greatly. Materials of this type are called *ferromagnetic*. They are various types of iron and steel, as well as alloys of nickel and

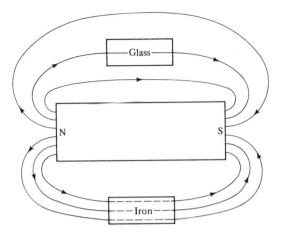

Figure 14-3 The effects of paramagnetic and ferromagnetic materials on flux paths. Glass is nonmagnetic and the lines of flux are unaffected by it. The iron has a high permeability and therefore less opposition to flux. The lines take this path despite the crowding.

cobalt. Certain ferromagnetic materials have as little as 10^{-5} of the opposition that air has to magnetic flux (Fig. 14-3).

When magnetizing force is applied to a ferromagnetic material, the material exhibits all the properties of a magnet. Some materials like soft iron and transformer steel lose magnetic properties very quickly when magnetizing force is removed. Such materials are used as part of *temporary magnets*.

Permanent magnets can retain magnetic properties for many years after being magnetized. The ferromagnetics most notable for this property are alloys of nickel and cobalt. A widely used commercial permanent magnet material is Alnico. The ability of a material to remain magnetized after magnetizing force is removed is called *retentivity*.

14-4. MOLECULAR THEORY OF MAGNETISM

Modern physics has determined that electrons have magnetic properties. Nearly two centuries ago it was learned that magnetic fields are associated with electric currents. In fact, one finds that *currents always produce magnetic fields*. There are theories of magnetism based upon the magnetic properties of electrons.

The molecular theory holds that all atoms and molecules have magnetic properties. Individual atoms and molecules are considered to behave as tiny bar magnets, called *domains*. Under this theory, ferromagnetic substances are those whose molecules reorient in terms of magnetism when influenced by magnetizing forces. It is thought that the domains align themselves so that the magnetic forces of each are added, which results in a material that exhibits magnetic properties. The domains of temporary magnets are considered to become randomly oriented when magnetizing force is removed. In the case of permanent magnets, the domains remain oriented so as to add to one another after magnetizing force is removed.

In paramagnetic and diamagnetic materials, magnetic orientation cannot occur despite the use of external magnetizing force. The magnetic domains

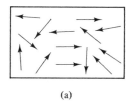 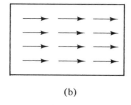

(a)   (b)

Figure 14-4 Non-magnetic and ferromagnetic materials when influenced by a magnetic field. (a) Non-magnetic material. The domains are randomly oriented. (b) Ferromagnetic material. The sketch shows the domains in supportive orientation.

of these substances are randomly oriented and remain so, with or without external magnetic force. These concepts are illustrated in Fig. 14-4.

14-5. FLUX DENSITY

So far we have considered magnetic fields in terms of flux. Flux cannot, by itself, be any indicator of the intensity or strength of a magnetic field. *Flux density* is an accurate means of describing the strength of a magnetic field. The symbol for flux density is B. The SI unit of flux density is the weber per square meter (Wb/m²).

$$B = \frac{\Phi}{A} \text{ Wb/m}^2 \tag{14-1}$$

where Φ is in webers and A is in square meters.

Other units for flux density are the gauss, a c.g.s. unit, and lines per square inch, an English unit.

14-6. CURRENT AND MAGNETIC FLUX

Whenever electric current flows in a conductor, we find a magnetic field surrounding the conductor. The field is circular and exhibits forces at right angles to the conductor. It may be visualized as concentric circles of magnetic force around the conductor, as in Fig. 14-5. The flux lines are complete

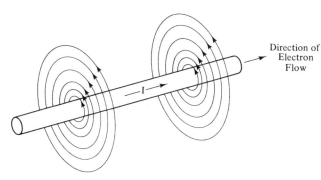

Figure 14-5 Magnetic flux around a current carrying conductor. To determine the direction of the flux, grasp the wire with the left hand, pointing the thumb in the direction of the current flow. The fingers point in the direction of the flux.

loops without identifiable poles, although the direction of the flux can be defined. We find the greatest concentration of flux at the surface of the conductor, with the flux density decreasing with increasing distance from the conductor. This reduction in flux density confirms the previous observation that flux lines of the same polarity repel each other. The flux around a current-carrying conductor may be demonstrated. We pass a wire through the center of a piece of cardboard and cause a current of about 60 A to flow in the wire. When we sprinkle iron filings onto the cardboard, they align themselves into concentric circles around the wire. The electromagnetic effect can also be observed with smaller currents by use of a compass. One finds that a compass needle will change position as the compass is moved around the wire. The compass needle moves so as to align itself in the direction of the concentric flux.

We have made several references to the direction of the concentric lines of force. The defined direction is found by using the fingers and thumb of the *left* hand. *Point the thumb in the direction of the electron current flow; the fingers then indicate the direction of the lines of force.* If, for example, we imagine a current direction in a wire that is at right angles to and into this page, then the flux is moving in a counterclockwise (ccw) direction. If the current direction is out of the page, then the lines of force are clockwise. This is illustrated in Fig. 14-6. If one uses *conventional* flow rather than electron flow, the same results are obtained. We simply use the right-hand fingers and thumb in place of the left.

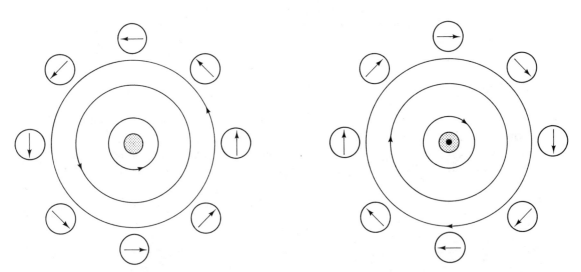

Figure 14-6 Using a compass to follow the direction of flux surrounding a conductor carrying current. In the left hand figure, the current direction is away from you. The flux is counterclockwise. The right hand figure has the current flowing towards you. Note the clockwise flux as indicated by the compass needle. Use the left hand rule to check the flux direction.

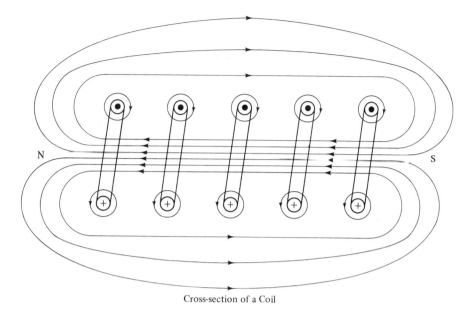

Cross-section of a Coil

Figure 14-7 The magnetic field established by current flow in a coil. We are assuming current flow (no circuit is shown); the top row of conductors carry current toward you, the lower ones carry current into the page. The flux within the coil is in one direction just as in a bar magnet. This is due to the flux of adjacent turns aiding each other. The result is flux that surrounds the coil, enabling the resulting field to have north and south poles.

When a current-carrying conductor is coiled into many loops, like a spiral, some of the flux from each turn of the spiral adds to flux produced by other turns as in Fig. 14-7. A common flux links the entire coil of wire and is concentrated along the axis of the coil. This concentration of flux is the result of the fields of individual turns combining into a single field of magnetic flux. The fields combine because they are all moving in the same direction within the coil.

If a ferromagnetic material like soft iron is used as core for the coil, the amount of flux is greatly increased. As a result, the flux density is also greatly increased. Such arrangements of coils and cores are used industrially for applications ranging from lifting scrap steel to small relays. It is common practice to refer to coils with ferromagnetic cores as *electromagnets*.

The magnetic field established by a current in a coil is the same as that of a bar magnet, including north and south poles.

The polarity and direction of the flux is found by another left-hand rule. *Let the fingers of the left hand indicate the direction of current in the coil; the thumb then points in the direction of the flux within the coil or core.* Remember, flux leaves the north pole of a magnet; therefore, the thumb points to the north pole of an electromagnet when the fingers are placed in the direction of current in the coil.

Sec. 14-6. Current and Magnetic Flux

14-7. PERMEABILITY

Different types of materials have varying effects upon the amount of flux in a magnetic field. *Permeability is a measure of the ability of a material to conduct magnetic lines of force.* The concept of permeability in magnetic systems is similar to the concept of conductivity in electric circuits. The symbol for permeability is the Greek letter μ (mu). The permeability of a vacuum (free space) is $4\pi \times 10^{-7}$.

Substances with permeabilities slightly greater than that of a vacuum are paramagnetic. Those substances with slightly less permeability than $4\pi \times 10^{-7}$ are diamagnetic. Ferromagnetic materials have permeabilities ranging from tens to thousands of times greater than that of a vacuum.

A more useful measure of permeability is *relative permeability* (μ_r). This quantity compares the permeability of a substance to the permeability of free space.

$$\mu_r = \frac{\mu}{\mu_0} \tag{14-2}$$

where μ = permeability of a substance
μ_0 = permeability of a vacuum

In computations with paramagnetic and diamagnetic materials, we shall use a relative permeability of 1. Table 14-1 lists the relative permeabilities of various ferromagnetic substances.

TABLE 14-1

Ferromagnetic	Relative Permeability
Nickel	50
Cast iron	1000
Cast steel	2000
Transformer steel	5000
Permalloy	30,000 to 80,000

We will find that the relative permeability of some ferromagnetic substances is not a constant, but varies with the flux density of the field within the material.

The product of relative permeability and the permeability of a vacuum results in the permeability of the substance in question.

$$\begin{aligned} \mu &= \mu_0 \mu_r \\ &= (4\pi \times 10^{-7})\mu_r \end{aligned} \tag{14-3}$$

EXAMPLE 14-1:
At a flux density of 600 mWb/m², a certain ferromagnetic has a relative permeability of 5000. Calculate the permeability of the material.

Solution:

$$\mu = 4\pi \times 10^{-7} \times \mu_r$$
$$= 12.56 \times 10^{-7} \times 5 \times 10^3$$
$$= 6.28 \times 10^{-3}$$

14-8. RELUCTANCE

Opposition to magnetic flux is called *reluctance* ($\mathcal{R}$). The unit of reluctance is the rel. The longer the flux path, the greater the reluctance. Decreasing the cross-sectional area of the flux path causes reluctance to increase. The permeability of the material making up the flux path is an important factor; the higher the permeability of the material within the path, the lower the reluctance.

$$\mathcal{R} = \frac{\ell}{\mu A} \tag{14-4}$$

where ℓ = length of the flux path in meters
μ = permeability of the flux path ($\mu_0 \mu_r$)
A = cross-sectional area of the flux path in square meters

EXAMPLE 14-2:
Find the reluctance of a piece of transformer steel that is rectangular, with a length of 0.152 m and cross-sectional dimensions of 0.051 m and 0.04 m, given that the relative permeability is 5.1×10^3.

Solution:
(a) Solve for the cross-sectional area:

$$A = 0.051 \times 0.04 = 2.04 \times 10^{-3} \text{ m}^2$$

(b) From Eq. (14-4):

$$\mathcal{R} = \frac{\ell}{\mu A} = \frac{\ell}{4\pi \times 10^{-7} \times \mu_r \times A}$$
$$= \frac{0.152}{12.56 \times 10^{-7} \times 5.1 \times 10^3 \times 2.04 \times 10^{-3}}$$
$$= 11.63 \times 10^3 \text{ rel}$$

EXAMPLE 14-3:
Given the same flux path dimensions as in Example 14-2, but with air in place of transformer steel for the flux path, solve for the reluctance.

Solution: The relative permeability of air is approximately 1. Therefore, the reluctance will be 5100 times that of the transformer steel.

$$\mathcal{R} = 11.63 \times 10^3 \times 5.1 \times 10^3$$
$$= 59.31 \times 10^6 \text{ rel}$$

Proof:

$$\mathcal{R} = \frac{\ell}{\mu A} = \frac{152 \times 10^{-3}}{12.56 \times 10^{-7} \times 2.04 \times 10^{-3}}$$
$$= 5.93 \times 10^7 \text{ rel}$$

14-9. MAGNETIC CIRCUIT

Magnetomotive force is the force that causes magnetic flux. It is similar to electromotive force in electric circuits. The only practical source of magnetomotive force is that produced by current in a coil of wire (electromagnetism).

Magnetomotive force is the product of the number of turns of a coil and the current in those turns. The symbol is F and the unit is ampere-turns (At).

$$F = NI \quad \text{At} \tag{14-5}$$

where N = number of turns of the coil
I = current in amperes

Magnetic circuits may be compared with electric circuits but with one precaution. *There are no magnetic insulators.* While we may have an electromotive force without an electric current simply by opening a circuit, whenever there is magnetomotive force, there will always be magnetic flux.

It is possible to compare magnetomotive force to electromotive force and to compare magnetic lines of force to electric current. A basic law of magnetic circuits relates flux to magnetomotive force, reluctance, and the physical properties of the flux path.

$$\Phi = \frac{NI}{\mathcal{R}} \quad \text{Wb} \tag{14-6}$$

If we substitute Eq. (14-4) for $\mathcal{R}$, we have

$$\Phi = \frac{\mu A N I}{\ell} \quad \text{Wb} \tag{14-7}$$

EXAMPLE 14-4:
A flux path has an average length of 9 in. The cross-sectional area of the path is 2.25 in. The relative permeability of the flux path is 3800. A 200-turn coil is wound upon the ferromagnetic core. A current of 130 mA is passed through the coil. (a) What is the flux in webers? (b) What is the flux density in webers/m²?

Solution:
(a) Change all dimensions to SI.

$$9 \text{ in.} = 9 \times 0.0254 = 0.229 \text{ m}$$
$$2.25 \text{ in.}^2 = 2.25 \times 6.45 \times 10^{-4} \text{ m}^2$$
$$= 1.45 \times 10^{-3} \text{ m}^2$$

(b) Solve for flux:

$$\Phi = \frac{\mu ANI}{\ell}$$

$$= \frac{12.56 \times 10^{-7} \times 3.8 \times 10^3 \times 1.45 \times 10^{-3} \times 200 \times 1.3 \times 10^{-1}}{2.29 \times 10^{-1}}$$

$$= 0.786 \times 10^{-3} \text{ Wb}$$

(c) Solve for flux density:

$$B = \frac{\Phi}{A} = \frac{0.786 \times 10^{-3}}{1.45 \times 10^{-3}}$$
$$= 0.542 \text{ Wb/m}^2$$

All our discussion up to this point has been based upon a uniform flux path. We have assumed that the flux path consisted of just one type of material with the same permeability throughout the path. This is true in some cases, but we must also consider applications with flux paths containing different substances and therefore different permeabilities. This latter case is the usual situation found in electromagnetic devices like transformers, motors, and generators.

Magnetizing force is the way that we measure magnetomotive force per unit length of the flux path. When a path contains materials of differing permeabilities, we determine the required magnetizing force for each part of the flux path. The total force required is the sum of the magnetizing forces required for each part of flux path. The symbol for magnetizing force is H, and the unit is ampere-turns per meter (At/m).

$$H = \frac{F}{\ell} = \frac{NI}{\ell} \quad \text{At/m} \tag{14-8}$$

where N is turns, I is current in amperes, and ℓ is length in meters.
Solving Eq. (14-8) for F, we have

$$NI = H\ell \quad \text{At} \tag{14-9}$$

Ampère's circuital law repeats what we have said in these paragraphs: *the total magnetomotive force needed in a magnetic circuit is the sum of the required forces for each section of the circuit as defined by the magnetizing force per section.*

$$NI_{\text{total}} = NI_1 + NI_2 + NI_3 + \cdots + NI_n \tag{14-10}$$
$$= H\ell_1 + H\ell_2 + H\ell_3 + \cdots + H\ell_n \tag{14-11}$$

Additional useful formulas are obtained from our previous work.

$$\Phi = \frac{\mu NIA}{\ell} \tag{14-7}$$

We obtain

$$NI = \frac{\Phi \ell}{\mu A} \tag{14-12}$$

Sec. 14-9. Magnetic Circuit

From our knowledge of flux density, we recognize the ratio of flux to area in Eq. (14-12):

$$NI = \frac{B\ell}{\mu} = \frac{B\ell}{12.56 \times 10^{-7} \times \mu_r} \qquad (14\text{-}13)$$

An important example of the need to know about magnetizing force is to be seen whenever we encounter an air gap in a flux path. In fact, we shall see that the magnetomotive force needed for a given flux density in a system that uses ferromagnetic material and a gap is used almost entirely to overcome the reluctance of the air gap.

EXAMPLE 14-5:

A relay has a ferromagnetic core that is 5 cm long. The relative permeability of the core is 4000. The movable portion of the relay (called the armature) is 3 mm away from the core when the relay is not operating. The average area of the flux path is $10^{-4} m^2$. A flux density of 40 mWb/m² is needed to operate the relay. (a) What is the magnetomotive force needed to establish the flux density in the core? (b) the specified flux density in the air gap? (c) the total magnetomotive force needed to operate the device? (d) If the coil has 300 turns, what current is needed to operate the relay? (e) Assuming that the air gap is zero when the armature closes, what minimum current is needed to hold the relay in operation?

Solution:

(a) $NI = \dfrac{B\ell}{\mu}$ (14-13)

$= \dfrac{40 \times 10^{-3} \times 5 \times 10^{-2}}{12.56 \times 10^{-7} \times 4 \times 10^{3}}$

$= 3.98 \times 10^{-1}$ At

(b) $NI = \dfrac{B\ell}{\mu}$ (14-13)

$= \dfrac{40 \times 10^{-3} \times 3 \times 10^{-3}}{12.56 \times 10^{-7}}$

$= 9.55 \times 10^{1}$ At

(c) $NI_t = NI_1 + NI_2$ (14-10)

$= 0.398 + 95.5$

$= 95.9$ At

(d) Divide the total ampere-turns by the number of turns to find the current needed to make the relay operate.

$$I = \frac{95.9 \text{ At}}{200 \text{ t}}$$
$$= 480 \text{ mA}$$

(e) With the air gap removed, we need only the magnetomotive force that sets up a flux density of 40 mWb/m², that is, 0.398 At. The holding current is found in the same manner as the operating current in part (d).

$$I = \frac{3.98 \times 10^{-1} \text{ At}}{200 \text{ t}}$$
$$= 2 \text{ mA}$$

Note: The example was devised to demonstrate that most of the magnetomotive force in a system is required because of the air gap. In a practical relay, the closing of the armature affects gap and cross section, and much more current than given in the solution is needed for holding the armature closed. Typically, holding current is in the range of 60% of operating current.

In this chapter we have explored the magnetic properties of coils and cores when current flows in the coil. In Chapter 15 we shall learn about the effects that magnetic fields have upon coils. We shall also study some applications of electromagnetism.

SUMMARY

1. The unit of magnetic flux is the weber (Wb). $1 \text{ Wb} = 10^8$ lines.
2. Magnetic materials may be paramagnetic, diamagnetic, or ferromagnetic.
3. Retentivity is the ability of a material to remain magnetized after magnetizing force has been removed.
4. Current flow always establishes magnetic flux.
5. Permeability is a measure of the ability of a material to conduct magnetic flux.
6. Reluctance ($\mathcal{R}$) is opposition to lines of magnetic flux.
7. Magnetomotive force is supplied by current flowing in a coil of wire. This is always stated in ampere-turns (At): $F = NI$ At.

PROBLEMS

14-1. A magnetic field has a flux of 10^{-4} Wb. How many lines of force are in the field?

14-2. A 0.16 mWb field is present in a flux path of 4×10^{-4} m^2 cross section. What is the flux density?

14-3. Suppose that the area of the flux path for Prob. 14-2 is tripled. What is the flux density?

14-4. The permeability of a substance is 2.14×10^{-3}. What is its relative permeability?

14-5. The relative permeability of a certain steel is 4800. What is the permeability of the steel?

14-6. Solve for the reluctance of a flux path that is 3 cm long, with a cross section of 8×10^{-4} m^2. The relative permeability of the material is 40×10^3.

14-7. Given the same ferromagnetic material as in Prob. 14-6, the length of the flux path is 1.5 cm, while the cross-sectional area is 16×10^{-4} m^2. What is the reluctance of the path?

14-8. A 450-turn coil carries a current of 30 mA. What is the magnetomotive force in ampere-turns?

14-9. Given that reluctance is 750 rels and that force equals 1.26 At, solve for flux in webers.

14-10. The length of a flux path is 4 cm. The cross-sectional area is 0.5×10^{-4} m^2. The relative permeability of the material is 1250. What is the flux if a magnetomotive force of 35 ampere-turns is supplied? What is the flux density of the magnetic path?

15

Applications of Electromagnetism

15-1. SOLENOIDS

A solenoid is a coil whose length is much greater than its diameter. The coil usually has a ferromagnetic core.

These electromagnetic devices have a wide variety of applications. They are used to sound door chimes in homes. Cars have starter solenoids whose function is to disengage the starter motor when the engine starts. Solenoids are used for control of the flow of high-pressure oil in jet aircraft. It is not our purpose to list all the uses of solenoids, but simply to illustrate the wide range of applications for these devices. All applications of solenoids require that some mechanical motion of the core take place. An understanding of the basic structure of a solenoid will make this mechanical motion clear.

The typical solenoid consists of a cylindrical coil wound on a hollow nonmagnetic coil form as in Fig. 15-1. Within the coil form is a movable ferromagnetic core. When no current is drawn by the coil, the core is only partly within the coil form. The method of mounting the solenoid and the application usually determine the means for keeping the movable core only partly within the coil. For example, in door chimes, a spring is used to keep

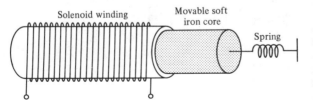

Figure 15-1 An example of solenoid construction. Not all solenoids have movable cores. In fact, a solenoid is simply a coil whose length is much greater than its diameter. But the name is used often to mean a device that performs some form of mechanical work.

the core in its rest position. When rated current passes through the coil, the core is drawn into the coil. The moving core provides the mechanical motion needed to strike a bar, sounding the chimes.

15-2. RELAYS

Relays are electromagnetic devices that operate by the movement of a *portion* of a ferromagnetic core to produce a desired result. In a relay, the movement of a portion of the core is used to operate switch contacts. Usually, there is no electrical connection between the circuit operating the relay and the circuit connected to the relay (switch) contacts. A sketch of a relay is shown in Fig. 15-2.

Relays are useful in industrial control circuits, as well as in other applications. Comparatively small currents used to energize the relay coil winding are capable of switching very large currents by use of the relay contacts. Sequenced operations in industry are often switched from step to step by relays.

Relay coils are frequently rated in terms of voltage and resistance or voltage and current. There are really two currents of interest: (1) the current needed to operate the relay, and (2) the minimum current needed to hold the relay in operation. It is also desirable to specify whether or not the relay is designed for ac or dc operation. The reasons for such a specification will be found in our study of inductive reactance (Chapter 19). Clearly, the current rating of the relay contacts must always be specified. Excessive currents will burn and pit the switch contacts.

A relay may have many sets of contacts, as in telephone relays, or simply one pair of contacts. Some relays are used to switch high currents by means of low-current control. Others are used to permit switching in many different circuits, all under the control of the single current through the relay coil. In still other uses, the relay is used as a protective device: if the current in the protected circuit goes beyond a certain limit, the relay acts to open the circuit or to reduce supply voltage to the circuit.

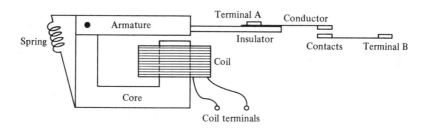

Figure 15-2 An elementary relay. When current flow in the coil reaches the level required to operate, the tension of the spring is overcome by the magnetic force pulling the armature to the body of the relay.

Relay contacts are defined as *normally closed* (n.c.) or *normally open* (n.o.), based upon the closed or open condition of the contacts when the relay coil is de-energized.

15-3. ELECTROMAGNETIC INDUCTION

Whenever a conductor is moved across a magnetic field, voltage is developed along the length of the conductor. Voltage is also developed along the length of a conductor whenever lines of flux are moved across a conductor. It is important that we understand that the relative motion between conductors and flux be such that lines of force are "cut" by conductors. No voltage is developed when the motion of the conductors is parallel to the direction of the lines of force so that no flux lines are cut. Voltages developed by flux cuttings are called *generated* emfs. This effect was first reported by Michael Faraday in England and Joseph Henry in the United States, in the same year (1831).

The amount of generated voltage depends upon the following:

1. Flux density (B): Recall that a magnetic field represents energy. The greater the flux density is, the greater the amount of energy that can be changed to electrical energy, that is, voltage.
2. The number of turns of the conductor that are cutting flux at any instant: The turns each act as a single conductor, developing a generated emf. The emfs of the turns are in series aiding and therefore add to each other. The result is a greater amount of voltage. In fact, if all other variables are held constant, the amount of generated voltage is exactly proportional to the number of turns. Doubling the turns will double the generated voltage.
3. The relative velocity between conductors and flux: The greater the relative velocity, the greater the amount of flux cuttings per second.

The direction of the conductors through the magnetic field determines the polarity of the generated voltage. A reversal of direction will result in a reversal of the polarity of the generated voltage.

We summarize these concepts with Faraday's law:

The voltage developed in a conductor as the result of relative motion between the conductor and a magnetic field is proportional to the rate of change of flux linkages or to the rate of flux cutting.

Flux linkages are simply another way of describing flux cuttings. Flux linkage is the product of flux and turns.

$$\text{Flux linkages} = N\Phi \qquad (15\text{-}1)$$

where N is the number of turns and Φ is the flux in webers.

Flux linkages are of value in the study of *induced* voltages. Whenever there is *change* in the flux surrounding a coil, voltages are induced in the coil. The flux is a form of energy; any change in this energy must result in a voltage in the coil. If we increase the flux surrounding a coil, there will be an induced voltage in the coil opposing this increase in flux. This induced voltage will prevent current through the coil from increasing; therefore, the induced voltage will be in series opposition to the applied voltage. On the other hand, if we decrease the flux around a coil, a voltage will be induced in the coil so as to keep the flux constant. In this case, the voltage will be in series aiding with the applied voltage so as to keep current in the coil constant.

Lenz's law summaries these ideas.

The induced voltage always has a polarity that is in such a direction as to oppose any changes in current through a coil and therefore any changes in the magnetic field surrounding a coil.

A flux change of one weber per second will induce a voltage of one volt per turn in a coil. The formula for average induced voltage is

$$V = N \frac{\Delta \Phi}{\Delta t} \qquad (15\text{-}2)$$

where N = number of turns
$\Delta \Phi$ = change in flux in webers
Δt = change in time in seconds

EXAMPLE 15-1:
A flux change of 0.07 Wb takes place in 5 ms. What is the voltage induced in a 200-turn coil?

Solution:

$$V = N \frac{\Delta \Phi}{\Delta t}$$
$$= 200 \times \frac{70 \times 10^{-3}}{5 \times 10^{-3}}$$
$$= 2800 \text{ V}$$

EXAMPLE 15-2
What is the induced voltage in the coil of Example 15-1 if the same flux change takes place in 2 s rather than 5 ms?

Solution:

$$V = 200 \times \frac{70 \times 10^{-3}}{2}$$
$$= 7 \text{ V}$$

The formula for generated voltage, if we assume a conductor moving at constant velocity through a uniform magnetic field, is

$$V = B\ell v \sin \theta \qquad (15\text{-}3)$$

where B = flux density in Wb/m²
ℓ = length of the conductor within the field in meters
v = velocity in m/s
θ = angle between lines of force and the direction of motion of the conductor

We may summarize the difference between generated and induced voltages: Generated voltages are developed as a result of mechanical motion between conductors and flux. Induced voltages are obtained whenever the *amount of flux around a coil changes.*

Suppose that a coil, consisting of many turns of wire, is arranged so that it can be rotated around its axis, as in Fig. 15-3. Let the ends of the coil of wire be connected to slip rings that press against contacts so that electrical connections are made to the coil even though the coil rotates. Now rotate this coil within a uniform magnetic field. At each instant, we shall make sure to rotate the coil at a constant speed in revolutions per minute (rpm). We find that the voltage available at the terminals of the coil is alternating and in the shape of a sine wave. The operation during one rotation of the coil (see Fig. 15-4) will help us understand some of the principles of generated emfs:

1. At an instant in time, during the rotation of the coil, the conductors are *parallel* to the lines of force of the magnetic field, and the generated voltage is zero. We shall define this position of the coil as 0°.

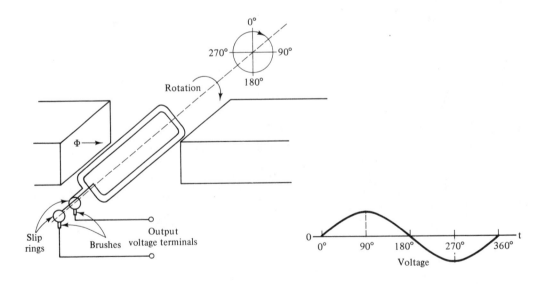

Figure 15-3 Generating an alternating voltage. Assume that the coil is rotating through a uniform magnetic field.

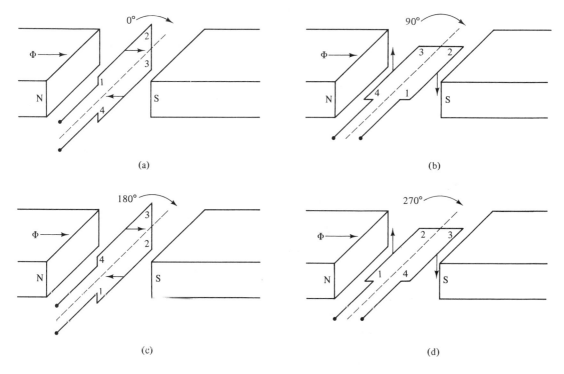

Figure 15-4 (a) The coil is at the 0° position. Conductors 1-2 and 3-4 are moving parallel to the flux and no lines of force are cut. There is no generated emf. (b) The coil has rotated through 90°. The conductors are generating maximum emf. The polarity of the induced emf's is such that V_{4-1} is positive. (c) The coil has rotated another 90°, so that it is at the 180° position. Once again, the conductors are moving parallel to the flux and no emf is generated. (d) Rotation through 90° brings the coil to the 270° position. Maximum flux cuttings take place at this instant. The output polarity is reversed with terminal 4 most negative. V_{4-1} is at the maximum negative point.

2. Following this instant, the conductors begin to cut lines of force and some voltage is generated. However, the conductors are at angles such that maximum rate of flux cutting has not been reached.
3. At 90° of rotation the conductors are at right angles to the flux, causing maximum rate of flux cutting and maximum output voltage.
4. As rotation proceeds beyond 90°, the angle between conductors and flux decreases, causing the rate of flux cutting to decrease, resulting in reduced output voltage.
5. At 180° (1/2 rotation) the conductors are once again parallel to the flux, and the output voltage is zero.
6. From 180° to 360° the amount of output voltage is the same as for the period from 0° to 180°. However, the polarity of the voltage is reversed because the conductors of the coil move through the magnetic field in directions opposite to those taken from 0° to 180°.

Clearly, one rotation of the coil produces one cycle of alternating voltage. Therefore, the frequency of the output voltage depends upon the number of revolutions per second of the coil. In fact, the frequency is equal to the number of revolutions per second (rps).

$$f = \text{rps} = \frac{\text{rpm}}{60} \tag{15-4}$$

where f is frequency in hertz (Hz).

An increase in rpm results in two changes; one is frequency, which increases as in Eq. (15-4). The other is in the output voltage, which also increases with increased rpm. This results from the higher rate of flux cutting that occurs because of the increased rpm.

On the other hand, if we hold rpm constant, but use a coil with fewer turns, we find that the generated voltage decreases. If, instead, we use more turns for the coil, generated voltage increases.

Similarly, changes in flux density directly affect the generated voltage. For a given coil rotating at a fixed rpm, the output voltage is directly proportional to flux density. The greater the flux density, the greater the output voltage.

All these effects upon generated voltage that we have considered are summarized by Eq. (15-3). While it is not common to use Eq. (15-3), we shall demonstrate some of the important aspects of generated emfs by the following example.

EXAMPLE 15-3:
The total length of conductor for a coil is calculated to be 45 m. The width of the coil is 0.23 m. The coil is rotated at a constant speed of 3600 rpm within a magnetic field whose flux density is 51.3 mWb/m². Calculate the peak value of the voltage available at the slip rings.

Solution:
1. To apply the formula, we must first determine the velocity of the coil in meters per second.
   (a) When rotating, the coil outlines a circle with a diameter equal to the width of the coil. The circumference of the path traced by the coil is π times the diameter.

   circumference = 3.14 × 0.23 = 0.722 m

   (b) Determine revolutions per second. This is so we may multiply the rps by the circumference to obtain meters per second, the velocity of the coil.

   $$\text{rps} = \frac{\text{rpm}}{60} = \frac{3600}{60} = 60 \text{ rps}$$

   (c) Velocity = rps × circumference
      = 60 × 0.722 = 43.33 m/s
2. In Eq. (15-3), $B = 0.0513$ Wb/m², $\ell = 45$ m, and we shall use 90° as the angle in order to obtain maximum output. The sine of 90° = 1.

$$v_{max} = 0.0513 \times 45 \times 43.33 \times 1$$
$$= 100 \text{ V}$$

At the rotational speed of 3600 rpm, the output is an ac voltage with a peak value of 100 V and a frequency of 60 Hz.

15-4. GENERATORS AND ALTERNATORS

If an electrical machine receives mechanical energy and supplies electrical energy, the machine is a generator. On the other hand, if the input to the machine is electrical energy and the output is mechanical energy, the machine is a motor. In this section we shall consider some elementary examples of electrical generators.

We have seen that an alternating voltage is developed when a coil is rotated within a magnetic field. This generated voltage is available at contacts riding on slip rings. These contacts are properly called *brushes* and are made of carbon to minimize wear and tear on the slip rings and to reduce friction.

Revolving a coil within a fixed magnetic field will always generate an alternating voltage. To obtain a dc output from the ac voltage, we use a split ring contactor, called a *commutator*. In the elementary single-coil generator, each end of the coil is connected to a commutator segment, as in Fig. 15-5. In our example the commutator has two segments because we have a single-coil generator. The brushes and commutator are arranged so that coil connections to the brushes switch once for each rotation. This switching (commutation) causes the voltage available at the terminals to be *pulsating dc*

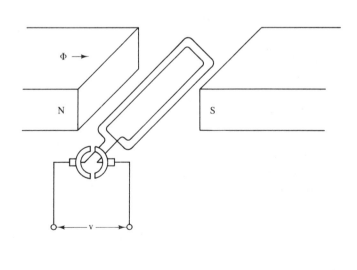

Figure 15-5 Generating a pulsating *dc* voltage.

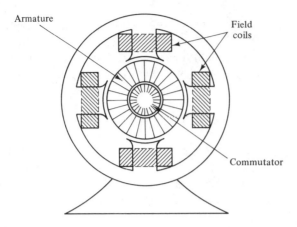

Figure 15-6 Elementary *dc* machine. This sketch is of a four-pole unit. The use of four poles results in less time between peak outputs (at a given speed); this means smoother operation with greater torque if a motor. If a generator, the average output voltage will be greater.

rather than ac. The brushes are located at the position that permits commutation to take place as the voltage drops to zero. This is done to minimize sparking at the brushes.

A single-coil generator is not practical. The output is too small and it pulsates from zero to some maximum level. Practical dc generators have many coils and several pairs of magnetic poles in order to develop a dc output with only a small amount of pulsation. For each coil, we must have a pair of commutator segments. A generator is sketched in Fig. 15-6.

The assembly of coils arranged around a high permeability core is called an *armature*. The magnetic fields are fixed around the armature in an assembly called a *yoke*. All dc generators are based upon a rotating armature and fixed magnetic fields. The armature coils are *always* connected to commutator segments in order to obtain dc at the brushes. The field for dc generators is derived from dc current flow in the field coil windings. We can control the amount of output voltage by varying the current in the field coils or by changing the speed of rotation of the armature. As long as the pole pieces are not in saturation, field current control is effective. If the pole pieces are saturated, the field flux remains relatively constant over a wide range of field current levels, and output control is poor.

Alternating-current generators are made differently from dc generators. Direct-current generators must have commutators to maintain a constant polarity of output. Because of this requirement, dc generators are limited in power and voltage capabilities. Compared to ac generators, dc generators must be made to produce very much lower voltages. They also handle lower load currents. With the ac generator, properly called an *alternator*, very large amounts of power are available at very high voltages. Because commutation is not needed, the output coils need not be rotated and are usually mounted in the frame of the machine. This set of coils is called the *stator windings*. Rotating within the stator is a set of electromagnets energized by dc current in order to produce a constant flux. In the alternator, the fields are rotated, and lines of flux are cut by the stator coils. The rotating unit is called the *rotor*.

Alternators are usually designed to produce three sets of ac voltages located 120° of rotation from each other. This is three-phase ac and is standard for the industry.

15-5. MOTORS

The function of any motor is to develop a rotational force called *torque*. In this section we shall discuss briefly some of the basic concepts of dc and ac motors.

Motors designed to be used on dc are very similar in construction to dc generators. Armature connections must be made by brushes and commutator. Armature resistance is very low, normally just a fraction of an ohm. The manner of connecting the field windings also determines the speed and torque characteristics of the motor.

Regardless of the type of motor, all develop torque in the same way. Torque is the result of interaction between the magnetic field developed by current in the armature conductors and the flux established by the field windings, as shown in Fig. 15-7.

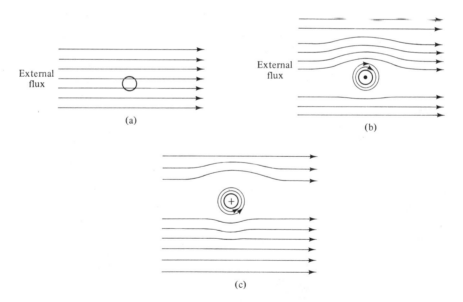

Figure 15-7 Motor action as a result of magnetic attraction and repulsion between the flux of an external field and the field around a wire carrying current. (a) The current through the conductor is zero. Therefore there is no flux from the conductor, and no interaction between magnetic fields. There is no force on the conductor. (b) The direction of current (electron flow) is out of the page. Conductor flux and external flux add and subtract, as shown resulting in a downward force on the wire. (c) Current direction is into the page. Flux is subtracted above the wire and added below the wire, resulting in an upward force on the wire.

Whenever a current-carrying conductor is placed within a magnetic field, a force will be developed that will move the conductor at right angles to the fixed magnetic field.

It should be of interest to note that this rule applies to generators as well as to motors. When current flows in the armature of a generator, "motor action" takes place. This action opposes the rotation of the generator armature. Therefore, as more current is supplied by a generator, more force is needed to rotate the armature.

Just as generators have motor action, motors develop generated emfs. The rotating armature coils cut lines of flux and generate a voltage that opposes the applied voltage. The counter emf reduces armature current and power losses. However, torque depends upon armature current, and reducing armature current also reduces torque.

The field strength of the poles has an important effect upon armature speed. As field flux increases, the motor slows down. This is caused by the increased amount of counter emf developed as armature conductors cut more lines of flux. The speed automatically reduces to the point where there is sufficient armature current to develop the torque needed to turn the load. On the other hand, a reduction in field flux increases the speed of a motor. With fewer lines of force, the counter emf is lower. This results in increased armature current and greater torque. The speed increases causing the counter emf to increase and reduce armature current to an amount that still permits the motor to turn the load.

There are three basic types of dc motors: series, shunt, and compound.

The *series* motor is arranged so that the same current flows through the armature and field windings. Because armature current must be carried by the field windings, they consist of a few turns of large wire. The series motor has very poor speed regulation (the variation in speed with load changes is great.) The series motor has very high torque, particularly starting torque. In fact, a series motor must *never* be started without a load. It will develop so high a speed that it may destroy itself.

The *shunt* motor is supplied with field current separately from armature current. The field windings are made of many turns of a wire that is much smaller than that used for the series motor. The separate field results in excellent speed regulation, but very poor torque characteristics.

The *compound* motor combines the characteristics of series and shunt motors. Compounds motors use shunt and series field windings. Armature current flows in the series field windings, which results in high torque characteristics. The shunt windings carry current separately from armature currents, which results in good speed characteristics. Figure 15-8 shows schematic diagrams of the dc motors.

There is a great variety of ac motors, far more than can be discussed in this text. Except for the series ac motor (often called the *universal* motor), most work on the same basic principles.

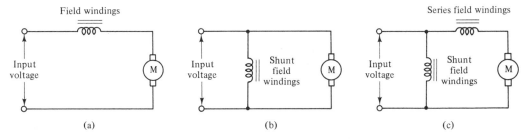

Figure 15-8 Schematic diagrams for *dc* motors: (a) Series motor, (b) shunt motor, (c) compound motor.

1. The rotor consists of an assembly of conductors placed in a high-permeability core. The rotor conductors are short circuited at both ends, resulting in what is called a *squirrel cage rotor*.
2. Currents and voltages are supplied to the rotor by means of electromagnetic induction from the stator.
3. The stator flux must rotate. The speed in rpm of the stator flux is known as *synchronous* speed. The rotor flux, established as a result of electromagnetic induction, attempts to revolve at the same speed as the stator flux. The rotor flux cannot revolve at the same speed as the stator flux because torque would not be developed. The difference between synchronous speed and rotor speed is called *slip*. *Torque is directly proportional to slip*. Increased load results in increased slip and therefore increased torque in order to turn the load.

One type of motor, known as the universal motor, operates on ac or dc. The motor is a low-horsepower unit, rarely as large as 1/3 hp. It is similar to a series dc motor except for the need for laminated cores. It is widely used in small appliances like vacuum cleaners, sewing machines, and mixers.

15-6. TRANSFORMERS

An extremely important application of electromagnetic induction is found in transformers. We shall begin a detailed study of them in Chapter 24. In this section we shall examine some basic principles applicable to all transformers.

We have already observed that a *change* in flux linkages results in an induced voltage. Also, we know that flux is established when there is current in a coil. Thus, in an ac circuit, where current is constantly changing, we find a constantly changing flux. Suppose that the flux of one coil links the turns of another coil. Flux linkages will then exist between the coils. Changes in current, resulting in changes in flux linkages, will cause induced voltages between the coils.

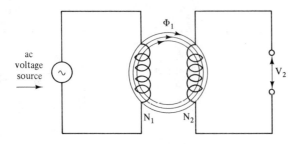

Figure 15-9 Mutually coupled coils.

When coils are arranged so that flux from one coil links the turns of the other, the coils are said to be mutually coupled (Fig. 15-9).

The amount of induced voltage is found by Eqs. (15-5) and (15-6). From coil 1 to coil 2,

$$v = kN_2 \frac{\Delta \Phi_1}{\Delta t} \qquad (15\text{-}5)$$

From coil 2 to coil 1,

$$v = kN_1 \frac{\Delta \Phi_2}{\Delta t} \qquad (15\text{-}6)$$

where Φ_1 = flux of coil 1
 Φ_2 = flux of coil 2
 N_1 = turns of coil 1
 N_2 = turns of coil 2
 k = coefficient of coupling

It is important to understand that k, the coefficient of coupling, is a way of describing the degree of coupling that exists between the coils. If *all* the flux of a coil linked *all* the turns of another coil, we would find that we have *unity* coupling and $k = 1$. Suppose that half of the flux of a coil links all the turns of another coil; then $k = 0.5$. If half the flux of a coil links half the turns of another coil, then $k = 1/2 \times 1/2 = 1/4 = 0.25$.

Mutually coupled coils are the basis for transformers, a very important and widely used device in alternating-current circuits.

Practical transformers range in size from extremely small, low-power units (fractions of a watt) to very large power transformers rated at hundreds of kilowatts. Practical values for the coefficient of coupling depend upon the transformer application. Transformers designed for radio frequency operation have very low values of k, typically from 0.01 to 0.2. On the other hand, transformers used for operation at power line and audio frequencies have values of k so close to unity that we consider k to be 1 for all practical purposes.

Transformers are useful for changing ac voltage and current levels, for isolating circuits, for load coupling, and for other purposes. We shall discuss these concepts in detail in Chapter 24.

15-7. DC METERS

Digital meters, which make use of modern electronic devices and provide digital readouts, have become popular in recent years. However, the most widely used instruments for measurement of current and voltage are still of the *analog* type: a scale and a pointer whose deflection across the scale depends upon the amount of voltage or current being measured. Analog meter movements are another example of an application of electromagnetism. While there are several different types of movements, we shall examine just one, the D'Arsonval movement.

A typical D'Arsonval meter movement is sketched in Fig. 15-10. The deflection of the pointer needle results from the torque developed between two magnetic fields in much the same manner as in electric motors.

One field is the fixed field of the magnet. The soft iron pole pieces direct this field across the core of the coil. The other is the field established by the coil when current flows. The amount of torque and therefore movement of the pointer depend upon the amount of current in the coil.

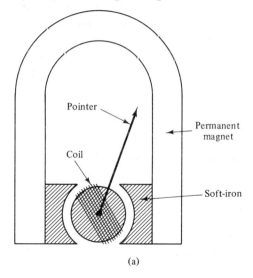

(a)

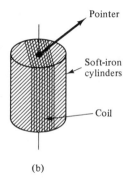

(b)

Figure 15-10 D'Arsonval meter movement.

The sensitivity of a meter is determined by its ability to read small currents. This sensitivity is given in terms of current needed for full-scale deflection of the pointer. A 1-mA movement requires 1 mA for full-scale deflection. A current of 100 µA will result in a deflection that is one-tenth of full-scale. Because very fine wire is used for the coil, the typical resistance of a 1-mA movement is approximately 50 Ω. To obtain a greater current sensitivity, we must use many more turns of fine wire. A 50-µA movement has a resistance on the order of 2000 Ω.

Remember that we required an ideal ammeter to have *zero* resistance. Clearly, if a 2-kΩ meter movement is used to make a current measurement, the resistance of the meter (R_m) will affect the circuit. In those circuits where R_m can cause serious error, we should make use of electronic instruments. Many of these instruments also use D'Arsonval movements for readout; however, the terminal resistance of the instrument more nearly approaches the ideal of zero.

Ammeters. We can increase the full-scale current value of a meter movement by means of a *shunt*. The shunt, a very low value of resistance, enables most current to bypass the meter movement. Properly selected, correct current measurements are readily made.

A meter shunt is simply one branch of a parallel circuit, with the meter resistance as the other branch. It follows then that the voltage across each branch must be the same. Given that

I_m = full-scale current sensitivity of the meter movement

I_T = full-scale current to be measured

I_{sh} = current in the shunt at full scale

R_{sh} = resistance of the shunt

R_m = resistance of the meter movement

then

$$R_{sh}(I_T - I_m) = R_m I_m$$

Therefore,

$$R_{sh} = R_m \frac{I_m}{I_T - I_m} \qquad (15\text{-}7)$$

EXAMPLE 15-4:
A 1-mA, 40-Ω movement is used to measure up to 500 mA full scale. What value of shunt resistance must be placed across the meter movement?

Solution:

$$R_{sh} = R_m \frac{I_m}{I_T - I_m}$$

$$= 40 \frac{1}{499}$$
$$= 0.080 \ \Omega$$

EXAMPLE 15-5:
Repeat Example 15-4, but with a 50-μA (0.05-mA) meter movement with $R_m = 2000 \ \Omega$.

Solution:
$$R_{sh} = 2 \ \text{k}\Omega \frac{0.05}{499.95}$$
$$= 0.200 \ \Omega$$

EXAMPLE 15-6:
Suppose that the meter of Example 15-5 is used to measure 5 mA (full scale). Solve for the resistance of the shunt.

Solution:
$$R_{sh} = 2 \ \text{k}\Omega \frac{0.05}{4.95}$$
$$= 20.202 \ \Omega$$

A multirange ammeter is illustrated in the circuit of Fig. 15-11.

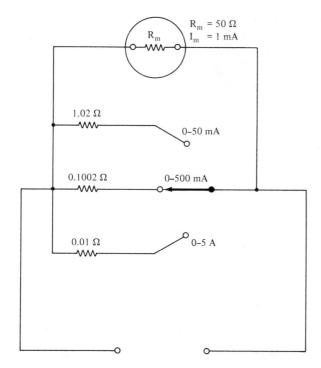

Figure 15-11 Multi-range ammeter.

Voltmeters. Recall that the ideal terminal resistance of a voltmeter is infinity. We approach this ideal only with electronic voltmeters. However, we shall discuss the use of a D'Arsonval meter movement as a voltmeter. For practical measurements, we must place resistors called multipliers in *series* with the meter movement. For example, if 10 V is to cause full-scale deflection of a 1-mA meter movement, we must have a total of 10 kΩ between the terminals. In the case of a 20-µA movement, we would need 500 kΩ between the terminals to make the same 10-V measurement. The selection of a multiplier resistance simply requires that we find the total full-scale resistance needed and then subtract the resistance of the meter movement. We shall make use of another indication of meter sensitivity to simplifly our approach.

Ohms/Volt. This convenient measure of meter sensitivity enables us to calculate multiplier resistors directly. The ohms/volt sensitivity of a meter movement is simply the reciprocal of the full-scale current sensitivity.

$$\frac{\text{ohms}}{\text{volt}} = \frac{1}{I_m} \qquad (15\text{-}8)$$

Therefore, a 1-mA meter has a sensitivity of 1000 Ω/V. For a full-scale capability of 50 V, we need a resistance of (50 V)(1000 Ω/V) = 50,000 Ω. Remember, the multiplier resistance is based upon *full-scale capability*, not actual voltage. Some typical Ω/V ratings are listed:

Current Rating	Ohms/Volt
1 mA	1,000
50 µA	20,000
20 µA	50,000
10 µA	100,000

A multirange voltmeter is sketched in Fig. 15-12.

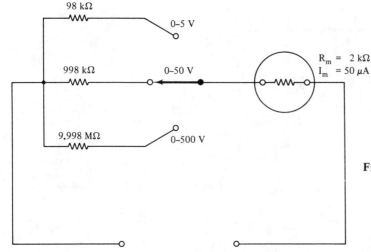

Figure 15-12 Multi-range voltmeter.

SUMMARY

1. A solenoid is simply a long coil.
2. Relays are used for electromechanical switching applications.
3. Voltages are induced whenever there is a change in the amount of flux linking a coil.
4. Whenever flux is "cut" by the turns of a coil, voltage is induced in the coil.
5. A generator is a machine that produces dc output voltage by rotating coils through magnetic fields. If the machine produces ac output, it usually referred to as an alternator.
6. Series dc motors have high starting torque and poor speed regulation.
7. Shunt motors have excellent speed regulation but poor starting torque.
8. Transformers are a nonmechanical way of transforming ac voltage and currents.
9. The most widely used dc meter movement is the D'Arsonval type.
10. Meter sensitivity is measured in ohms/volt or full-scale movement current.
11. Voltmeters use multiplier resistances to extend voltage range; ammeters require shunt resistances to extend current range.

PROBLEMS

15-1. The flux linkages between a coil and its magnetic field are equal to 2.5. Suppose that the current through the coil is reduced by 50%. What are the new flux linkages.

15-2. A flux change, taking place in 0.25 s, results in an induced voltage of 4 V. What is the induced voltage if the same amount of flux change takes place in 25 ms?

15-3. With a flux density of 75 mWb/m^2, a generated voltage reaches a peak value of 45 V. What is the peak voltage if the flux density changes to 100 mWb/m^2?

15-4. A speed of 1200 rpm results in a peak generated voltage of 100 V. What is the peak output voltage at 1800 rpm?

15-5. Two coils are mutually coupled. If all the flux of one coil links all the turns of the other coil, what is the coefficient of coupling?

15-6. Suppose that half the flux of one coil links three-fourths the turns of another coil. What is the coefficient of coupling between the coils?

15-7. The coefficient of coupling between two coils is 0.8. A flux change of 80 μWb takes place in 500 μs in the first coil. If the second coil consists of 75 turns, what is the induced voltage in the second coil?

15-8. Restate the following meter sensitivities in terms of ohms/volt.
  (a) 1 mA   (c) 100 μA
  (b) 50 μA   (d) 20 μA

15-9. A meter with an internal resistance of 2100 Ω requires 50 μA for full-scale deflection. Calculate the multiplier resistances for each of the following full-scale ranges: 0–0.5, 5, 50, and 500 V.

15-10. Calculate the shunts needed to use a 100-μA meter as an ammeter with ranges of 0–500 μA, 5 mA, and 50 mA. The internal resistance of the movement is 1100 Ω.

16

Inductance

16-1. INTRODUCTION

Inductance is that property of a circuit that *opposes changes in current*. This opposition is the result of energy exchange between a current-carrying coil and its surrounding magnetic field.

In Chapter 14 we learned that an electric current always has a magnetic field associated with it. Also, we know that the amount of magnetic flux is directly proportional to current. When the current in a coil is constant, the flux around the coil is constant, and there is no change in flux linkages. Recall from Eq. (15-1) that flux linkage is simply the product of flux and turns. From our studies in Chapter 15 we learned that, whenever there is a *change* in the flux surrounding a coil, there is a voltage *induced* in the coil that opposes the *change* in flux. The induced voltage cannot prevent the change in flux. Instead, we find that the rate of change of current and, therefore, rate of change of flux is decreased by inductance.

When current attempts to increase, the flux surrounding a coil begins to increase, which results in a change in flux linkages. Several physical processes are taking place: First, the increasing flux represents energy taken from the circuit. Second, as the flux attempts to increase, flux linkages change, which results in an induced voltage. Third, the polarity of the induced voltage *opposes* the applied voltage. Eventually, the current increases to a level determined by voltage and resistance. The time required to reach maximum current is determined by the inductance of the coil and the resistance of the circuit. At this maximum level of current, which we shall assume is constant, there is no effect of inductance in the circuit. Should we attempt to change the current, the effect of inductance is immediately present.

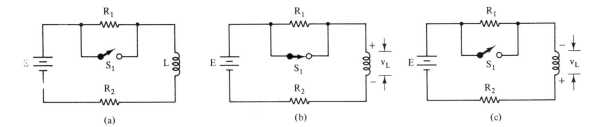

Figure 16-1 Inductance opposes current change. (a) Switch S_1 is open and current is assumed to have reached a steady value. Flux linkages are constant and there is no induced emf. For all practical purposes, all of E is dropped across R_1 and R_2. (b) Switch S_1 is closed. Current attempts to rise. This results in an expanding magnetic field, and *increasing* flux linkages. The changing flux linkages result in an induced emf across the coil. The induced emf has a polarity that is in *series-opposition* to the source voltage. The current eventually increases to a new steady value, equal to E/R_2. (c) When S_1 is opened, the current should decrease to the same level as in step (a). However the changing flux linkages cause induced emf's that attempt to maintain current. The polarity of the induced emf is *series-aiding*.

When current attempts to decrease, similar physical processes take place. The decreasing flux returns energy to the circuit. The resultant changing flux linkages produce an induced voltage. In this case, the induced voltage has a polarity that *aids* the applied voltage. Once again, we find that current eventually decreases to a level determined by Ohm's law. The effects of inductance as a result of current changes are shown in Fig. 16-1.

The effect of inductance in electrical circuits is similar to the effect of inertia in a mechanical system. The flywheel attached to the driveshaft of an automobile engine prevents very rapid changes in engine rpm because of its inertia, or opposition to change in motion.

16-2. INDUCED VOLTAGES

When we studied electromagnetic induction in Chapter 15, we found that induced voltages depended upon relative motion between a coil and a magnetic field. These were defined as generated voltages and were determined by the *rate* of flux cutting (Faraday's law). We also learned that changes in flux linkages resulted in induced voltages proportional to the *rate of change* of flux linkages.

$$V = N \frac{d\Phi}{dt} \tag{16-1}$$

where $d\Phi/dt$ is the instantaneous rate of change of the flux linking the turns (N) of a coil. Thus, a gradual change in flux results in much less induced voltage as compared to that caused by an abrupt change in flux as shown in Fig. 16-2.

Because the flux surrounding a coil is proportional to the current in the

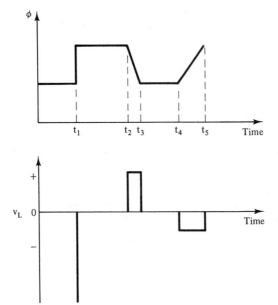

Figure 16-2 Induced voltage is a function of the rate of change of flux.

coil, induced voltages are obtained whenever current changes. Induced voltage is often called *counter emf*, because the polarity of the induced voltage opposes the change of flux. We summarize this rule in Lenz' law.

The polarity of an induced voltage is always that which attempts to establish a current that opposes any change in the flux surrounding a conductor or coil.

16-3. UNIT OF INDUCTANCE

A unit of inductance, representing a proportionality factor, enables us to relate induced voltage to the rate of change of current and flux. The unit is the *henry* (H), named for the American physicist, Joseph Henry. The symbol for inductance is L. We may write induced voltage in terms of inductance and rate of change of current.

$$V = L \frac{di}{dt} \qquad (16\text{-}2)$$

where di/dt = instantaneous rate of change of current
L = inductance in henrys (H)

EXAMPLE 16-1:
At some instant, the rate of change of current through a 20-mH coil is 15 ma/μs. What is the amount of the instantaneous induced voltage?

Solution:

$$V = L \frac{di}{dt}$$

given that $L = 20 \times 10^{-3}$ and that $di/dt = 15 \times 10^{-3}/10^{-6} = 15 \times 10^3$:

$$V = 20 \times 10^{-3} \times 15 \times 10^3$$
$$= 300 \text{ V}$$

EXAMPLE 16-2:
Suppose that the rate of change of current in a 7-mH coil is 20 μA/ns. Solve for the induced voltage.

Solution:

$$V = L \frac{di}{dt}$$
$$= 7 \times 10^{-3} \frac{20 \times 10^{-6}}{10^{-9}}$$
$$= 140 \text{ V}$$

The voltage in Eq. (16-2) is exactly the same as the induced voltage calculated by Eq. (16-1). Setting them equal to each other, we have

$$N \frac{d\Phi}{dt} = L \frac{di}{dt}$$

Solving for L,

$$L = N \frac{d\Phi}{di} \qquad (16\text{-}3)$$

where $\quad N =$ turns of the coil
$d\Phi/di =$ instantaneous rate of change of flux with respect to current.

We can derive a formula for inductance that permits us to solve for inductance without knowing $d\Phi/di$. We make use of some of the material we learned in Chapter 14.

$$\Phi = \frac{\mu NIA}{\ell} \text{ Wb} \qquad (14\text{-}7)$$

We write Eq. (14-7) in terms of instantaneous flux and current.

$$d\Phi = \frac{\mu NA}{\ell} di$$

Substituting into Eq. (16-3), we have

$$L = \frac{\mu AN^2}{\ell} \text{ H} \qquad (16\text{-}4)$$

where $\mu = \mu_0 \mu_r$
$A =$ area in square meters
$N =$ turns
$\ell =$ length in meters

Note a very important point: *Inductance is proportional to the square of the turns of a coil.*

Practical values of inductance range from nanohenries to henries.

$$1 \text{ nanohenry (nH)} = 10^{-9} \text{ H}$$
$$1 \text{ microhenry } (\mu\text{H}) = 10^{-6} \text{ H}$$
$$1 \text{ millihenry (mH)} = 10^{-3} \text{ H}$$

EXAMPLE 16-3:
A 200-turn coil has a diameter of 3 cm and a length of 9 cm. The core is air, with $\mu = 12.56 \times 10^{-7}$ H/m.

Solution:
1. Solve for area.

$$A = \frac{\pi d^2}{4} = \frac{3.14(3 \times 10^{-2})^2}{4}$$
$$= 7.069 \times 10^{-4} \text{ m}^2$$

2. Solve for inductance by Eq. (16-4).

$$L = \frac{\mu A N^2}{\ell}$$
$$= \frac{(12.56 \times 10^{-7})(7.069 \times 10^{-4})(200^2)}{9 \times 10^{-2}}$$
$$= 3.946 \times 10^{-4} \text{ H}$$
$$= 0.395 \text{ mH}$$

Self-inductance. There are two kinds of inductance, self and mutual. Self-inductance refers to the effects due to flux changes surrounding a coil or conductor, when that flux is the result of the current carried by the conductor or coil. Inductance and self-inductance have the same meaning. Common practice is to simply use the word inductance.

16-4. MUTUAL INDUCTANCE

The flux of one coil may link the turns of another coil. In this case, the coils are mutually coupled and we say that *mutual* inductance exists between the coils. Mutually induced voltages are found as follows:

$$V_1 = M \frac{di_2}{dt} \tag{16-5}$$

$$V_2 = M \frac{di_1}{dt} \tag{16-6}$$

where di_2/dt = instantaneous rate of change of current in coil 2
di_1/dt = instantaneous rate of change of current in coil 1
V_1 = instantaneous voltage induced in coil 1 by current change in coil 2

V_2 = instantaneous voltage induced in coil 2 by current change in coil 1

M = mutual inductance in henrys

We shall find that mutual inductance depends upon the self-inductance of the coils and the *coefficient of coupling* between the coils.

The coefficient of coupling (k) is the ratio of the actual flux linkages between mutually coupled coils to the maximum theoretically possible flux linkages. Maximum theoretical flux linkages occur when every line of flux from each coil links every turn of the other coil. This is defined as unity coupling. At this degree of coupling, $k = 1$. Practical values of k range from very low amounts to those approximating 1. The type of core material and the arrangement of the coils on the core determine the coefficient of coupling.

We can represent mutually induced voltages by Faraday's law:

$$v_1 = kN_1 \frac{d\Phi_2}{dt} \quad (16\text{-}7)$$

$$v_2 = kN_2 \frac{d\Phi_1}{dt} \quad (16\text{-}8)$$

Solving Eqs. (16-5) and (16-7) for M, we have

$$M = kN_1 \frac{d\Phi_2}{di_2} \quad (16\text{-}9)$$

Solving Eqs. (16-6) and (16-8) for M, we have

$$M = kN_2 \frac{d\Phi_1}{di_1} \quad (16\text{-}10)$$

Mutual inductance is the same amount going from coil 1 to coil 2 as it is going from coil 1 to coil 2. We must take more steps before we get a workable formula for M.

1. Multiplying Eqs. (16-9) and (16-10),

$$M^2 = k^2 N_1 N_2 \left(\frac{d\Phi_2}{di_2}\right)\left(\frac{d\Phi_1}{di_1}\right)$$

2. Rewriting to represent L_1 and L_2,

$$M^2 = k^2 N_1 \left(\frac{d\Phi_1}{di_1}\right) N_2 \left(\frac{d\Phi_2}{di_2}\right)$$

3. $N_1 \dfrac{d\Phi_1}{di_1} = L_1$

   $N_2 \dfrac{d\Phi_2}{di_2} = L_2$

4. $M^2 = k^2 L_1 L_2$

5. Solving for M,

$$M = k\sqrt{L_1 L_2} \qquad (16\text{-}11)$$

EXAMPLE 16-4:
Two coils are mutually coupled with $k = 0.4$. Given that $L_1 = 150$ mH and $L_2 = 200$ mH, what is the mutual inductance?

Solution:

$$\begin{aligned} M &= k\sqrt{L_1 L_2} \\ &= 0.4\sqrt{(150 \times 10^{-3})(200 \times 10^{-3})} \\ &= 21.91 \text{ mH} \end{aligned}$$

The coefficient of coupling is found by solving Eq. (16-11) for k.

$$k = \frac{M}{\sqrt{L_1 L_2}} \qquad (16\text{-}12)$$

EXAMPLE 16-5:
Two coils are mutually coupled. The inductance of coil 1 is 300 mH. Coil 2 has an inductance of 250 mH. What is the coefficient of coupling if the mutual inductance is 100 mH?

Solution:

$$\begin{aligned} k &= \frac{100 \times 10^{-3}}{\sqrt{(300 \times 10^{-3})(250 \times 10^{-3})}} \\ &= 0.365 \end{aligned}$$

Mutual inductance is the basis of operation of transformers, which are extremely useful devices. Audio and power transformers (Chapter 24) have high permeability flux paths with the coefficient of coupling assumed to be 1 (unity coupling). Transformers used at radio frequencies (Sec. 24-7) cannot make use of high-permeability flux paths and, as a result, have much lower coefficients of coupling.

16-5. INDUCTORS

All circuits have inductance. This is so because any wire has some self-inductance, even if only a fraction of a nanohenry. However, when we require a circuit to have a specific amount of inductance, we use a component designed for the purpose. This component is often called a coil, a choke, or an inductor. In a sense, we may consider that an inductor adds inductance to a circuit in the same way that a capacitor adds capacitance. There is a great difference that we must also consider. Inductors are made of coils of wire. In addition to inductance, we have the resistance of the wire and capacitance between adjacent turns of the wire. At certain frequencies we find that the capacitance between the turns has a greater effect upon a circuit than the inductance does.

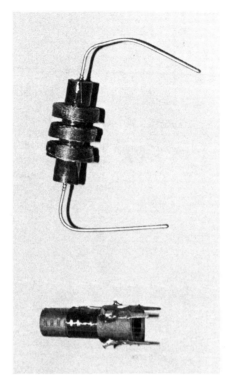

Figure 16-3

We identify inductors in terms of inductance and current capacity. Inductors for use at power and audio frequencies make use of laminated ferromagnetic cores. This permits high amounts of inductance in fairly small packages.

At higher frequencies, as in radio, we cannot use laminated ferromagnetic cores. At most, we can use powdered iron in a ceramic binder. For very low values of inductance, we use a few turns of wire with air as the core. Typical small inductors are shown in Fig. 16-3.

Inductors are used as part of tuning circuits, as well as in filter circuits. These items are covered in later chapters. In many applications, the inductor is used to minimize alternating currents while permitting direct current flow. When used in this application, we call the inductor a *choke*. We have radio-frequency chokes (RFC) and audio-frequency chokes (AFC).

16-6. SERIES AND PARALLEL INDUCTORS

As in any series circuit, current is the same in all parts of the circuit, and the sum of the voltage drops is equal to the applied voltage. If no mutual inductance exists between inductors in series, as in Fig. 16-4, total inductance is the sum of the inductances.

For inductors in series, $M = 0$:

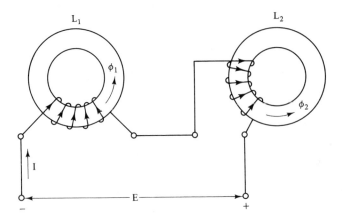

Figure 16-4 Inductors in series, without mutual coupling. The fluxes of the inductors are isolated from each other.
$L_T = L_1 + L_2$

$$e = v_1 + v_2 + v_3$$

From Eq. (16-2),

$$L_T \frac{di}{dt} = L_1 \frac{di}{dt} + L_2 \frac{di}{dt} + \cdots + L_n \frac{di}{dt}$$

Dividing by di/dt,

$$L_T = L_1 + L_2 + \cdots + L_n \qquad (16\text{-}13)$$

With inductors in parallel, we have voltage as the common quantity, while currents can differ. Total current is the sum of branch currents. Inductors in parallel add just as resistors in parallel do.

Solve Eq. (16-2) for v/L:

$$v = L \frac{di}{dt}$$

and

$$v/L = di/dt$$

Solving for branch current rates:

$$\frac{di_T}{dt} = \frac{v}{L_T} \qquad \frac{di_2}{dt} = \frac{v}{L_2}$$

$$\frac{di_1}{dt} = \frac{v}{L_1} \qquad \frac{di_n}{dt} = \frac{v}{L_n}$$

Adding, we have

$$\frac{di_T}{dt} = \frac{di_1}{dt} + \frac{di_2}{dt} + \cdots + \frac{di_n}{dt}$$

Substituting v/L,

$$\frac{v}{L_T} = \frac{v}{L_1} + \frac{v}{L_2} + \cdots + \frac{v}{L_n}$$

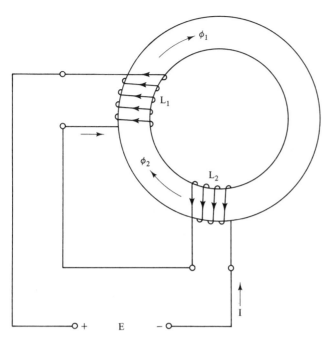

Figure 16-5 Inductors in series, fluxes aiding.
$$L_T = L_1 + L_2 + 2M$$

Solving for L_T, we have

$$L_T = \frac{1}{\frac{1}{L_1} + \frac{1}{L_2} + \cdots + \frac{1}{L_n}} \tag{16-14}$$

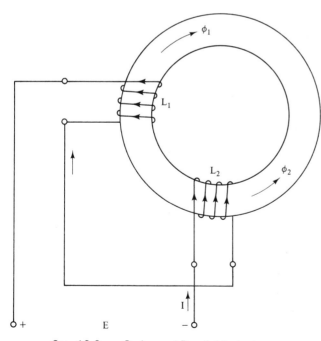

Figure 16-6 Inductors in series, fluxes opposing. With opposing fluxes, the total inductance is less than the sum of the inductors.
$$L_T = L_1 + L_2 - 2M$$

Sec. 16-6. Series and Parallel Inductors

Inductors in series can have mutual inductance between them. If the magnetic field polarity of each coil is the same so that they *aid* each other, as in Fig. 16-5, the inductors are in *series aiding*. *Series-opposing* inductors occur when the flux directions of the coils are different, as in Fig. 16-6, so that they oppose each other. With series-aiding inductors, the total inductance is *greater* than the sum of the inductances. With series-opposing inductors, the total inductance is *less* than the sum of the inductances.

$$L_T = L_1 + L_2 \pm 2M \qquad (16\text{-}15)$$

EXAMPLE 16-6:

Given the series inductors shown in Fig. 16-7, (a) determine if the mutual coupling is aiding or opposing, and (b) calculate the total inductance.

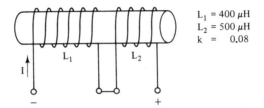

$L_1 = 400\ \mu H$
$L_2 = 500\ \mu H$
$k = 0.08$

Figure 16-7

Solution:
(a) By use of the left-hand rule, we find that the north pole of L_1 is at the right side. In a similar manner, we find that L_2 establishes a flux of the same polarity. With fluxes aiding, the mutual inductance is aiding.
(b) Solve for M.

$$M = k\sqrt{L_1 L_2}$$
$$= 0.08\sqrt{(400 \times 10^{-6})(500 \times 10^{-6})}$$
$$= 35.8\ \mu H$$

Solve for L_T.

$$L_T = L_1 + L_2 + 2M$$
$$= (400 + 500 + 2 \times 35.8)\ \mu H$$
$$= 971.6\ \mu H$$

EXAMPLE 16-7:

Given the series inductors of Fig. 16-8, solve for total inductance, given that $L_1 = L_2 = 50$ mH and $k = 0.1$.

Solution:
1. The current flowing through L_1 establishes a counterclockwise flux.
2. The current in L_2 establishes a counterclockwise flux. Therefore, the coils are series aiding.
3. Solve for M.

$$M = k\sqrt{L_1 L_2}$$
$$= 0.1\sqrt{(50 \times 10^{-3})(50 \times 10^{-3})}$$
$$= 5\ mH$$

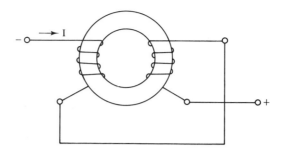

Figure 16-8

4. $L_T = L_1 + L_2 + 2M$
    $= 50 \text{ mH} + 50 \text{ mH} + 10 \text{ mH}$
    $= 110 \text{ mH}$

EXAMPLE 16-8:
Given the series inductors as in Example 16-7, but with the wiring change shown in Fig. 16-9, solve for total inductance.

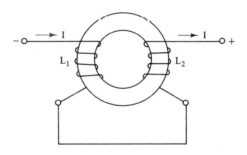

Figure 16-9

Solution:
1. The flux of L_1 is counterclockwise.
2. The flux of L_2 is clockwise. Therefore, the coils are series opposing.
3. $M = 5$ mH (same value as in Example 16-7).
4. $L_T = L_1 + L_2 - 2M$
    $= 50 \text{ mH} + 50 \text{ mH} - 10 \text{ mH}$
    $= 90 \text{ mH}$

We rarely find mutual inductance between coils connected in parallel. If such a situation exists, we use the following formula:

$$L_T = \frac{L_1 L_2 - M^2}{L_1 + L_2 \pm 2M} \qquad (16\text{-}16)$$

With aiding fluxes, we use the negative sign for M in the denominator. For opposing fluxes the positive sign is used.

16-7. ENERGY STORAGE

We have referred to inductance and capacitance as energy-storage properties of a circuit. This is true, but some clarification is needed to avoid misunderstanding.

Remember that we said the charge on a capacitor is stored in the dielectric. We can remove a charged capacitor from a circuit. There will still be energy storage in the capacitor. If the leakage current is very low, as in a mica capacitor, the capacitor remains charged for many hours. No similar situation is found with inductance. There is no way to sustain a magnetic field when current drops to zero.

An inductor takes energy from a circuit as current attempts to increase the flux. When current decreases, the magnetic field returns energy to the circuit. In both instances the effect is observed as a counter emf, always of a polarity that attempts to maintain constant current and flux.

The solution for energy storage in joules makes use of the following formula:

$$W = \frac{LI^2}{2} \text{ J} \qquad (16\text{-}17)$$

In this formula, I is a steady-state current. We have calculated the energy stored in the magnetic field. A change in current results in a change in energy,

$$\Delta W = \frac{L(\Delta I)^2}{2} \qquad (16\text{-}18)$$

where ΔI is the difference between the original and new current. When the new current is less than the original, energy is returned to the circuit. When the new current is greater, energy is taken from the circuit.

SUMMARY

1. Inductance opposes changes in current.
2. Induced voltage is often referred to as counter emf.
3. The unit of inductance is the henry (H). Other practical units are millihenrys (mH), microhenrys (μH), and nanohenrys (nH).
4. Mutual inductance occurs when the flux of one coil links the turns of another coil.
5. Inductors in series add directly.
6. The mutual inductance of series or parallel inductors may be aiding or opposing. This depends upon the flux directions established by the individual coils.
7. Inductance stores energy in the form of a magnetic field.

PROBLEMS

16-1. Change the following to millihenrys:

        0.76 H   1.2 H
        0.005 H   680 μH
        0.015 H   0.02 H

16-2. Change the following to microhenrys:

        1.5 mH   165 mH
        0.03 mH   0.85 mH
        0.008 mH   10 mH

16-3. In the sketch of Fig. 16-10, the current rises at a rate of 1.5 mA/μs. Solve for the amount and polarity of induced voltage.

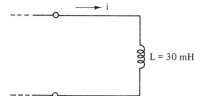

Figure 16-10

16-4. Repeat Prob. 16-3, but with current decreasing at 2 mA/μs.

16-5. A radio-frequency coil is wound of 20 turns of stiff copper wire. The coil is free standing, with air as the core. The length is 4 cm and the diamter is 2.5 cm. Calculate the inductance of the coil.

16-6. A choke coil consists of 150 turns wound upon a high-permeability core. Given that $\mu_r = 3540$, coil length is 5 cm, and cross-sectional area is 5×10^{-4} m², solve for inductance.

16-7. Two unconnected coils are mutually coupled. Given that $L_1 = 350$ mH, $L_2 = 200$ mH, and $k = 0.6$, solve for mutual inductance.

16-8. Two coils are wound upon the same high-permeability core, as in an audio transformer. Assuming that $k \simeq 1$, what is the mutual inductance, given that $L_1 = 3$ H and $L_2 = 4$ H?

16-9. Solve for the coefficient of coupling between coupled coils, given the data $L_1 = 40$ μH, $L_2 = 80$ μH, and $M = 11.3$ μH.

16-10. Solve for total inductance in the circuit of Fig. 16-11.

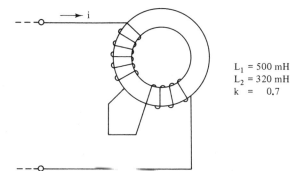

$L_1 = 500$ mH
$L_2 = 320$ mH
$k = 0.7$

Figure 16-11

16-11. In the circuit of Fig. 16-12, solve for the total inductance.

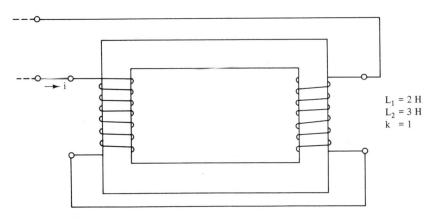

Figure 16-12

16-12. Two coils, each equal to 50 mH, are connected in parallel with $k = 0.15$ (aiding). Calculate the total inductance.

16-13. Solve for total inductance in each circuit of Fig. 16-13. Assume that there is zero mutual inductance between coils.

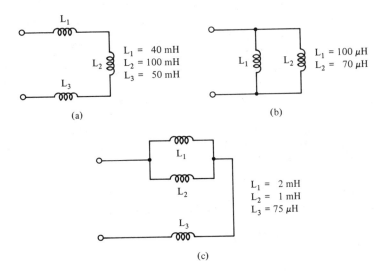

Figure 16-13

16-14. Calculate the energy stored in the magnetic field of a coil when $I = 300$ mA and $L = 2$ H.

16-15. Suppose that the current in Prob. 16-14 decreases to 250 mA. How much energy is returned to the circuit?

17

Transient Response and Time Constants

17-1. TRANSIENT VOLTAGES AND CURRENTS

With dc voltages and currents, we refer to quantities that are either constant, change gradually with respect to time, or pulsate at some given rate, but in one direction.

With ac voltages and currents, we refer to quantities that vary above and below zero, so the average value of the wave is zero. Regardless of wave shape, the voltage or current has *periodicity*. That is, we have a specific time interval for a cycle. At the end of the period of a cycle, the cycle repeats.

In the case of *transient* voltages and currents, there need not be a fixed time interval between events. A transient voltage or current may appear occasionally and randomly. On the other hand, the portion of a wave that we handle as a transient may be part of a periodic wave. *Be aware that we do use transient analysis formulas with pulse wave forms even though these waves have periodicity.* We treat the leading and falling edges of these pulses as transients. In fact, it is common to treat any abrupt change of a wave as a transient.

Transients occur in ac as well as dc circuits. They are caused in many ways; for example, they occur whenever circuits switch on or off. An example is sketched in Fig. 17-1. Transients are generated whenever there is any abrupt change in operating conditions. In this chapter we shall work with switched circuits or pulse waves to analyze the effect of transients in *RC* and *RL* circuits.

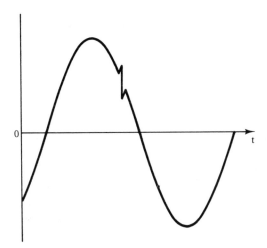

Figure 17-1 The ac wave is a periodic wave. However, a transient of voltage appears in the positive half-cycle. This transient could have been the result of some other circuit coming on or being cut-off.

17-2. RESISTANCE–CAPACITANCE (*RC*) CIRCUITS WITH DC SOURCES

Capacitors can be used in dc circuits as coupling devices. In this application, a capacitor is used to *prevent* the transfer of dc voltage from one part of a circuit to another. At the same time, it must permit (couple) the transfer of changing or ac voltages from one part of the circuit to another. This is a common usage in electronic amplifier circuits. We must use dc voltages to properly operate an amplifer. The desired output is the signal voltage, not the dc component. The capacitor couples the signal to the output load resistor while preventing dc from appearing in the output. When properly

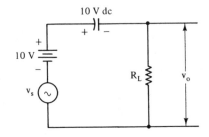

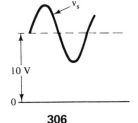
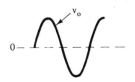

Figure 17-2 Capacitive coupling.

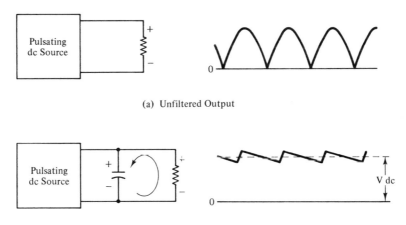

(a) Unfiltered Output

(b) Shunt Capacitance Filtering

Figure 17-3 The use of a capacitor across the load results in a higher average voltage, with less variation from max to min.

selected in comparison to the resistance of a circuit, the capacitor acts as a short circuit for ac and an open circuit for dc. Figure 17-2 is an illustration of capacitive coupling. In this circuit, the charge on the capacitor is equal to battery voltage. The only voltage across the load is signal voltage. This voltage is the result of the capacitor charging and discharging as signal voltage adds to and subtracts from the 10-V dc level.

Capacitors are also used to maintain average levels of voltage across resistors when current in the remainder of the circuit falls. In this application, a capacitor is in parallel with the load resistor. The capacitor charges to the maximum voltage developed across the resistor. When current attempts to drop, the capacitor discharges through the resistor, causing the drop-off in voltage to be very gradual. With sufficient capacitance, in relation to the resistance, any drop-off in voltage between pulses of charging current can be made to be as small as 1/1000 of the maximum voltage. Figure 17-3 illustrates this application of changing a pulsating dc voltage into a very much "smoother" dc voltage.

17-3. VOLTAGE RISE IN AN *RC* CIRCUIT

Consider the circuit of Fig. 17-4. Assume that charge on the capacitor is zero before the switch is closed. At the instant the switch is closed, the charge on the capacitor is zero. Therefore, all the source voltage is across the resistor and current is *maximum*.

$$i = \frac{E}{R}, \quad t = 0$$

Immediately after the switch closes, the capacitor *begins* to charge. The voltage across the capacitor is in *series opposition* to the source voltage.

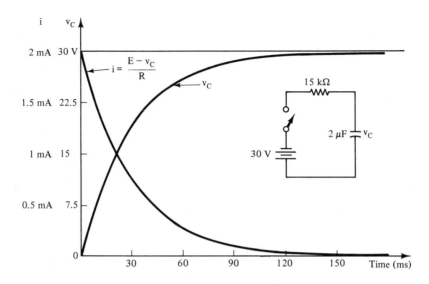

Figure 17-4 As the voltage across the capacitor increases, current in the circuit decreases. Eventually, current approximates zero, as the voltage across the capacitor equals, for all practical purposes, the applied voltage. In this graph, current begins at 2 mA; (30 V/15 kΩ). Note that the voltage across the capacitor never quite reaches 30 V.

The voltage across the resistor decreases by the amount of voltage across the capacitor.

$$v_R = E - v_C, \quad t > 0$$

With less voltage across the resistor, current decreases. Thus the *rate* of charge of the capacitor decreases.

The voltage across the capacitor continues to rise as charge continues to accumulate. With each increase in v_C, v_R decreases, which results in reduced current. While this explanation may make the process appear to occur in pieces, this is a continuous process that ends only when the capacitor is fully charged.

When the charge on the capacitor is such that the voltage across the capacitor is approximately equal to the supply voltage, the current is zero.

From Kirchhoff's voltage law, we know that

$$E = v_R + v_C \tag{17-1}$$

and

$$E = iR + v_C \tag{17-2}$$

Substituting,

$$i = C \frac{dv_C}{dt}$$

we have

$$E = RC\frac{dv_C}{dt} + v_C \tag{17-3}$$

This equation can only be solved as a differential equation and is beyond the scope of math needed by any electronics technician. However, we can make very good use of the solution to Eq. (17-3). We solve for v_C:

$$v_C = E(1 - e^{-t/RC}) \tag{17-4}$$

where E = applied voltage
R = resistance in the charging circuit
t = time that the capacitor charges
e = the base of natural logarithms, 2.718
C = capacitance in the charging circuit

With scientific calculators available to us, we need only know how to use e, not understand where it comes from. We shall explain how to solve problems with e^{-x} using the calculator, but for our purposes in this section, simply refer to Table 17-1 to get the values of e^{-x} and $(1 - e^{-x})$ that we need.

In the table, we treat t/RC as the exponent x. The value of x at any instant is the ratio of time in seconds to the product RC.

$$x = \frac{t}{RC}$$

TABLE 17-1

$t/RC = x$	e^{-x}	$1 - e^{-x}$
0	1	0
0.25	0.7788	0.2212
0.5	0.6065	0.3935
1.0	0.3679	0.6321
2.0	0.1353	0.8647
3.0	0.04979	0.95021
4.0	0.01832	0.98168
5.0	0.006738	0.99326
10.0	0.0000454	0.99995

EXAMPLE 17-1:
In the circuit of Fig. 17-5, after the switch is closed, what is the voltage across the capacitor at (a) 2.5 ms, (b) 5 ms, (c) 10 ms, (d) 30 ms, and (e) 50 ms?

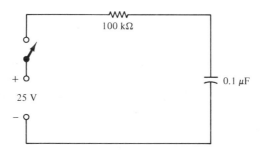

Figure 17-5

Solution:

$$v_C = E(1 - e^{-t/RC})$$

1. We begin by solving for RC.

$$RC = 1 \times 10^5 \times 0.1 \times 10^{-6} = 0.1 \times 10^{-1}$$
$$= 10 \times 10^{-3}$$

2. t/RC is solved at each step.
3. The formula is solved for v_C by use of Table 17-1.

(a) $t = 2.5$ ms:

$$\frac{t}{RC} = \frac{2.5}{10} = 0.25 = x$$

$$v_C = 25(1 - e^{-0.25})$$
$$= 25 \times 0.2212 = 5.53 \text{ V}$$

(b) $t = 5$ ms:

$$\frac{t}{RC} = x = \frac{5}{10} = 0.5$$

$$v_C = 25(1 - e^{-0.5})$$
$$= 25 \times 0.3935 = 9.84 \text{ V}$$

(c) $t = 10$ ms:

$$\frac{t}{RC} = 1 = x$$

$$v_C = 25(1 - e^{-1})$$
$$= 25 \times 0.6321 = 15.8 \text{ V}$$

(d) $t = 30$ ms:

$$\frac{t}{RC} = 3$$

$$v_C = 25(1 - e^{-3})$$
$$= 25 \times 0.9502 = 23.76 \text{ V}$$

(e) $t = 50$ ms:

$$x = 5$$

$$v_C = 25(1 - e^{-5})$$
$$= 25 \times 0.9933 = 24.8 \text{ V}$$

17-4. MEANING OF THE *RC* TIME CONSTANT

The time constant is a way of describing the time needed to charge or discharge a capacitor. Time constant is the time in seconds that causes the fraction t/RC to equal 1. In the time equal to one time constant, a capacitor will charge by 63% of the available charging voltage. We can see this from Table 17-1. At $t/RC = 1$, we have e^{-x} equal to 0.3679. This decimal repre-

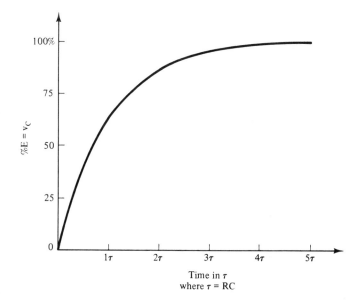

Figure 17-6 Capacitor charge as a percent of applied voltage. In approximately 5 time constants the capacitor charges fully.

sents the fraction of the *remaining* voltage. The capacitor has charged to 0.6321 of the applied voltage. At $x = 2$, we find that the capacitor has charged to 0.8647 of the applied voltage. If you take 63.21% of 0.3679 and *add* the result to 0.6321, you will get 0.8647.

Clearly, the value of time that will result in one time constant must equal the product of resistance and capacitance. We use the symbol τ for time constant.

$$\tau = RC \qquad (17\text{-}5)$$

1. (a) With R in megohms and C in microfarads, τ is in seconds.
2. (b) With R in kilohms and C in microfarads, τ is in milliseconds (ms).
3. (c) With R in kilohms and C in nanofarads, τ is in microseconds (μs).
4. (d) With R in megohms and C in picofarads, τ is in microseconds (μs).
5. (e) With R in kilohms and C in picofarads, τ is in nanoseconds (ns).

Figure 17-6 is a graph of %E versus t/τ. Voltage is not given because the graph of a charging capacitor is always of the same shape. A capacitor is assumed to be fully charged at five time constants.

17-5. DISCHARGE IN AN *RC* CIRCUIT

Consider the circuit of Fig. 17-7. Assume that S_1 has been closed for at least five time constants (5τ). Now we open S_1 and close S_2. The capacitor must discharge through the resistor R_2. At the instant S_2 closes, v_C is maximum

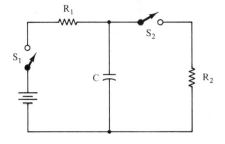

Figure 17-7 With switch S2 open and S1 closed, the capacitor charges through resistor R_1. When we open S1 and close S2, the capacitor discharges through resistor R_2.

and the discharge current is maximum. As the voltage across the capacitor decreases, current decreases, reducing the *rate* of discharge.

The voltage across resistor R_2 and the capacitor are always equal to each other.

$$v_{R2} = v_C$$
$$iR_2 = v_C$$

In a capacitive circuit, $i = C(dv_C/dt)$; therefore,

$$v_C = RC\frac{dv_C}{dt}$$

If we solve this equation for v_C, we have

$$v_C = V_{C(max)} e^{-t/RC} \qquad (17\text{-}6)$$

where $V_{C(max)}$ was the voltage across the capacitor at the start of discharge.

Given that the voltage across R_2 is equal to v_C, we can solve for the discharge current.

$$i = \frac{v_C}{R} = \frac{V_{C(max)}}{R} e^{-t/RC} \qquad (17\text{-}7)$$

Figure 17-8 illustrates the relationship between t/τ and the discharge of a capacitor. In charging a capacitor, we assume that it is fully charged in a

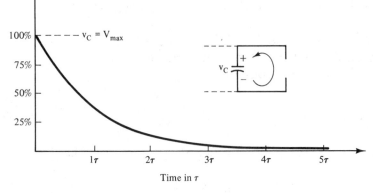

Figure 17-8 Discharge of a capacitor through a shunt resistance. After five time constants the discharge is virtually complete.

time equal to, or greater than, five time constants. Similarly, we find that for all practical purposes a capacitor is fully discharged after five time constants.

For $v_C = E$ (charging), and $v_C = 0$ (discharging),

$$t \geqslant 5\tau \qquad (17\text{-}8)$$

where $\tau = RC$.

EXAMPLE 17-2:

In the circuit of Fig. 17-9, what value of R is needed if we wish to cause a delay of 2 s between the closing of the switch and full charge of the capacitor?

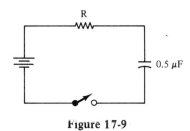

Figure 17-9

Solution:

1. $t = 5\tau$

$$\tau = \frac{t}{5} = \frac{2}{5} = 0.4$$

2. $\tau = RC$

$$R = \frac{\tau}{C}$$

$$= \frac{0.4}{0.5 \times 10^{-6}} = 0.8 \times 10^6$$

$$= 800 \text{ k}\Omega$$

EXAMPLE 17-3:

A 0.1-μF capacitor is charged to 40 V. If a 10-kΩ resistor is switched across the capacitor and the charging source is removed, how much time in milliseconds is needed for capacitor voltage to drop to zero?

Solution:

$$t = 5\tau$$

$$\tau = RC = 10 \times 10^3 \times 0.1 \times 10^{-6}$$
$$= 1 \times 10^{-3} = 1 \text{ ms}$$

$$t = 5\tau = 5 \text{ ms}$$

17-6. PULSE RESPONSE OF *RC* CIRCUITS

An important area of application for *RC* circuits deals with pulsed inputs. These pulse voltages may vary between zero and some level, or they may vary between equal and opposite voltage levels. The first type are typical of

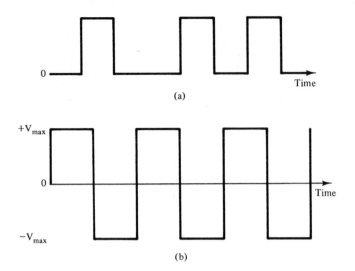

Figure 17-10 (a) Pulse waveform, typical of some digital processes. (b) Square wave, typical of signal generator output.

those found in digital systems. Signal generators are a typical source of pulse signals that vary between equal positive and negative levels. Also, we find in digital systems that pulses need not have periodicity, although the duration of the pulse may be fixed by the design characteristics of the system. On the other hand, pulses obtained from signal sources always have periodicity (Fig. 17-10).

Let us examine the response of one type of *RC* circuit to pulse inputs. *The circuit is called a differentiator, provided that the time constant of the circuit is much less than the time of a pulse.* An example of a differentiator is shown in Fig. 17-11. The output voltage is developed across the resistor.

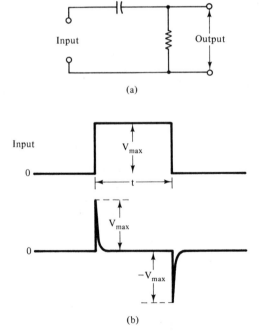

Figure 17-11 Input and output waveforms of a differentiator.

314   Chap. 17   Transient Response and Time Constants

Because the time constant of the circuit is a small fraction of pulse duration, the capacitor charges and discharges fully. The resulting output is a pair of differentiated pulses, a pair of spikes of voltage. The positive spike occurs on the leading edge of the input signal, with the negative pulse occurring when the input drops to zero.

The rising leading edge of the input pulse is equivalent to the closing of a switch in a charging circuit. When the input pulse drops to zero, it is equivalent to short circuiting the input and allowing the capacitor to discharge. Note that the output voltage is a peak-to-peak value that is twice that of the input voltage. A detailed explanation of differentiator operation is given in Fig. 17-12.

When the time constant of the *RC* circuit is a more appreciable portion of the pulse duration, the output voltage decreases. It is no longer a sharp spike of voltage, although we find abrupt changes in output that occur when pulse voltage rises and falls. Such operation is shown in Fig. 17-13. The resulting output is stable after approximately five cycles of input.

When the output is taken across a capacitor, as in Fig. 17-14, the circuit can operate as an *integrator*. For the circuit to operate properly, the time constant must be very much greater than the time of a pulse. In this kind of circuit, the voltage across the capacitor is triangular if the pulse input is rectangular.

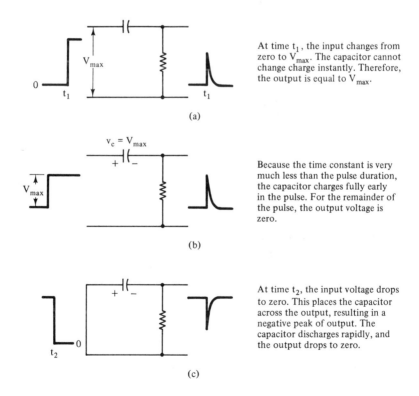

At time t_1, the input changes from zero to V_{max}. The capacitor cannot change charge instantly. Therefore, the output is equal to V_{max}.

(a)

Because the time constant is very much less than the pulse duration, the capacitor charges fully early in the pulse. For the remainder of the pulse, the output voltage is zero.

(b)

At time t_2, the input voltage drops to zero. This places the capacitor across the output, resulting in a negative peak of output. The capacitor discharges rapidly, and the output drops to zero.

(c)

Figure 17-12 Differentiator action.

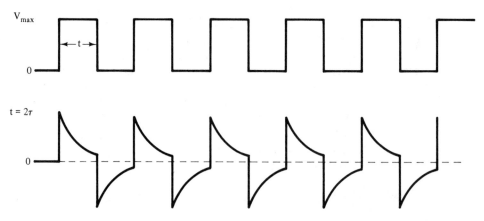

Figure 17-13 In this case, the time constant is one-half of pulse duration.

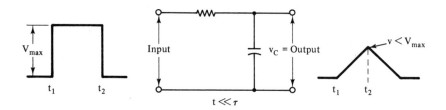

Figure 17-14 An integrator circuit. The circuit functions in this way because the time constant is much greater than pulse duration.

A special application of integrator circuits is found when the space between pulses is less than the time of a pulse. Because the discharge time is less than the charge time, charge accumulates with each input pulse, which results in a voltage across the capacitor that approaches the peak value of the input. An example of an integrator application is found in television and is shown in Fig. 17-15. The vertical synchronizing pulse consists of six short-duration pulses separated by very short off periods. The synch pulse used by

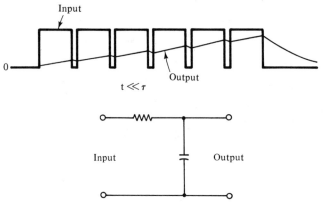

Figure 17-15 The use of an integrator circuit in television.

316   Chap. 17   Transient Response and Time Constants the TV receiver is the integrated pulse taken at the output of the integrator circuit. The system has high immunity to interference because of the relatively long time constant of the circuit.

17-7. CURRENT RISE IN LR CIRCUITS

Just as capacitance offers opposition to change in voltage, inductance opposes current change. Consider the circuit of Fig. 17-16. With the switch open, current is zero. At the instant that the switch closes, current attempts to flow. There is an induced voltage across the coil that is in opposition to applied voltage. At the instant the switch closes, the *rate of change* of current is maximum, and the induced voltage is nearly equal to the applied voltage. The rate of change of current decreases with time, which allows current to reach a maximum value that is determined by Ohm's law.

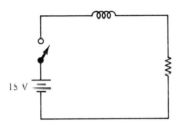

Figure 17-16 An elementary LR circuit.

17-8. L/R TIME CONSTANT

The instantaneous value of current can be found only by the use of advanced mathematics.

$$E = v_R + v_L$$
$$= iR + L\frac{di}{dt}$$

Solving for i, we have

$$i = \frac{E}{R}(1 - e^{-Rt/L}) \qquad (17\text{-}9)$$

where E = applied voltage
R = resistance in series with L
t = time that the current is allowed to rise
e = base of natural logarithms

Note that the "shape" of Eq. (17-9) is the same as that of Eq. (17-4).

The values of e^{-x} in Table 17-1 are used in the solution of Eq. (17-9). The time constant, τ, of the circuit is that unit of time that causes x to equal 1. When t is equal to L/R, we have one time constant.

$$\frac{R}{L} \times \frac{L}{R} = 1$$

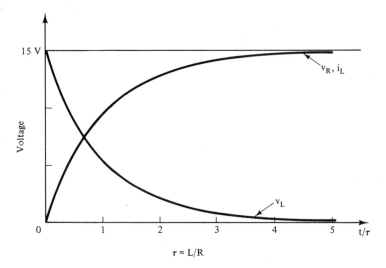

Figure 17-17 The rise of current in an inductive circuit.

Therefore,

$$\tau = \frac{L}{R} \qquad (17\text{-}10)$$

Current reaches its maximum at approximately five time constants.

The voltage across the inductor is maximum at the instant of "charge" and decreases with time. If the resistance of the inductor winding is very small compared to the resistance of the circuit, we can assume that the voltage across the coil eventually becomes zero, for all practical purposes. On the other hand, the voltage across the resistor rises as current increases and becomes equal to applied voltage after five time constants (5τ). These processes are shown in Fig. 17-17.

EXAMPLE 17-4:
A 50-mH inductor is used in a timing circuit. What value of resistance must be placed in series with the inductor for $\tau = 5$ μs?

Solution:

$$\tau = \frac{L}{R}$$

$$R = \frac{L}{\tau} = \frac{50 \times 10^{-3}}{5 \times 10^{-6}} = 10 \times 10^3$$

$$= 10 \text{ k}\Omega$$

EXAMPLE 17-5:
What is the time constant of an *RL* circuit if $L = 500$ mH and $R = 100$ kΩ?

Solution:

$$\tau = \frac{L}{R} = \frac{500 \times 10^{-3}}{100 \times 10^3}$$

$$= 5 \times 10^{-6}$$

$$= 5 \text{ μs}$$

17-9. DISCHARGE IN AN *RL* CIRCUIT

Suppose that we replace source voltage in an *RL* circuit with a short circuit. Current must drop, but the effect of the energy stored in the magnetic field surrounding the inductor causes a time delay before current drops to zero. At the instant current *begins* to fall, the voltage induced in the coil equals the original source voltage, with a polarity that maintains the original current direction. At this instant, we have

$$v_L = -E \qquad (17\text{-}11)$$

As the flux decreases, induced emf decreases, which results in reduced current. The current at any instant is found by Eq. (17-12).

$$i = \frac{E}{R} e^{-Rt/L} \qquad (17\text{-}12)$$

After 5τ, current is assumed to be zero.

17-10. PULSE RESPONSE OF AN *LR* CIRCUIT

Resistance–capacitance differentiator and integrator circuits are widely used. While resistance–inductance circuits can be used for similar purposes, it is rarely done. We find *LR* circuits have greater cost, greater weight, and interfere with other circuits. The interference is the result of induced emfs into adjacent circuits by the changing flux of the inductor.

It is important that we understand the reaction of an *LR* circuit to pulses. In future studies you will find the need to *prevent* sharp spikes of voltage across inductors.

Assume that the pulse of voltage shown in Fig. 17-18 is applied to the circuit shown. We shall assume that the time constant is much less than the duration of the pulse.

$$\tau \ll t_2 - t_1$$

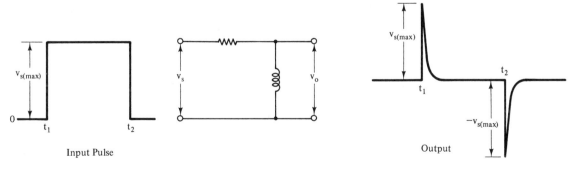

Figure 17-18 L/R differentiator circuit.

At time t_1, the applied voltage switches from zero to its maximum value. The voltage across the inductor is a spike equal to the applied voltage. This equals the applied voltage in order to cause the current to be zero. With the very short time constant, current reaches its maximum value in a fraction of the pulse period. The voltage across the inductor drops to zero when the current is a constant.

At time t_2, the source voltage switches back to zero. This is the same as shorting the input terminals. The inductor is in *parallel* with the resistor. The collapsing magnetic field attempts to maintain current. The resulting output is a differentiated pulse (spike) of voltage with a peak value equal to V_s but of opposite polarity.

EXAMPLE 17-6:
What is the voltage wave form across the inductor of Fig. 17-19 if pulse duration is 10 ms? Sketch the resultant wave.

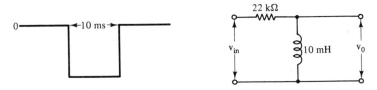

Figure 17-19

Solution:
(a) Solve for the time constant.

$$\tau = \frac{L}{R} = \frac{10 \times 10^{-3}}{22 \times 10^3}$$
$$= 0.455 \times 10^{-6}$$
$$= 0.455 \ \mu s$$

(b) Pulse duration is very much greater than the time constant. Therefore, the circuit will work as a differentiator. A negative spike is developed at the start of the pulse and a positive spike at the end of the pulse.
(c) A sketch of v_L shows the output wave (Fig. 17-20).

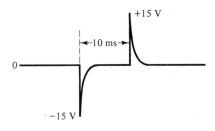

Figure 17-20

17-11. USING THE SCIENTIFIC CALCULATOR

In solutions for instantaneous values of current, voltage, and time, we make use of scientific calculators. We need to evaluate e^{-x}, or given an e^{-x}, we must find x. The actual process is simple. We find difficulties because there is no uniformity in the names assigned to the function key we must use. In many calculators, particularly those made by Texas Instruments, we cannot find an $\boxed{e^x}$ key. Of course, these splendid calculators make e^x solutions; we simply use a different procedure.

Solutions for e^{-x}:
The key we must eventually use may be identified in several different ways.
1. $\boxed{\ln x}$ (natural log of x)
2. $\boxed{\ln}$ (natural log; x is understood)
3. $\boxed{e^x}$

Note that, while we need to work with e^{-x}, all calculators are designed to perform the e^x function. We must make x negative by keying the *change sign* when we enter the exponent. Do not use the minus sign; remember that this instructs the calculator to subtract!

For those calculators with $\boxed{\ln}$ or $\boxed{\ln x}$ keys, we must invert the function. The calculator has an $\boxed{\text{INV}}$ or $\boxed{\text{ARC}}$ key. The names mean the same thing. To get e^x or e^{-x}, we begin by entering x or $-x$, then $\boxed{\text{INV}}$, then $\boxed{\ln}$ or $\boxed{\ln x}$.

EXAMPLE 17-7:
Solve for (a) $e^{-1.5}$, (b) $e^{-0.5}$, (c) e^{-5}.

Solution:
(a) Enter 1.5, press $\boxed{+/-}$, press $\boxed{\text{INV}}$, press $\boxed{\ln x}$. The display should read
$$0.2231301601$$
We would round off to 0.223.
(b) Enter 0.5, press $\boxed{+/-}$, press $\boxed{\text{INV}}$, press $\boxed{\ln x}$. The result is
$$0.607 \text{ (rounded)}$$
(c) Enter 5, press $\boxed{+/-}$, press $\boxed{\text{INV}}$, press $\boxed{\ln x}$.
$$e^{-5} = 0.00674$$

EXAMPLE 17-8:

Solve for $e^{-0.05}$, e^{-2}, $e^{-2.8}$ on a calculator with an $\boxed{e^x}$ key.

Solution:

1. Enter 0.05, press $\boxed{+/-}$, press $\boxed{e^x}$.
$$e^{-0.05} = 0.951$$
2. Enter 2, press $\boxed{+/-}$, press $\boxed{e^x}$.
$$e^{-2} = 0.135$$
3. Enter 2.8, press $\boxed{+/-}$, press $\boxed{e^x}$.
$$e^{-2.8} = 0.0608$$

Solutions for time in RL and RC circuits require that we make a solution for x. In these cases we know the value of e^{-x}. We need to know what value of x resulted in the amount represented by e^{-x}. To solve for x, we need to take the natural logarithm of the number.
Given that

$$e^x = y$$

then

$$x = \ln y$$

For calculators with an $\boxed{\ln}$ or $\boxed{\ln x}$ key, we simply enter y and press $\boxed{\ln x}$. For calculators with an $\boxed{e^x}$ key, we must invert the function. We enter y, press $\boxed{INV}$ or $\boxed{ARC}$, then press $\boxed{e^x}$.

For values of e^x that are less than 1, the result is negative. This will always be our case, and this simply means that we have made a solution for $-x$, given e^{-x}.

Once we have x, we solve for time, t: In RC circuits,

$$t = (x)RC \qquad (17\text{-}13)$$

In RL circuits,

$$t = (x)\frac{L}{R} \qquad (17\text{-}14)$$

EXAMPLE 17-9:

We find that e^{-x} is 0.56. Solve for x.

Solution:

Enter 0.56, press $\boxed{\ln x}$. The display when rounded is

$$-0.580$$

In a differently keyed calculator we would enter 0.56, press $\boxed{\text{INV}}$, press $\boxed{e^x}$.
$$x = -0.580$$

We shall put our knowledge to use in solving problems in transient analysis. In the following examples we shall make solutions in the traditional manner, followed by calculator procedure.

EXAMPLE 17-10:
Given the circuit and input voltage of Fig. 17-21, what is the voltage across the capacitor 5 ms after time t_1?

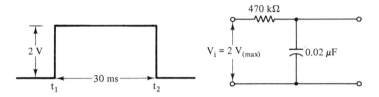

Figure 17-21

Solution:
1. Solve for τ.
$$\tau = RC = 470 \text{ k}\Omega \times 0.02 \text{ }\mu\text{F}$$
$$= 9.4 \text{ ms}$$

2. Solve the formula:
$$v_C = V_{max}(1 - e^{-t/RC})$$
(a) We treat $e^{-t/RC}$ as e^{-x}, where $x = t/\tau$.
(b) $\dfrac{t}{\tau} = \dfrac{5 \text{ ms}}{9.4 \text{ ms}} = 0.532$
(c) $e^{-0.532} = 0.5874$
(d) $v_C = 2(1 - 0.5874)$
$= 0.825$ V

Calculator Procedure:
1. 5 $\boxed{\text{EE}}$ 3 $\boxed{+/-}$ $\boxed{\div}$ $\boxed{(}$ 470 $\boxed{\text{EE}}$ 3 $\boxed{\times}$ .02 $\boxed{\text{EE}}$ 6 $\boxed{+/-}$ $\boxed{)}$ $\boxed{=}$
Display: 5.319 ... -01
2. $\boxed{+/-}$ $\boxed{\text{INV}}$ $\boxed{\ln x}$
Display: 5.875 ... -01
3. $\boxed{+/-}$ $\boxed{+}$ 1 $\boxed{=}$
Display: 4.1252 ... -01
4. $\boxed{\times}$ 2 $\boxed{=}$
Display: 8.25 ... -01

$$v_C = 0.825 \text{ V}$$

EXAMPLE 17-11:

Given the circuit and pulse input of Example 17-10, what is the voltage across the capacitor at time t_2?

Solution:
    1. The time constant is known.

$$\tau = 9.4 \text{ ms}$$

    2. Solve the equation:

$$v_C = 2(1 - e^{-t/\tau})$$

    (a) $\dfrac{t}{\tau} = \dfrac{30 \text{ ms}}{9.4 \text{ ms}} = 3.191$
    (b) $e^{-3.191} = 0.0411$
    (c) $v_C = 2(1 - 0.0411)$
           $= 1.918$ V

Calculator Procedure:
1. 30 [EE] 3 [+/−] [÷] [(] 470 [EE] 3 [×] .02 [EE] 6 [+/−] [)] [=]
   Display: 3.191 ...   00
2. [+/−] [eˣ] or [+/−] [INV] [lnx]
   Display: 4.111 ...   −02
3. [+/−] [+] 1 [=]
   Display: 9.5889 ...   −01
4. [×] 2 [=]
   Display: 1.9177 ...   00

$$v_C = 1.918 \text{ V}$$

EXAMPLE 17-12:

In the circuit of Example 17-10, what is the voltage across the capacitor 5τ after time t_2?

Solution:

$$\begin{aligned} v_C &= V_{C(\text{max})} e^{-t/\tau} \\ &= 1.918(e^{-5}) \\ &= 1.918 \times 0.00673 \\ &= 0.0129 \text{ V} \end{aligned}$$

Calculator Procedure:
1.918 [×] 5 [+/−] [INV] [lnx] [=]
Display: 0.01292 ...

$$v_C = 0.0129 \text{ V}$$

EXAMPLE 17-13:

In the circuit of Fig. 17-22, what is the time required for the voltage across the capacitor to reach (a) one-half the pulse amplitude and (b) the full pulse amplitude?

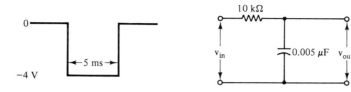

Figure 17-22

Solution:
(a) Solve for the time constant, τ.

$$\tau = RC = 10 \text{ k}\Omega \times 0.005 \text{ }\mu\text{F} = 50 \text{ }\mu\text{s}$$

Then we must solve the charging formula for t:

$$v_C = V_{max}(1 - e^{-t/\tau})$$

Dividing by V_{max},

$$\frac{v_C}{V_{max}} = 1 - e^{-t/\tau}$$

Transposing,

$$e^{-t/\tau} = 1 - \frac{v_C}{V_{max}}$$

$$-t/\tau = \ln\left(1 - \frac{v_C}{V_{max}}\right)$$

$$t = \tau \left| \ln\left(1 - \frac{v_C}{V_{max}}\right) \right|$$

$$t = 50 \times 10^{-6} \times \left| \ln\left(1 - \frac{2}{4}\right) \right|$$
$$= 50 \times 10^{-6} \times |\ln 0.5|$$
$$= 34.66 \times 10^{-6}$$
$$= 34.66 \text{ }\mu\text{s}$$

Calculator Procedure:
(a) We begin by solving for x:

$$2 \boxed{\div} 4 \boxed{=} \boxed{+/-} \boxed{+} 1 \boxed{=} \boxed{\ln x}$$

Display: $-.6931\ldots$
Because we are interested in the magnitude of x, we use the change sign key: $\boxed{+/-}$

$$|x| = 0.6931$$

$\boxed{\times}$ 10 $\boxed{\text{EE}}$ 3 $\boxed{\times}$ .005 $\boxed{\text{EE}}$ 6 $\boxed{+/-}$ $\boxed{=}$
Display: $3.4657\ldots$ -05

$$t = 34.66 \text{ }\mu\text{s}$$

Sec. 17-11. Using the Scientific Calculator

(b) The time needed to fully charge the capacitor is 5τ.
$$t = 5 \times 50 \times 10^{-6}$$
$$= 250\ \mu s$$

EXAMPLE 17-14:
What is the time constant of the circuit of Fig. 17-23?

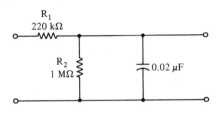

Figure 17-23

Solution: We apply Thevenin's theorem, using the capacitor as the load. The Thevenin resistance is the parallel combination of R_1 and R_2 as in Fig. 17-24.

$$220\ k\Omega\ \|\ 1\ M\Omega = 180\ k\Omega$$
$$\tau = RC = 180 \times 10^3 \times 2 \times 10^{-8}$$
$$= 360 \times 10^{-5} = 3.6 \times 10^{-3}$$
$$= 3.6\ ms$$

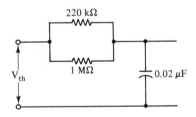

Figure 17-24

Calculator Procedure:
$R_1 \| R_2$:

$$220\ \boxed{EE}\ 3\ \boxed{\times}\ 1\ \boxed{EE}\ 6\ \boxed{\div}$$
$$\boxed{(}\ 220\ \boxed{EE}\ 3\ \boxed{+}\ 1\ \boxed{EE}\ 6\ \boxed{)}\ \boxed{=}$$

Display: 1.803 ...   05

$$R_{Th} = 180\ k\Omega$$

τ:

$$180\ \boxed{EE}\ 3\ \boxed{\times}\ .02\ \boxed{EE}\ 6\ \boxed{+/-}\ \boxed{=}$$

Display: 3.6   −03

$$\tau = 3.6\ ms$$

EXAMPLE 17-15:
Given the circuit of Fig. 17-25, solve for (a) the time constant and (b) the voltage across the resistor 300 μs after the switch closes.

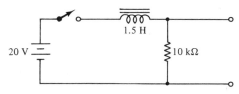

Figure 17-25

(a) $\tau = \dfrac{L}{R} = \dfrac{1.5}{10 \times 10^3}$

$= 0.15$ ms $= 150$ μs

(b) Solve for the current with $t = 300$ μs.
Given:

$$i = \frac{E}{R}(1 - e^{-t/\tau})$$

$$= \frac{20}{10 \times 10^3}(1 - e^{-2})$$

$$= 1.73 \text{ mA}$$

The voltage drop across the resistor is found by Ohm's law.

$$v_R = iR$$
$$= 1.73 \text{ mA} \times 10 \text{ k}\Omega$$
$$= 17.3 \text{ V}$$

Calculator Procedure:

(a) $\tau = 1.5$ ÷ 10 EE 3 =
Display: 1.5   −0.4

$\tau = 150 \times 10^{-6} = 150$ μs

(b) Solution for v_R:
300 EE 6 +/− ÷ 150 EE 6 +/− =
Display: 2   00
+/− INV lnx
Display: 1.3533 . . .   −01
+/− + 1 =
Display: 8.6466 . . .   −01
× 20 ÷ 10 EE 3 =
Display: 1.72932 . . .   −03
× 10 EE 3 =
Display: 1.72932 . . .   01

$$v_R = 17.29 \text{ V}$$

SUMMARY

1. A capacitor cannot change charge in zero time.
2. The voltage across a capacitor rises gradually in a nonlinear manner in resistance-capacitance (RC) circuits.
3. The time constant (τ) of a resistance-capacitance circuit is R times C.
4. A capacitor is assumed to be fully charged after 5τ.
5. It takes 5τ to fully discharge a capacitor.
6. Current in inductance-resistance (LR) circuits rises gradually and nonlinearly.
7. The time constant (τ) of an LR circuit is R divided by L.
8. It takes 5τ to fully establish and to fully discharge the flux around an inductor in an LR circuit.
9. A differentiator circuit is a circuit with a time constant that is very much less than the time of an input pulse.
10. An integrator circuit is a form of circuit where the time constant is very much greater than the time of an input pulse.

PROBLEMS

17-1. Calculate the time constant (τ) for each of the following RC combinations:
10 kΩ, 100 μF 220 kΩ, 0.5 μF
47 kΩ, 0.1 μF 470 kΩ, 5 nF
5.6 kΩ, 200 pF 15 kΩ, 50 pF

17-2. A 15-V source charges a 2-μF capacitor through 1 MΩ of resistance. Assuming zero initial voltage across the capacitor, how much time is needed for $v_C \simeq 15$ V?

17-3. A 0.05-F capacitor is charged to 40 V. If it is discharged through a 2.2-kΩ resistor, how much time is needed for the voltage across the capacitor to be approximately zero?

17-4. In a series RC circuit, $C = 0.3$ μF and $R = 27$ kΩ. Solve for v_C at 10 μs, 2 ms, 10 ms, 20 ms and 100 ms after the instant the switch to a 24-V battery is closed.

17-5. Given the data found in Prob. 17-4, what is the voltage across R at each of the specified time intervals?

17-6. In the circuit of Problem 17-4, $C = 15$ μF and $R = 100$ kΩ. Solve for v_C and v_R at 10 ms, 100 ms, 1.5 s, 4 s, and 10 s.

17-7. In the circuit of Prob. 17-4, $C = 100$ pF and $R = 15$ kΩ. Solve for v_C and v_R at 10 ns, 100 ns, 1.5 μs, 4 μs, and 10 μs.

17-8. In the circuit of Fig. 17-26, $C = 0.6$ μF and $R = 10$ kΩ. Solve for the voltage across the parallel circuit 50 μs, 3 ms, 10 ms, and 50 ms after the instant that the switch is closed. The initial voltage across the capacitor is 18 V.

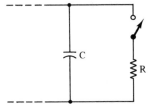

Figure 17-26

17-9. In the circuit of Fig. 17-26, $C = 15$ nF and $R = 4$ kΩ. Solve for the voltage across the RC circuit 0.5 μs, 30 μs, 100 μs, and 500 μs after the switch is closed. Initial $v_C = 18$ V.

17-10. In the circuit of Fig. 17-26, $C = 2$ μF and $R = 300$ kΩ. Solve for the voltage across the RC circuit 5 ms, 300 ms, 1 s, and 5 s after the switch is closed. Initial $v_C = 18$ V.

17-11. Given the circuit of Fig. 17-27, how long after the switch closes at point a does v_C reach 20 V?

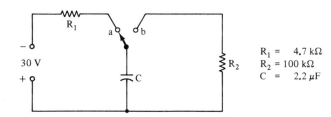

$R_1 = 4.7$ kΩ
$R_2 = 100$ kΩ
$C = 2.2$ μF

Figure 17-27

17-12. Given the data of Prob. 17-11, assume that at the instant $v_C = 20$ V the switch is moved from point a to point b. (a) How much time is required for v_C to drop to 8 V? (b) What is the minimum amount of time needed for v_C to approximate zero?

17-13. Solve for the time constant of the circuit of Fig. 17-28.

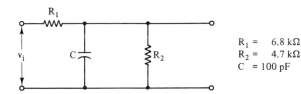

$R_1 = 6.8$ kΩ
$R_2 = 4.7$ kΩ
$C = 100$ pF

Figure 17-28

17-14. A positive pulse with an amplitude of 20 V is applied to the circuit of Fig. 17-28. (a) What is the minimum duration of the pulse in order that $v_C = 6$ V? (b) What is the minimum duration of pulse for $V_C \approx 20$ V?

17-15. Sketch the output voltage of the circuit of Fig. 17-29, given the input pulse.

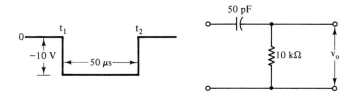

Figure 17-29

17-16. The input pulse duration in Prob. 17-15 is changed so that $t_2 - t_1 = 1$ μs. Can the circuit be an effective differentiator for this input? Explain.

17-17. Calculate maximum output amplitude for the RC integrator circuit of Fig. 17-30, given the input pulse.

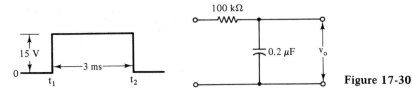

Figure 17-30

17-18. Suppose that the pulse input for the circuit of Fig. 17-30 has duration of 30 ms. Will the circuit work effectively as an integrator? Explain.

17-19. Given a RC differentiator circuit with $\tau = 0.6$ μs and a pulse wave input of 8 V. Each cycle of input is at +8 V for 10 μs and at 0 V for 15 μs. Sketch the input and output waves and specify output voltage.

17-20. Calculate time constants for the following L/R combinations:

10 μH, 100 Ω	50 μH, 10 kΩ
50 μH, 10 Ω	100 mH, 50 Ω
100 mH, 10 kΩ	250 mH, 75 Ω

18

Alternating Voltages and Currents

18-1. FREQUENCY AND PERIOD

In our studies of types of voltage (Chapter 3), we introduced some of the principles of alternating (ac) voltages. We begin our present studies with a review of basic concepts.

All alternating voltages and currents have *periodicity*. This simply means that the time of a cycle for a given ac voltage or current is constant. Regardless of wave shape, the time needed to go from one point on a wave to the *same* point on the next wave is the same. We refer to this time as the *period* (T). *Period is the time of one cycle.* Various waves are shown in Fig. 18-1.

We define a cycle as one complete alternation of an ac wave. The number of cycles per second is the *frequency* of an ac wave. The unit of frequency is the hertz (Hz). Typical values of frequency depend upon the application of the ac voltage and current.

The electrical power furnished by the utilities is ac. In the United States, as in much of the world, this is at a frequency of 60 Hz. In other countries, notably England, ac power is supplied at 50 Hz.

Audio (sound) signals range in frequency from 20 Hz to 20 kHz. The lower frequencies represent deep bass sounds, while the higher frequencies represent treble sounds.

Video frequencies in commercial TV range from 30 Hz to 4 MHz. The high frequencies represent very fine picture details, while the very low frequencies represent the average brightness level of a scene.

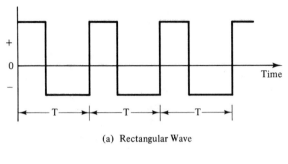

(a) Rectangular Wave

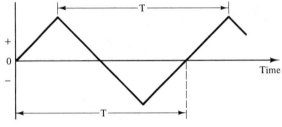

(b) Triangular Wave

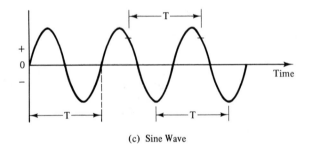

(c) Sine Wave

Figure 18-1 Examples of alternating voltages and currents.

Radio-frequency waves can be at any frequency, from as low as 10 kHz to over 100 GHz (10^{11} Hz).

Because period defines the time of one cycle and frequency the number of cycles per second, these quantities, as shown in Fig. 18-2, are related as follows:

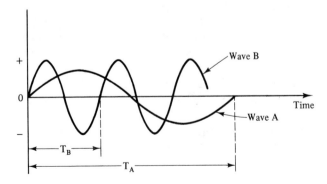

Figure 18-2 Sine waves of different frequencies illustrate that period and frequency are reciprocal relationships.

$$T = \frac{1}{f} \tag{18-1}$$

$$f = \frac{1}{T} \tag{18-2}$$

EXAMPLE 18-1:
What is the period of 60 Hz ac?

Solution:

$$\begin{aligned} T &= \frac{1}{f} \\ &= \frac{1}{60} = 0.0167 \text{ s} \\ &= 16.7 \text{ ms} \end{aligned}$$

EXAMPLE 18-2:
The period of a wave is 1 ms. What is the frequency of the wave?

Solution:

$$\begin{aligned} f &= \frac{1}{T} \\ &= \frac{1}{1 \times 10^{-3}} = 10^3 \text{ Hz} \\ &= 1 \text{ kHz} \end{aligned}$$

EXAMPLE 18-3:
Your favorite FM station operates at 103.2 MHz. What is the period of one cycle?

Solution:

$$\begin{aligned} T &= \frac{1}{f} \\ &= \frac{1}{103.2 \times 10^6} = 0.00969 \times 10^{-6} \text{ s} \\ &= 9.69 \text{ ns} \end{aligned}$$

EXAMPLE 18-4:
The period of an S-band radar wave is 0.5 ns. What is the frequency of the wave?

Solution:

$$\begin{aligned} f &= \frac{1}{T} \\ &= \frac{1}{0.5 \times 10^{-9}} = 2 \times 10^9 \text{ Hz} \\ &= 2 \text{ GHz} \end{aligned}$$

18-2. TYPES OF AC WAVES

Alternating voltages and currents may have any wave shape; some are shown in Fig. 18-3. The only constant requirement is that they have periodicity. The amount of voltage or current, called *amplitude*, may vary within a few cycles, as in audio and video signals, or the amplitude may be constant from cycle to cycle as in electrical power systems. Some types of waveforms are the following:

1. *Pulse wave forms*. In these waves, voltage or current switches abruptly between constant levels. Typically, it switches from a maximum positive level to an equal but negative maximum level. These abrupt changes are called *steps*. A change from maximum negative to maximum positive is a positive step. A negative step takes place when the change is from maximum positive to maximum negative. A form of pulse wave used in digital systems requires that the step changes take place from zero level to maximum positive. There are no negative swings of voltage. The wave is pulsating dc, but has the same shape as the wave described previously.
2. *Triangular waves*. In these waves the *rate of rise and fall are constant and equal*. This results in a straight line rise over time from maximum negative to maximum positive, followed by a similar linear fall back to maximum negative.
3. *Sawtooth waves*. These waves are similar to triangular waves, but with differing rates of rise and fall. Typically, the rate of rise is very

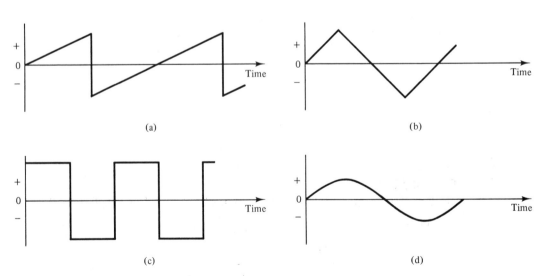

Figure 18-3 The sine wave is the basis for the analysis of all other waveforms. Various ac waveforms: (a) sawtooth wave, (b) triangle wave, (c) square wave, (d) sine wave.

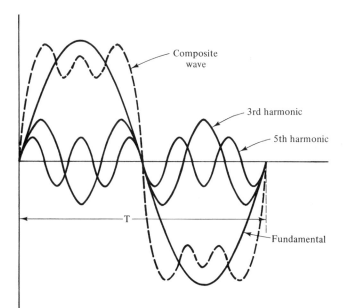

Figure 18-4 A complex wave consists of a fundamental sine wave and harmonics of the fundamental.

much less than the rate of fall. This results in a wave form with a gradual but constant increase to maximum positive level, followed by a rapid but constant change to maximum negative level. These waves are commonly used as time bases in certain types of test equipment.

4. *Sine waves.* These are a special category of wave form. It can be shown that *any* periodic wave, regardless of shape, consists of combinations of sine waves, variations of sine waves, and in some cases, dc. By variations of sine waves, we mean waves that are sine wave in shape, but do not meet the mathematical definition of a sine wave. All waves that have the shape of a sine wave are referred to as *sinusoidal*.

It is beyond the scope of this textbook to do any mathematical analysis of nonsinusoidal waves, but we should understand the meaning of *harmonic*. The relationship between a fundamental frequency and harmonics is shown in Fig. 18-4.

A harmonic is a sinusoidal wave that is at a frequency that is a whole number multiple of the fundamental frequency of the nonsinusoidal wave.

EXAMPLE 18-5:
A sawtooth wave has a period of 0.2 ms. (a) What is the fundamental frequency of the wave? (b) What is the frequency of the second harmonic? (c) What is the frequency of the fifth harmonic?

Solution:

(a) $f = \dfrac{1}{T}$

$= \dfrac{1}{0.2 \times 10^{-3}} = 5 \times 10^3$ Hz

$= 5$ kHz

This is the fundamental frequency.

(b) The second harmonic is at $2 \times f$.

$$2f = 2 \times 5 \text{ kHz} = 10 \text{ kHz}$$

(c) The frequency of the fifth harmonic is $5 \times f$.

$$5f = 5 \times 5 \text{ kHz} = 25 \text{ kHz}$$

18-3. SINE WAVE

Up to this point, we have considered a complete cycle only in terms of time. With sine waves it is useful to consider a cycle in terms of degrees. One complete cycle is equal to 360°. This enables us to consider all sine waves in the same way regardless of frequency. A cycle begins at 0° and ends at 360° (see Fig. 18-5).

At 0°, the value of a sine wave is zero. The wave is crossing the axis at this instant and is in a positive direction. Immediately after 0°, the wave has some positive value.

At 90°, a sine wave reaches its maximum positive amplitude. For angles between 90° and 180°, the values are positive and decreasing.

At 180°, a sine wave is zero. The wave is crossing the axis at this instant, but in a negative direction.

At 270°, a sine wave is at its maximum negative amplitude. For angles between 270° and 360°, the values are negative and decreasing.

At 360°, equal to 0°, the cycle repeats.

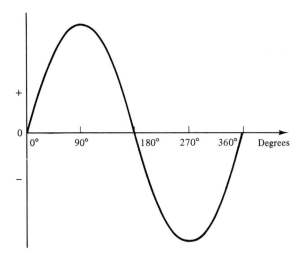

Figure 18-5 In this sketch of a sine wave time is represented by degrees.

For any angle, which actually represents a fraction of a cycle, we can define the relationship between peak amplitude and the amplitude at a particular part of a cycle.

18-4. EQUATION OF A SINE WAVE

The equation of a sine wave of voltage is given in Eq. (18-3). The formula for current is of the same form.

$$v = V_{max} \sin \theta \qquad (18\text{-}3)$$

Similarly,

$$i = I_{max} \sin \theta \qquad (18\text{-}4)$$

where I_{max} = peak amplitude
$\sin \theta$ = sine of any angle
i = instantaneous value of current angle θ

Note that we use lowercase letters to represent instantaneous values of current and voltage, except for peak (maximum) values.

EXAMPLE 18-6:

A sine wave of voltage has a peak amplitude of 15 V. What is the instantaneous value of the voltage at (a) 30°, (b) 45°, (c) 120°, (d) 210°, and (e) 270°? Use Table 18-1 to obtain values for the sines of the angles.

Solution:

$$v = V_{max} \sin \theta$$

(a) $v = 15 \sin 30°$
$= 15 \times 0.5 = 7.5$ V
(b) $v = 15 \sin 45°$
$= 15 \times 0.707 = 10.61$ V
(c) $v = 15 \times \sin 120°$
$= 15 \times 0.866 = 12.99$ V
(d) $v = 15 \times \sin 210°$
$= 15 \times -0.5 = -7.5$ V
(e) $v = 15 \times \sin 270°$
$= 15 \times -1 = -15$ V

EXAMPLE 18-7:

At 30° a current has an instantaneous value of 30 mA. What is the peak value of the current?

Solution:
1. $i = I_{max} \sin \theta$
Therefore,

$$I_{max} = \frac{i}{\sin \theta}$$

TABLE 18-1

SINES OF SELECTED ANGLES

Angle (°)	Sine
0	0
30	0.5
45	0.707
60	0.866
90	1.0
120	0.866
135	0.707
150	0.5
180	0
210	−0.5
225	−0.707
240	−0.866
270	−1.0
300	−0.866
315	−0.707
330	−0.5
360	0

2. $\sin 30° = 0.5$

3. $I_{max} = \dfrac{30 \text{ mA}}{0.5} = 60 \text{ mA}$

While the equation of a sine wave is accurate, it is a general expression for a sine wave of any frequency. In Sec. 18-9, we shall consider another equation for sine waves that includes frequency.

18-5. PHASOR REPRESENTATION OF A SINE WAVE

Another way to understand the unique nature of a sine wave is by the use of a special form of vector, called a *phasor*. A phasor is a rotating radius, whose length represents the peak value of a sine wave. The defined direction of rotation is counterclockwise (ccw). The phasor completes one revolution in a time equal to the period of a given sine wave, as shown in Fig. 18-6.

As the phasor rotates through a cycle, the distance from the tip of the phasor to the horizontal axis represents instantaneous values of voltage or current. One can see from the illustration that, for angles between 0 and 180°, instantaneous values are above the axis. Therefore, they are positive. Instantaneous values below the axis are negative. These are generated between 180° and 360°.

We do not use phasor drawings to calculate values. However, the concept of phasors is very valuable in understanding phase shifts and phase

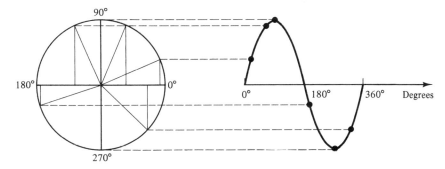

Figure 18-6 The generation of a sine wave by a rotating radius. The radius represents the maximum value of the wave.

differences in ac circuits. Compare the illustrations of sine waves that are shifted in phase with those of phasors. The ease with which phasor sketches may be understood is obvious.

18-6. PHASE ANGLES AND PHASE DIFFERENCES

So far we have considered sine waves that go through 0 at 0°, and in a positive direction. We say that these sine waves "begin" at 0°. In electricity and electronics it is common to have sine waves that do not begin at 0°. Such sine waves have been *shifted in phase* and have what is called a *phase angle*.

It is important to understand that we can only think of a phase angle for a wave provided that we have some reference for comparison. All sinusoidal waves have the same general form: therefore, we need to be able to compare waves to a reference wave or some reference point. In the illustrations, this reference point is 0°.

If the phase-shifted wave has a positive value at the 0° reference point, we consider the wave to have a leading phase angle. Positive angles between 0° and 180° are leading phase angles.

If the instantaneous value of a wave is negative at the 0° reference point, the wave has a lagging phase angle. Negative angles between 0° and 180° or positive angles between 180° and 360° represent lagging phase angles.

If we visualize a phasor, a leading phase angle causes it to be moved farther in a ccw direction. A lagging phase angle results in clockwise displacement of the phasor. Study Figs. 18-7 and 18-8 very carefully. It can be difficult to visualize lead or lag on sine curves. With phasor sketches, one has little, if any, trouble in doing so. This is shown in Fig. 18-9.

The general equation for a sine wave including phase shift is given in Eqs. (18-5) and (18-6).

$$v = V_{max} \sin(\theta \pm \alpha) \quad (18\text{-}5)$$

$$i = I_{max} \sin(\theta \pm \alpha) \quad (18\text{-}6)$$

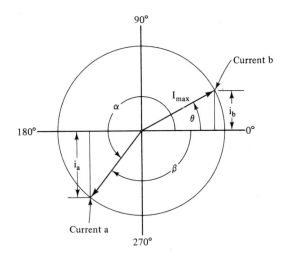

Figure 18-7 Using phasors to illustrate phase differences between two currents.

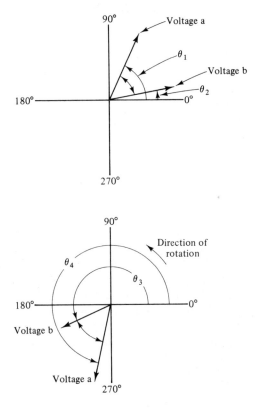

Figure 18-8 Voltages out of phase are clearly shown by use of phasors. Voltage a leads voltage b by the angular difference between the phasors.

Chap. 18 Alternating Voltages and Currents

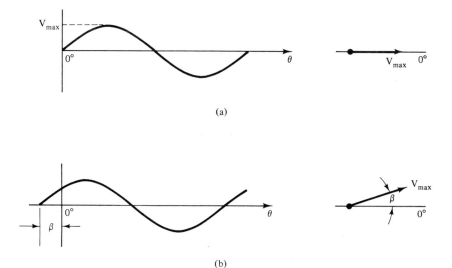

(a)

(b)

(c)

Figure 18-9 (a) A sine wave without phase shift. (b) A sine wave with a leading phase shift. (c) A sine wave with a lagging phase shift.

EXAMPLE 18-8:
A current leads a reference voltage by 30°. The peak amplitude of the current is 100 mA. (a) Write the equation of the current. (b) What is the instantaneous value of the current at 15° of the voltage? (c) What is the instantaneous value of the current at 180°? (d) For each solution, sketch the phasor diagram.

Solution:
(a) $i = I_{max} \sin(\theta \pm \alpha)$
   $= 100 \sin(\theta + 30°)$ mA   (Fig. 18-10)

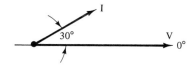

Figure 18-10

Sec. 18-6.   Phase Angles and Phase Differences

(b) $i = 100 \sin(15 + 30°)$ mA (Fig. 18-11)
   $= 100 \sin 45°$
   $= 100 \times 0.707$
   $= 70.7$ mA

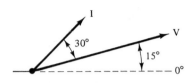

Figure 18-11

(c) $i = 100 \sin(180 + 30°)$ mA (Fig. 18-12)
   $= 100 \sin 210°$
   $= 100 \times -0.5$
   $= -50$ mA

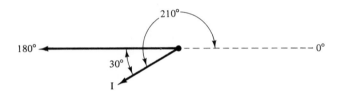

Figure 18-12

EXAMPLE 18-9:
A voltage lags a reference voltage by 60°. The peak value of the voltage is 40 V. (a) Write the equation of the voltage. (b) What is the value of the voltage when the reference voltage is at 0°? (c) What is the value of the voltage when the reference voltage is at 180°?

Solution:
(a) $v = V_{max} \sin(\theta \pm \alpha)$
   $v = 40 \sin(\theta - 60°)$ V
(b) $v = 40 \sin(0 - 60°)$ V
   $= 40 \sin -60°$
   $= 40 \times -0.866$
   $= -34.64$ V
(c) $v = 40 \sin(180 - 60°)$ V
   $= 40 \sin 120°$
   $= 40 \times 0.5$
   $= 20$ V

18-7. EFFECTIVE VALUE OF A SINE WAVE

For ac voltages and currents, power dissipation in a resistor is a continuously varying quantity. With purely resistive circuits there is no phase shift between current and voltage. The power dissipated in a resistor is the product of current and voltage at each instant. The average power over a cycle is used to determine the *effective* value of a sine wave. This quantity is also often called the *root-mean-square* (rms) value of a sine wave.

The effective value of a sine wave is that amount that is *numerically equal* to a dc voltage or current in terms of power. Thus ac voltage of 100 V_{rms} will produce exactly the same amount of power in a resistance as 100 V_{dc}. The relationship between effective and peak values are given in Eqs. (18-7) and (18-8).

$$V_{rms} = \frac{V_{max}}{\sqrt{2}} = 0.707 V_{max} \qquad (18\text{-}7)$$

$$I_{rms} = \frac{I_{max}}{\sqrt{2}} = 0.707 I_{max} \qquad (18\text{-}8)$$

If we know effective values, we can solve for peak values of sine waves.

$$V_{max} = \sqrt{2}\, V_{rms} = 1.414 V_{rms} \qquad (18\text{-}9)$$
$$I_{max} = \sqrt{2}\, I_{rms} = 1.414 I_{rms} \qquad (18\text{-}10)$$

EXAMPLE 18-10:
What is the peak value of an ac voltage that will develop the same amount of power as 120 V_{dc}?

Solution:
1. The effective value of the ac voltage must be equal to the dc voltage.

$$V_{rms} = V_{dc} = 120 \text{ V}$$

2. Assuming that the ac voltage is a sine wave, the peak value is found:

$$\begin{aligned}V_{max} &= 1.414\, V_{rms}\\ &= 1.414 \times 120\\ &= 169.7 \text{ V}\end{aligned}$$

EXAMPLE 18-11:
The peak value of a current is 30 mA. What is the effective value of the current?

Solution:

$$\begin{aligned}I_{rms} &= 0.707 I_{max}\\ &= 0.707 \times 30 \text{ mA}\\ &= 21.2 \text{ mA}\end{aligned}$$

EXAMPLE 18-12:
A dc current develops 80 W in a 20-Ω load. (a) What is the dc current? (b) What is the peak value of an ac current that will dissipate the same amount of power?

Solution:
(a) $P = I^2 R$
Therefore,
$$I = \sqrt{P/R}$$
$$= \sqrt{80/20} = \sqrt{4}$$
$$= 2 \text{ A}_{dc}$$

(b) I_{rms} must equal I_{dc} for the power dissipations to be the same.
$$I_{rms} = I_{dc} = 2 \text{ A}$$
$$I_{max} = 1.414 I_{rms}$$
$$= 1.414 \times 2 \text{ A}$$
$$= 2.828 \text{ A}$$

So far, we have been careful to identify effective values by using rms as a subscript. For all future references to effective values, we shall make use of a widely used convention. When an alternating current or voltage is given as an effective value, we need only use a capital letter.

$$I = I_{rms} = I_{eff}$$
$$V = V_{rms} = V_{eff}$$

The use of lowercase letters indicates instantaneous values. We usually use these for a current or voltage at some instant. For all other values of ac current or voltage, we shall use identifying subscripts.

We normally specify ac voltages and currents as effective values. To show phase shift, we simply add angle notation to the effective value. As an example, a voltage with an effective value of 35 V and an angle of lag of 45° would be written

$$\dot{V}_a = 35\underline{/-45°} \text{ V}$$

Leading phase shifts are represented by positive angles. For example,

$$\dot{I}_b = 4\underline{/22°} \text{ mA}$$

The overdot notation above V_a and I_b is used to show that these are phasor quantities. It is common practice to use the "dot" or to print the phasors as boldface letters.

18-8. AVERAGE VALUES

A pure ac wave must have an average value of zero. For example, the positive half of a sine wave is exactly equal to its negative half. If we take the average of each half, they are equal but of opposite sign. The overall average is zero.

The average value of a wave is the amount that would be read by a *dc meter*. Clearly, an ac wave will have an average only if it has a dc component. Yet we are interested in what is called the average value of a sine wave.

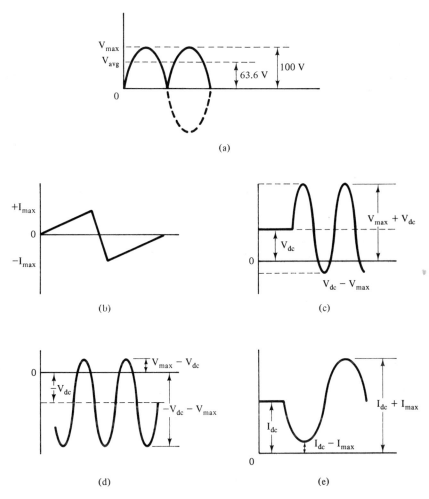

Figure 18-13 Average values of various voltages and currents.

It can be shown by calculus that the average value of a sine wave from 0° to 180°, *taken over half the cycle*, is related to the peak value of the wave.

$$V_{avg} = 0.636 V_{max} \qquad (18\text{-}11)$$

$$I_{avg} = 0.636 I_{max} \qquad (18\text{-}12)$$

Taking the average from 180° to 360°, we obtain the negative of the value from 0° to 180°.

Figure 18-13 (a) shows a rectified sine wave (changed to pulsating dc). In this case, the average value of the entire wave is found by Eqs. (18-11) and (18-12). Also shown are examples of ac waves with and without dc components. Study them carefully.

18-9. PERIODIC FUNCTIONS

Up to this point we have described instantaneous values of a sine wave as functions of the sine of an angle. While this method of solution is certainly valid, it does not take the *frequency* of the sine wave into consideration.

Consider the cycle of a phasor. One complete revolution of the phasor corresponds to one cycle of ac. The time to complete one revolution is the period of the ac wave. Thus, a high-frequency wave completes more revolutions per second than a lower-frequency wave. Also, the time required to complete one rotation of a phasor is greater the lower the frequency.

One rotation of a phasor equals 360° or 2π *radians*. A radian is an angle that results in an arc on the circumference of a circle that is exactly equal to the length of the radius of the circle, as shown in Fig. 18-14.

Therefore, each rotation of a phasor moves a distance of 2π radians. We define velocity as distance divided by time. In this instance, we are interested in the *angular velocity* of a phasor in radians per second. The symbol for angular velocity is ω. If we substitute $1/f$ for T, we obtain

$$\omega = 2\pi f \quad \text{rad/s} \qquad (18\text{-}13)$$

If we multiply angular velocity by time in seconds, we obtain radians. We may write the equation of a sine wave using angular velocity. This allows us to take the frequency of the sine wave into the solution of an instantaneous current or voltage. The equations for voltage or current are referred to as periodic functions.

$$i = I_{max} \sin \omega t \qquad (18\text{-}14)$$

$$v = V_{max} \sin \omega t \qquad (18\text{-}15)$$

where $\omega = 2\pi f$ rad/s
t = time in seconds

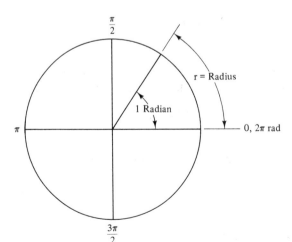

Figure 18-14 One radian is the angle that intercepts an arc on the circumference of a circle that is exactly equal in length to a radius.

EXAMPLE 18-13:
Given the following list of frequencies, solve for angular velocity in radians per second.

Solution:

Frequency	$\omega = 2\pi f$
60 Hz	377
200 Hz	1257
1 kHz	6280
10 MHz	6.28×10^7

EXAMPLE 18-14:
A sine wave of voltage has a peak amplitude of 10 V. The frequency of the wave is 200 kHz. Write the periodic function of the voltage.

Solution:

$$v = V_{max} \sin \omega t$$
$$= 10 \sin (6.28 \times 200 \times 10^3) t$$
$$= 10 \sin 1.256 \times 10^6 t$$

We have another consideration relating to periodic functions, *phase shift*. If we take phase shift into the equation, we write the periodic function of a sine wave of voltage as

$$v = V_{max} \sin (\omega t \pm \phi) \qquad (18\text{-}16)$$

where ϕ is the phase shift in degrees and ωt is in radians.

We may change from radians to degrees by using the following conversions. There are 360° in 2π radians. Therefore,

$$1 \text{ rad} = \frac{360°}{2\pi} = 57.3°$$

We multiply radians by 57.3° to get degrees; and we divide degrees by 57.3° to get radians. For example, 60° equals 1.05 radians, and 1.57 radians is approximately equal to 90°.

Periodic functions are useful in derivations and in the understanding of other important ideas in ac circuits. Of particular importance is the concept of angular velocity. You will find much use for this quantity in our studies of ac circuits. For the solution of problems involving sine waves, we usually work with Eqs. (18-3) and (18-4). However, we shall do problems in periodic functions after the following section.

Sec. 18-9. Periodic Functions

18-10. USING A SCIENTIFIC CALCULATOR

In this section we shall make use of the calculator for solutions involving sine functions. You need not know trigonometry in order to *use* the calculator. All scientific calculators can handle problems involving the basic trigonometric functions: sine, cosine, and tangent. Also, all scientific calculators can be switched from degrees to radian measure, so we can make solutions of sine curves in radians without the need to convert radians into degrees. To change from degrees (the normal mode of operation of the calculator) to radians, we must press the $\boxed{\text{DRG}}$ key. The calculator display must remind you that it is working with radians and will do so with some kind of identifier listed it its instruction book, or it will display "deg," "rad," or "grad." The function keys identified as $\boxed{\sin}$, $\boxed{\cos}$, and $\boxed{\tan}$ are for trigonometric functions of a right angle. In this chapter we shall be concerned only with the sine of an angle.

To find the sine of an angle, enter the angle and then press $\boxed{\sin}$. The display shows the value of the sine of the angle. This sequence is also true for any other trig functions. Always enter the angle first, and then press the proper function key.

If we know the value of the sine of an angle, but do not know the angle, we use the inverse of the function to find the angle. Various makers identify the inverse operation in several different ways. Many use $\boxed{\text{INV}}$ as a separate key; others use the same key but identify it by $\boxed{\text{ARC}}$. Many others use a color-coded second function key. You operate this key, first, and then the sine key to get the inverse sine function. Some identify the operation as a second function, with $\sin^{-1}$ printed directly above the $\boxed{\sin}$ key.

EXAMPLE 18-15:
Using the calculator, determine the sines of the following angles: 0.5°, 10°, 31.5°, 48°, −45°, 92°, 189°, 260°, 345°, and −15°.

Solution:
For each angle, simply enter the angle and press $\boxed{\sin}$. You do not need to clear the calculator between entries.

Angle	Sine	Angle	Sine
0.5°	0.00873	92°	0.9994
10	0.1736	189	−0.1564
31.5	0.5225	260	−0.9848
48	0.7431	345	−0.2588
−45	−0.7071	−15	−0.2588

For negative angles, enter the angle, then $\boxed{+/-}$.

EXAMPLE 18-16:
Given the following values of sines of angles, what are the angles? Given: 0.76, 0.51, 0.5, −0.5, −1, −0.32.

Solution:
For each positive value, simply enter the value, then $\boxed{\text{INV}}$, $\boxed{\text{sin}}$. For negative values, enter the value, then $\boxed{+/-}$, $\boxed{\text{INV}}$, $\boxed{\text{sin}}$.

Value	Angle
0.76	49.46°
0.51	30.66
0.5	30
−0.5	−30 (same as 330°)
−1.	−90 (same as 270°)
−0.32	−18.66 (same as 341.34°)

EXAMPLE 18-17:
A sine wave has a peak amplitude of 60 V. What is the instantaneous value of the voltage at 130°?

Solution:
$$v = V_{max} \sin \theta$$
$$= 60 \sin 130°$$
$$= 46 \text{ V}$$

Solution by calculator requires the following sequence of steps:

$$130 \; \boxed{\text{sin}} \; \boxed{\times} \; 60 \; \boxed{=} \; 45.96$$

or

$$60 \; \boxed{\times} \; 130 \; \boxed{\text{sin}} \; \boxed{=} \; 45.96$$

When working with functions of angles, the angle must always be entered *before* the trig function key is operated.

Calculations of peak values, when instantaneous value and the angle are known, may be made. From the equation of a sine wave, we have

$$V_{max} = \frac{v}{\sin \theta}$$

The calculator sequence is

Enter v, $\boxed{\div}$ enter θ, press $\boxed{\text{sin}}$ $\boxed{=}$

EXAMPLE 18-18:
At 48°, the instantaneous value of a current is 0.32 A. What is the peak amplitude of the current?

Solution:

$$I_{max} = \frac{i}{\sin \theta}$$
$$= \frac{0.32}{\sin 48°}$$
$$= 0.431 \text{ A}$$

Calculator Solution:

.32, $\boxed{\div}$ 48 $\boxed{\sin}$ $\boxed{=}$ 0.4306...

We can use the calculator to solve problems in periodic functions. All scientific calculators have a means of switching from degrees to radians. This is the $\boxed{\text{DRG}}$ key. This change must be made before making any radian measure computations involving trigonometric functions. Be sure that you return the calculator to the degree mode before you attempt to work other types of trig problems! To return to degree mode, the $\boxed{\text{DRG}}$ key must be stroked twice. The reason is that the calculator is programmed to progress from degrees to radians to gradians with each stroke of the $\boxed{\text{DRG}}$ key.

We calculate 2π as either $2 \times \pi$ or simply use 6.28 as an approximation.

EXAMPLE 18-19:
(a) What is the angular velocity of a 1.5-kHz sinusoidal signal? (b) Write the equation of the wave if $V_{max} = 30$ mV.

Solution:
(a) $\omega = 2\pi f$
$= 2 \times \pi \times 1.5 \times 10^3$
$= 9.425 \times 10^3$ rad/s
(b) $v = V_{max} \sin \omega t$
$= 30 \sin (9.425 \times 10^3 t)$

EXAMPLE 18-20:
In the sine-wave voltage of Example 18-19, what is the voltage at time $t = 150 \mu s$?

Solution:

$$v = 30 \sin (9.425 \times 10^3 \times 150 \times 10^{-6})$$
$$= 30 \sin 1.414 \text{ rad}$$
$$= 29.63 \text{ V}$$

Calculator Solution:

30 sin 1.414 rad = v

(a) Place calculator in radian mode.
(b) Enter 30, $\boxed{\times}$ 1.414 $\boxed{\sin}$ $\boxed{=}$ 29.63 ...

SUMMARY

1. The frequencies associated with ac depend upon the application of the voltage and current.
2. Frequency and period are inversely related: $T = 1/f$.
3. Any ac wave can be shown to consist of sinusoidal waves made up of a fundamental frequency signal and harmonics of the fundamental.
4. A cycle of a sine wave is 360°, beginning with 0°.
5. The phase shift of a sine wave can be leading or lagging.
6. The effective value of a sine wave is the value that produces the same heating effect as a numerically equal amount of dc. This is often called the rms value.
7. The peak value of a voltage or current is greater than the effective value. $V_{max} = 1.414 V_{rms}$.
8. The average value of any ac wave is zero, unless dc is added to the wave.
9. Rather than use degrees to describe a sinusoidal wave, we can use radians. This is of value in understanding frequency effects.
10. The angular velocity (ω) of a sinusoidal wave is stated in radians per second: $\omega = 2\pi f$ rad/s.

PROBLEMS

18-1. Calculate the period of each of the following:

10 Hz	25 Hz
100 Hz	2 kHz
40 kHz	200 kHz
1 MHz	250 MHz
500 MHz	4 GHz

18-2. The period of a signal is 3 μs. (a) What is the time of two cycles of this signal? (b) What is the time of one half-cycle?

18-3. Given the period, calculate frequency:

20 ms	0.5 ms
2 ms	100 μs
400 μs	50 ns
100 ns	10 ns
2.5 μs	1.25 ns

18-4. The significant harmonics of a certain wave are the third, fifth, and seventh. If the fundamental frequency is 3.5 kHz, determine the frequencies of the harmonics.

18-5. The second harmonic of a wave has a period of 50 μs. What is the fundamental frequency of the wave?

18-6. A sine wave of current has a peak amplitude of 25 mA. (a) What is the instantaneous value of the current at 135° of the cycle? (b) What is the instantaneous value at 225° of the cycle?

18-7. A sine wave of voltage has a peak amplitude of 80 V. Determine the amplitudes at 30°, 150°, 210°, and 330°.

18-8. A sine wave of voltage has an instantaneous amplitude of 190.5 V at 60° of the cycle. What is the peak value of the voltage?

18-9. The equation of a certain voltage is

$$v = 150 \sin (\theta + 25°) \text{ mV}$$

Solve for v, given the following values of θ: 0°, 20°, 40°, 180°, and 300°.

18-10. A sine wave of current with a peak amplitude of 40 mA lags a reference voltage by 50°. Write the equation of the current.

18-11. Given the phasor diagram of Fig. 18-15, (a) write the equation of current b using current a as the reference; (b) write the equation of current a with b as the reference.

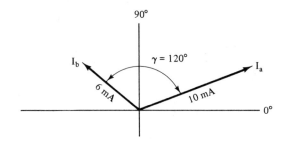

Figure 18-15

18-12. A certain parallel circuit has differing branch currents. With voltage as the reference, the currents are

$$i_1 = 35 \sin (\theta + 28°) \text{ mA}$$
$$i_2 = 12 \sin (\theta - 45°) \text{ mA}$$

(a) What is the phase angle between the currents?
(b) Sketch a phasor diagram of the currents.

18-13. A voltage has the equation $v = 48 \sin (\theta - 30°)$ mV. Starting with $\theta = 0°$ as the reference, how many degrees of the cycle must take place in order that $v = 48$ mV?

18-14. The peak amplitude of a certain sine wave of current is 3 A. What is the effective value of the current?

18-15. The current of Prob. 18-14 flows in a 25-Ω resistor. How much power is dissipated in the resistor?

18-16. A sine wave of voltage develops 40 W in a 30-Ω resistor. (a) What is the effective value of the voltage? (b) What is the peak value of the voltage?

18-17. A 24-V battery is used with a portable heater. We wish to replace the battery with power from an ac source. (a) What must be the effective value of the ac source voltage? (b) What is the peak value of this voltage?

18-18. A sine wave of current, with a peak value of 40 mA, flows in a 10-kΩ resistor. What is the power dissipated in the resistor?

18-19. A certain sine wave of voltage develops in 5 W in a 100-Ω resistor. What is the effective value of the voltage?

18-20. Solve for the average value of the voltage of Prob. 18-19 over the portion of the cycle from 0° to 180°.

18-21. A sine wave of current has a peak amplitude of 140 mA. What is the average value of the current from 180° to 360° of the cycle?

18-22. A certain voltage is given as $v = 167 \sin 377t$. (a) solve for the frequency of the voltage. (b) Solve for the effective value of the voltage.

18-23. Given the voltage of Prob. 18-22, what is the instantaneous value at time $t = 9.1$ ms?

18-24. The equation of a certain current is $i = 38 \sin 6280t$ mA. Circuit voltage, with $v_{max} = 50$ V, leads the current by $35°$. Write the periodic function of the voltage with current as the reference.

18-25. A complex wave is made up of the following voltages:

$$v_c = 3 \sin 37{,}680t \text{ mV}$$

$$v_b = 12 \sin 18{,}840t \text{ mV}$$

$$v_a = 48 \sin 9420t \text{ mV}$$

(a) Are the voltages harmonically related? (b) If so, identify each of the frequencies.

19

Resistance, Inductance, and Capacitance in AC Circuits

19-1. RATE OF CHANGE OF A SINE WAVE

We now begin our studies of alternating-current circuits. In these circuits we shall study the effects of inductance and capacitance in addition to those of resistance. We shall find that inductance and capacitance have effects that depend upon the frequency of voltage and current. Mathematically, we are concerned with the *rate of change* of current and voltage. Because our circuit analysis is based upon sine waves, we need to understand the rate of change of a sine wave.

To determine the instantaneous rate of change of any quantity, we must use differential calculus, a math beyond the scope of this text. However, we can make use of the result obtained by calculus. We can also use a practical approach to visualize the graph of rate of change of a sine wave.

There are two points on a sine wave where the rate of change is zero. These are at 90° and 270°. We find that there are three points where the rate of change is maximum. These are at 0°, 180°, and 360°. If we plot the rate of change curve, we find that we have a sinusoid, but with an angle of lead of 90°, as shown in Fig. 19-1. A curve that is sinusoidal, but leading a sine curve by 90°, represents a *cosine wave*.

If we consider the effect of frequency upon the rate of change, it should be clear that rate of change is proportional to frequency. As frequency increases, the period decreases. Thus changes must take place in less time, resulting in a greater rate of change as in Fig. 19-2.

Finally, we examine the mathematics. Given that

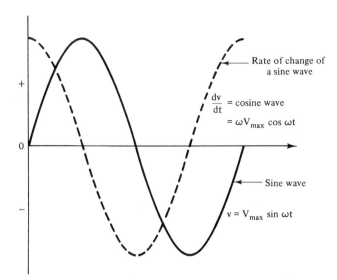

Figure 19-1 The rate of change of a sine wave.

$$i = I_{max} \sin \omega t$$

where $\omega = 2\pi f$, then

$$\frac{di}{dt} = \omega I_{max} \cos \omega t \qquad (19\text{-}1)$$

where di/dt = instantaneous rate of change of current.

Similarly,

$$v = V_{max} \sin \omega t$$

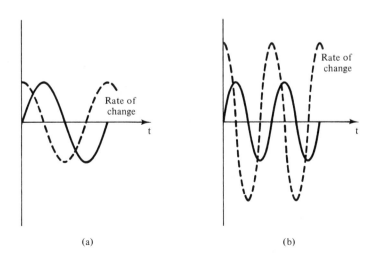

Figure 19-2 The rate of change of a sine wave depends upon the frequency of the wave.

Sec. 19-1. Rate of Change of a Sine Wave

and
$$\frac{dv}{dt} = \omega V_{max} \cos \omega t \quad (19\text{-}2)$$

In both equations we can substitute $\sin(\omega t + 90°)$ for $\cos \omega t$ and $2\pi f$ for ω.

$$\frac{di}{dt} = 2\pi f I_{max} \sin(2\pi ft + 90°)$$

$$\frac{dv}{dt} = 2\pi f V_{max} \sin(2\pi ft + 90°)$$

Note that the value of the rate of change is directly related to frequency, as we had predicted.

19-2. RESISTANCE IN AC CIRCUITS

The effects of resistance are the same in ac circuits as they are in dc circuits. When we use Ohm's law, we must be sure to use the same kinds of values for voltage and current.

$$I_{max} = \frac{V_{max}}{R} \quad (19\text{-}3)$$

$$\mathbf{I} = \frac{\mathbf{V}}{R} \quad (19\text{-}4)$$

where $\mathbf{V}$ and $\mathbf{I}$ are effective values.

$$i = \frac{V_{max} \sin(\omega t \pm \alpha)}{R} \quad (19\text{-}5)$$

$$i = \frac{V_{max} \sin \theta}{R} \quad (19\text{-}6)$$

In a similar manner, we find voltage:

$$V_{max} = I_{max} R \quad (19\text{-}7)$$
$$\mathbf{V} = \mathbf{I}R \quad (19\text{-}8)$$
$$v = R(I_{max} \sin \omega t) \quad (19\text{-}9)$$
$$v = R(I_{max} \sin \theta) \quad (19\text{-}10)$$

To find resistance,

$$R = \frac{V_{max} \sin \theta}{I_{max} \sin \theta} = \frac{\mathbf{V}}{\mathbf{I}} = \frac{v}{i} \quad (19\text{-}11)$$

EXAMPLE 19-1:

A current with a peak value of 30 mA flows through a 470-Ω resistor. (a) What is the peak value of the voltage drop across the resistor? (b) What is the effective value of voltage across the resistor?

Solution:
(a) $V_{max} = I_{max} R$
$= 30 \times 10^{-3} \times 470$
$= 14.1$ V
(b) $V = 0.707 V_{max}$
$= 0.707 \times 14.1$
$= 9.97$ V

Alternate solution:

$$V = IR$$
$$= 0.707 \times 30 \times 10^{-3} \times 470$$
$$= 9.97 \text{ V}$$

EXAMPLE 19-2:

The voltage drop across a 2.2-kΩ resistor has a peak value of 28 V. (a) What is the peak value of current in the resistor? (b) What is the current when the voltage is at an angle of 38°?

Solution:

(a) $I_{max} = \dfrac{V_{max}}{R}$

$= \dfrac{28}{2.2 \times 10^3}$

$= 12.73 \times 10^{-3}$ A
$= 12.73$ mA

(b) $i = \dfrac{v}{R}$

$v = V_{max} \sin 38°$
$= 28 \sin 38°$
$= 17.24$ V

Thus

$$i = \dfrac{17.24}{2.2} \times 10^3$$
$$= 7.84 \text{ mA}$$

Alternate solution:

$$i = I_{max} \sin 38°$$
$$= 12.73 \sin 38° \text{ mA}$$
$$= 7.84 \text{ mA}$$

EXAMPLE 19-3:

The voltage across a 220-Ω resistor is 30 V_{rms}. What is the peak value of current?

Solution:

1. $I_{rms} = \dfrac{V_{rms}}{R}$

   $= \dfrac{30}{220}$

   $= 0.136$ A

2. $I_{max} = 1.414 I_{rms}$

   $= 1.414 \times 0.136$

   $= 0.192$ A

Alternate solution:

$$V_{max} = 1.414 V_{rms}$$
$$= 1.414 \times 30$$
$$= 42.4 \text{ V}$$

$$I_{max} = \dfrac{V_{max}}{R}$$
$$= \dfrac{42.4}{220}$$
$$= 0.193 \text{ A}$$

EXAMPLE 19-4:

A current of 75 mA (rms) flows through a resistor. The drop across the resistor is 135 V_{rms}. (a) Solve for the resistance. (b) Demonstrate that the resistance is the same if peak values of current and voltage are used in the solution. (c) Demonstrate that the value of the resistance is the same if we use instantaneous values at 250° of the cycle.

Solution:

(a) $R = \dfrac{V_{rms}}{I_{rms}} = \dfrac{135}{0.075}$

   $= 1800 \ \Omega$

(b) $V_{max} = 1.414 V_{rms}$

   $= 1.414 \times 135 = 190.89$ V

   $I_{max} = 1.414 I_{rms}$

   $= 1.414 \times 0.075 = 0.10605$ A

   $R = \dfrac{V_{max}}{I_{max}}$

   $= \dfrac{190.89}{0.10605}$

   $= 1800 \ \Omega$

(c) $v = 190.89 \sin 250°$

   $= -179.38$ V

   $i = 0.10605 \sin 250°$

   $= -0.09965$

   $R = \dfrac{v}{i} = \dfrac{-179.38}{-0.09965}$

   $= 1800 \ \Omega$

Most types of resistors offer constant amounts of resistance over a very wide range of frequencies. Therefore, the voltage drop across a resistor is dependent upon the current in the resistor and is not affected by frequency. *There is no phase shift between current and voltage in a resistance.*

Power dissipation is usually computed using effective values of current and voltage. This results in what is often called *average power*.

$$P = V_{rms} I_{rms} \qquad (19\text{-}12)$$

$$= I_{rms}^2 R \qquad (19\text{-}13)$$

$$= \frac{V_{rms}^2}{R} \qquad (19\text{-}14)$$

Because the current and voltage are in phase in a resistor, the products of current and voltage are positive over an entire cycle. The instantaneous values of power dissipation result in a power curve that is always positive and is at twice the frequency of current and voltage as in Fig. 19-3.

So far we have considered average, peak, effective, and instantaneous values of voltage and current. We have one more value to learn. *Peak-to-peak* (p-p) values of voltage and current are used when we observe wave forms with an oscilloscope. We shall learn that most ac meters are not accurate at frequencies beyond a few kilohertz. The peak-to-peak value of a sine wave is simply twice its peak value (Fig. 19-4).

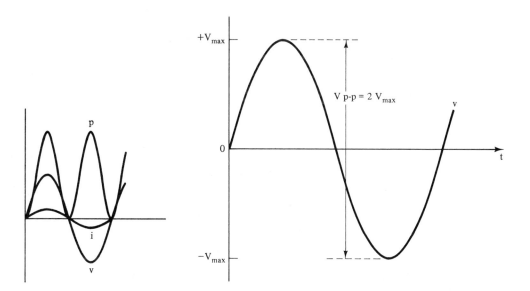

Figure 19-3 A curve of power dissipation in a resistor. Using instantaneous values of voltage and current products results in a double frequency curve.

Figure 19-4 A peak to peak measurement of a sine wave of voltage. Peak to peak measurements are very useful with all types of waveforms.

Sec. 19-2. Resistance in ac Circuits

$$V_{p\text{-}p} = 2V_{max} \tag{19-15}$$
$$= 2.828 V_{rms} \tag{19-16}$$
$$V_{rms} = \frac{V_{p\text{-}p}}{2.828} \tag{19-17}$$

EXAMPLE 19-5:

A current develops 56.56 $V_{p\text{-}p}$ across a 1-kΩ resistor. (a) What is the effective value of the voltage? (b) Solve for the effective value of the current. (c) What is the power dissipated in the resistor?

Solution:

(a) $V_{rms} = \dfrac{V_{p\text{-}p}}{2.828} = \dfrac{56.56}{2.828}$
$= 20$ V

(b) $I_{rms} = \dfrac{V_{rms}}{R} = \dfrac{20}{10^3}$
$= 20$ mA

(c) $P = I_{rms}^2 R$
$= (20 \times 10^{-3})^2 \times 1000$
$= 400 \times 10^{-6} \times 1000$
$= 400$ mW $= 0.4$ W

19-3. INDUCTANCE IN AC CIRCUITS

Inductance is that property of a circuit that opposes *changes* in current. When current is a steady, nonchanging amount, inductive effects are not present in a circuit.

Recall that the voltage across an inductor is proportional to the inductance and the rate of change of current.

$$v_L = L \frac{di}{dt} \tag{16-2}$$

When current is a sine wave, the rate of change curve of the current is a cosine wave, which is a sine wave shifted by $+90°$.

$$i = I_{max} \sin 2\pi ft$$

$$\frac{di}{dt} = 2\pi f I_{max} \sin(2\pi ft + 90°) \tag{19-18}$$

This is exactly the same as Eq. (19-1), but written so that we may see the effect of frequency upon the amount of the rate of change. Also, we have written the amount of phase shift, rather than simply identifying the curve as a cosine wave.

When we substitute Eq. (19-18) into the formula for voltage across an inductor, we have

$$v_L = 2\pi fL \times I_{max} \sin(2\pi ft + 90°) \qquad (19\text{-}19)$$

A voltage drop is always the product of current and opposition. In this instance, the opposition is represented by $2\pi fL$, and is called *inductive reactance*. Reactance is a form of opposition that is considerably different from resistance. It is frequency dependent, that is, changing with frequency; it always causes a 90° shift between current and voltage; and, as we shall find, it does not dissipate power.

The symbol for reactance is X. The units are in ohms. Inductive reactance is then

$$X_L = 2\pi fL = \omega L \qquad (19\text{-}20)$$

EXAMPLE 19-6:
Calculate the inductive reactance of a 100-mH coil at 100 Hz, 1 kHz, 10 kHz, 100 kHz, 1 MHz, and 5 MHz.

Solution:

$$X_L = 2\pi fL$$

$f = 100$ Hz: $X_L = 6.28 \times 100 \times 0.1$
$\phantom{f = 100 \text{ Hz: } X_L }= 62.8 \ \Omega$

$f = 1$ kHz: $X_L = 6.28 \times 10^3 \times 0.1$
$\phantom{f = 1 \text{ kHz: } X_L }= 628 \ \Omega$

$f = 10$ kHz: $X_L = 6.28 \times 10^4 \times 0.1$
$\phantom{f = 10 \text{ kHz: } X_L }= 6.28 \text{ k}\Omega$

$f = 100$ kHz: $X_L = 6.28 \times 10^5 \times 0.1$
$\phantom{f = 100 \text{ kHz: } X_L }= 62.8 \text{ k}\Omega$

$f = 1$ MHz: $X_L = 6.28 \times 10^6 \times 0.1$
$\phantom{f = 1 \text{ MHz: } X_L }= 628 \text{ k}\Omega$

$f = 5$ MHz $X_L = 6.28 \times 5 \times 10^6 \times 0.1$
$\phantom{f = 5 \text{ MHz } X_L }= 3.14 \times 10^6$
$\phantom{f = 5 \text{ MHz } X_L }= 3.14 \text{ M}\Omega$

Let us restate the effects of inductance in ac circuits.

1. The voltage across a *purely* inductive circuit element leads current by 90°, as shown in Fig. 19-5.
2. Inductance offers opposition to current that is proportional to the frequency of the current and the amount of inductance (Fig. 19-6). The opposition is called inductive reactance, X_L.

We stated earlier that inductance does not dissipate power. Instead, we find an energy exchange taking place between the magnetic field around an inductor and the current source. When current rises, energy is stored in the magnetic field. When current decreases, the magnetic field returns energy to the circuit. Figure 19-7 shows how the volt-ampere products are positive and

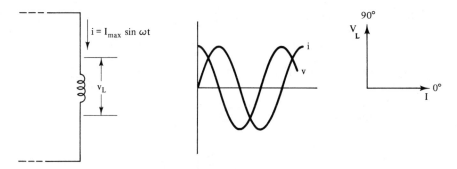

Figure 19-5 Inductance causes current to lag voltage. In a pure inductance the voltage across the coil will lead current by 90°.

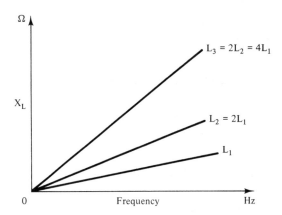

Figure 19-6 Inductive reactance is proportional to the frequency of signal and the amount of inductance of an inductor.

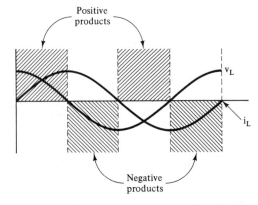

Figure 19-7 Energy exchange between an inductor and a source takes place because current and voltage are 90° out of phase.

negative. Positive values show energy storage, while negative values indicate energy returned to the circuit.

In practical inductors, some energy is dissipated. This is due to the resistance of the wire with which the coil is wound.

19-4. CAPACITANCE IN AC CIRCUITS

Capacitance is the circuit property that opposes changes in voltage. It does so by charging and discharging with voltage changes. Recall that current flow in a capacitor is limited to those instances when a capacitor is charging or discharging. In an ac circuit, voltage is constantly changing. It is reasonable then to expect that with higher frequencies, because they represent greater amounts of change per second, we will find greater amounts of current. Also, we learned that the quantity of charge depended upon the amount of capacitance in farads ($Q = CV$). We should then expect to find that the opposition offered by a capacitor is inversely proportional to both the frequency of the applied voltage and the capacitance. This is indeed the case.

If we assume a sinusoidal voltage across a capacitor, we may then solve for current in a capacitor. From our studies of transient response of capacitance, we have

$$i_C = C \frac{dv_C}{dt}$$

where dv_C/dt is the instantaneous rate of change of the voltage. Substituting for dv_C/dt,

$$i_C = C \times \omega V_{max} \sin(\omega t + 90°) \qquad (19\text{-}21)$$

We find that the current leads the voltage by 90° and is in phase with the rate of change of voltage, as in Fig. 19-8.

If we divide Eq. (19-21) by ωC, we have the general form of a voltage drop.

$$\frac{1}{\omega C} i_C = V_{max} \sin(\omega t + 90°)$$

The opposition offered by the capacitor is capacitive reactance, X_C.

$$X_C = \frac{1}{\omega C} = \frac{1}{2\pi f C} \quad \text{ohms} \qquad (19\text{-}22)$$

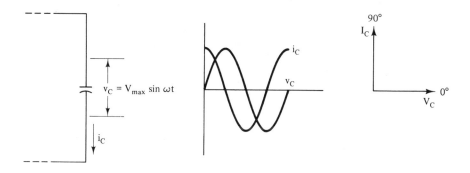

Figure 19-8 Current in a capacitor is a function of the rate of change of voltage.

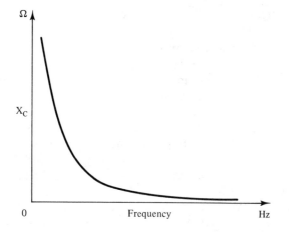

Figure 19-9 Capacitive reactance decreases as frequency increases.

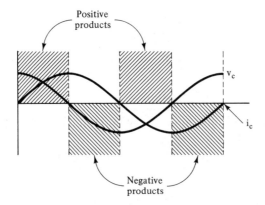

Figure 19-10 The current in a capacitor leads voltage across it by 90°, resulting in an energy exchange between a source and the electric field of the capacitor.

Note that the reactance decreases as frequency increases, as shown in Fig. 19-9. Also, capacitive reactance decreases as capacitance increases.

As with pure inductance, pure capacitance cannot dissipate power. Once again, there is an energy exchange between the capacitor and the source (Fig. 19-10). When voltage rises, capacitance stores energy in the form of electric charge. When voltage decreases, energy is returned to the circuit by the discharge of the capacitor.

EXAMPLE 19-7:
Given a 0.01-μF capacitor, solve for X_C at 200 Hz, 1 kHz, 5 kHz, 40 kHz, and 1 MHz.

Solution:

$$X_C = \frac{1}{2\pi fC}$$

$f = 200$ Hz: $X_C = \dfrac{1}{6.28 \times 200 \times 0.01 \times 10^{-6}}$
$= 79.6$ kΩ

$f = 1 \text{ kHz}: \quad X_C = \dfrac{1}{6.28 \times 10^3 \times 0.01 \times 10^{-6}}$
$= 15.9 \text{ k}\Omega$

$f = 5 \text{ kHz}: \quad X_C = \dfrac{1}{6.28 \times 5 \times 10^3 \times 0.01 \times 10^{-6}}$
$= 3.18 \text{ k}\Omega$

$f = 40 \text{ kHz}: \quad X_C = \dfrac{1}{6.28 \times 40 \times 10^3 \times 0.01 \times 10^{-6}}$
$= 398 \ \Omega$

$f = 1 \text{ MHz}: \quad X_C = \dfrac{1}{6.28 \times 10^6 \times 0.01 \times 10^{-6}}$
$= 15.9 \ \Omega$

We restate the effects of capacitance in ac circuits.

1. Current in a capacitor leads voltage across it by 90° (voltage lags by 90°).
2. Capacitive reactance is inversely proportional to frequency and inversely proportional to capacitance.

Figure 19-11 illustrates the phase relations in a series RLC circuit.

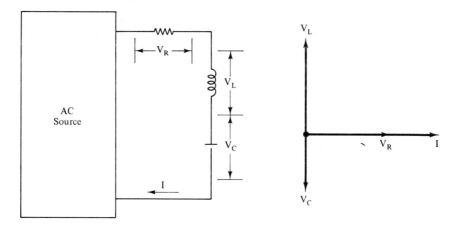

Figure 19-11 Phase relations in an R, L, C series circuit. Using current as the reference, we find that V_R is in phase with current, V_C lags current by 90°, and V_L leads current by 90°.

19-5. BASIC RIGHT TRIANGLE TRIGONOMETRY

Our studies of ac circuits shall include use of some of the basic relations of the sides and angles of right triangles. Therefore, before we begin studying

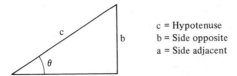

Figure 19-12

practical circuits, we review some of the math related to right triangles. We begin our review with some definitions (Fig. 19-12).

1. *Hypotenuse:* The side opposite the right (90°) angle is called the hypotenuse.
2. *Sine:* The sine of an angle is the ratio of the side opposite the angle to the hypotenuse.

$$\sin \theta = \frac{\text{opposite}}{\text{hypotenuse}} = \frac{b}{c}$$

   Values of sines vary between 0 and ±1, depending upon the angle.

3. *Cosine:* The cosine of an angle is the ratio of the adjacent side to the hypotenuse.

$$\cos \theta = \frac{\text{adjacent}}{\text{hypotenuse}} = \frac{a}{c}$$

   Values of cosines vary between ±1 and 0, but lead sines by 90°.

4. *Tangent:* The tangent of an angle is the ratio of the side opposite the angle to the side adjacent the angle.

$$\tan \theta = \frac{\text{opposite}}{\text{adjacent}} = \frac{b}{a}$$

   Values of tangents vary from 0 at 0°, to 1 at 45°, to infinity at 90°, from negative infinity to −1 at 135°, to 0 at 180°, to infinity at 270°, and from negative infinity to 0 at 360°.

In our work we shall frequently use the tangent function to find an angle. We shall calculate the value of the tangent of an angle and then find the angle. We state this in a formula form as

$$\theta = \tan^{-1} \frac{b}{a}$$

The use of $\tan^{-1}$ is simply a means of stating, "is the angle whose tangent equals...". The equation reads, "theta is the angle whose tangent equals $b \div a$."

We also have another notation for the same expression. Both forms are widely used.

$$\theta = \arctan \frac{b}{a}$$

Arctan has exactly the same meaning as $\tan^{-1}$.

We make use of the sine and cosine functions to find instantaneous values. We also make use of these functions to find resistance and reactance in circuits. Therefore, we find these functions used in several ways.

To find a or b,

$$b = c \sin \theta$$

$$a = c \cos \theta$$

To find c when a or b and θ are known,

$$c = \frac{b}{\sin \theta} = \frac{a}{\cos \theta}$$

EXAMPLE 19-8:

The right triangle of Fig. 19-13 has the following sides: $a = 20$, $b = 30$, and $c = 36.06$. Solve for (a) the included angle θ, (b) $\sin \theta$, (c) $\cos \theta$.

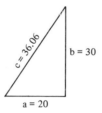

Figure 19-13

Solution:

(a) $\theta = \tan^{-1} \dfrac{b}{a}$

$\quad = \tan^{-1} \dfrac{30}{20} = \tan^{-1} 1.5$

$\quad = 56.31°$

(b) $\sin \theta = \sin 56.31°$

$\quad = 0.832$

(c) $\cos \theta = \cos 56.31°$

$\quad = 0.555$

EXAMPLE 19-9:

We are given the hypotenuse and an included angle in Fig. 19-14. Solve for the sides a and b, given that $c = 50$ and $\theta = -40°$.

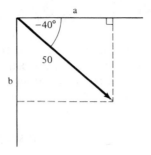

Figure 19-14

Solution:

$$a = 50 \cos -40°$$
$$= 50 \times 40 \boxed{+/-} \boxed{\cos} \boxed{=}$$
$$= 38.3$$

$$b = 50 \sin -40°$$
$$= 50 \times 40 \boxed{+/-} \boxed{\sin} \boxed{=}$$
$$= -32.14$$

Check: In a right triangle, the square of the hypotenuse equals the sum of the squares of the other sides.

$$50^2 = 38.3^2 + (-32.14)^2$$
$$2500 = 1467 + 1033$$
$$2500 = 2500$$

19-6. POLAR AND RECTANGULAR FORMS

All vector quantities must represent magnitude and direction. The phasor is simply a radius vector. As the radius rotates, we find right triangles generated whose sides change with each instant of rotation. However, the hypotenuse of these right triangles is a constant quantity.

Whenever we use the hypotenuse and the included angle, we are representing a vector in polar form.

$$\mathbf{C} = \dot{C} = C\underline{/\pm \theta} \qquad (19\text{-}23)$$

A vector may also be represented by its component parts, the sides a and b of the right triangle. The values of these sides depend upon the hypotenuse and the included angle. We show the magnitudes of a and b on the coordinate axes.

$$C\underline{/\pm \theta} = a \pm jb \qquad (19\text{-}24)$$

The letter j in the b term is used to indicate that this term is on an axis that is rotated 90° from the other axis. A positive j represents 90° of counterclockwise (ccw) rotation. A negative j represents a clockwise rotation of 90°.

19-7. RECTANGULAR TO POLAR CONVERSION

A phasor is properly represented by either form of the vector. However, we shall find that multiplication and division are very much more convenient when working with the polar form of a phasor.

Given $a + jb$,

$$\theta = \tan^{-1} \frac{b}{a} \qquad (19\text{-}25)$$

$$c = \frac{a}{\cos \theta} = \frac{b}{\sin \theta} \qquad (19\text{-}26)$$

EXAMPLE 19-10:
A current is stated as $12 - j15$ mA. What is the polar form of this current?

Solution:
1. Sketch the values (Fig. 19-15).

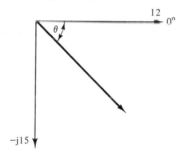

Figure 19-15

2. $\theta = \tan^{-1} \frac{-15}{12}$
   $= \tan^{-1}(-1.25)$
   $= -51.34°$
   Enter 15 [+/−] [÷] 12 [=] −1.25, [inv] [tan]
   −51.340 . . .

3a. $I = \dfrac{12}{\cos -51.34°}$
   $= 19.21$ mA
   Enter 12 [÷] 51.34 [+/−] [cos] [=]
   19.209 . . .

3b. $I = \dfrac{-15}{\sin -51.34°}$
   $= 19.21$ mA

Enter 15 [+/−] [÷] 51.34 [+/−] [sin] [=]

19.209 . . .

Finally, we write the polar form of the phasor.

$$I = 19.21 \underline{/-51.34°} \text{ mA}$$

EXAMPLE 19-11:

The rectangular form of a voltage is given as $-206.7 + j75.2$. State the voltage in polar form.

Solution:
1. Sketch the problem (Fig. 19-16).

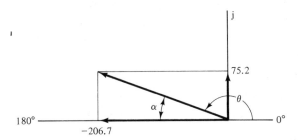

Figure 19-16

2. $\theta = \tan^{-1} \dfrac{75.2}{-206.7}$
   $= \tan^{-1}(-0.3638)$
   $= 160°$

   Enter 75.2 [÷] 206.7 [+/−] [=] −0.3638, [inv] [tan]

   −19.99 . . .

Study the sketch of the problem. Note that the right triangle includes the angle α. The angle θ is measured from 0° to the phasor c. To find θ, we subtract $|\alpha|$ from 180°.

$$\theta = 180° - |\alpha| = 180 - 20°$$
$$= 160°$$

3. $V = \dfrac{-206.7}{\cos 160°} = \dfrac{75.2}{\sin 160°}$

Calculator operations:
Enter 206.7 [+/−] [÷] 160 [cos] [=]

219.96 . . .

Enter 75.2 [÷] 160 [sin] [=]

219.87 . . .

The polar form of the vector is

$$V = 220\underline{/160°} \text{ V}$$

19-8. POLAR TO RECTANGULAR CONVERSION

Addition and subtraction of phasors *require* that they be in rectangular form. The only exceptions are when polar values are in phase or exactly 180° out of phase with each other. The relations between polar and rectangular forms are shown in Fig. 19-17.

From the math we have,

$$a = c \cos \theta$$
$$b = c \sin \theta$$

Therefore,

$$C\underline{/\pm\theta} = C \cos \theta \pm jC \sin \theta \tag{19-27}$$

For angles between 0° and 90° both terms are positive.

For angles between 90° and 180°, the "real" term is negative and the *j* term positive.

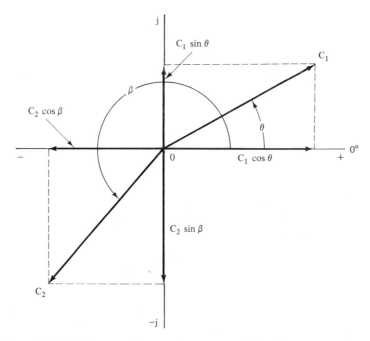

Figure 19-17 Polar quantities may be represented by their component parts. These quantities, located on the coordinate axes, are the rectangular form of the polar phasor.

For angles between 180° and 270° both terms are negative.

For angles between 270° and 360° (−90° and 0°), the "real" term is positive and the j term negative.

EXAMPLE 19-12:
A current is given as $3.5\underline{/50°}$ A. Convert the phasor into rectangular form.

Solution:
$$3.5\underline{/50°} = 3.5 \cos 50° + j3.5 \sin 50° \text{ A}$$
$$= 2.25 + j2.68 \text{ A}$$

Calculator operations:
Enter 3.5 $\boxed{\times}$ 50 $\boxed{\cos}$ $\boxed{=}$ 2.25
Enter 3.5 $\boxed{\times}$ 50 $\boxed{\sin}$ $\boxed{=}$ 2.68

EXAMPLE 19-13:
What is the rectangular form of $85\underline{/110°}$ mV?

Solution:
$$85\underline{/110°} \text{ mV} = 85 \cos 110° + j85 \sin 110° \text{ mV}$$
$$= -29.1 + j79.9 \text{ mV}$$

Calculator operations:
Enter 85 $\boxed{\times}$ 110 $\boxed{\cos}$ $\boxed{=}$ −29.07 . . .
Enter 85 $\boxed{\times}$ 110 $\boxed{\sin}$ $\boxed{=}$ 79.87 . . .

EXAMPLE 19-14:
A voltage is given as $20\underline{/240°}$ V. What is the rectangular form of the voltage?

Solution:
$$20\underline{/240°} \text{ V} = 20 \cos 240° + j20 \sin 240°$$
$$= -10 - j17.32 \text{ V}$$

Calculator operations:
Enter 20 $\boxed{\times}$ 240 $\boxed{\cos}$ $\boxed{=}$ −10 . . .
Enter 20 $\boxed{\times}$ 240 $\boxed{\sin}$ $\boxed{=}$ −17.32 . . .

EXAMPLE 19-15:
The polar form of a current is $100\underline{/-35°}$ mA. Write the rectangular form of the current.

Solution:
$$100\underline{/-35°} \text{ mA} = 100 \cos -35° + j100 \sin -35° \text{ mA}$$
$$= 81.91 - j57.36 \text{ mA}$$

Calculator operations:
Enter 100 $\boxed{\times}$ 35 $\boxed{+/-}$ $\boxed{\cos}$ $\boxed{=}$ 81.91 . . .
Enter 100 $\boxed{\times}$ 35 $\boxed{+/-}$ $\boxed{\sin}$ $\boxed{=}$ 57.357 . . .

19-9. PHASOR ARITHMETIC

Addition and Subtraction. To add and subtract phasors we must use rectangular form. Simply add or subtract like terms.

Multiplication. We may multiply phasors in either form. However, it is very much more convenient to work with the polar form. Simply multiply magnitudes and *add* angles.

Division. As with multiplication, we find that division of one phasor by another is best done with the polar form. Divide numerator by denominator, and *subtract* the angle of the denominator from the angle of the numerator. The subtraction is algebraic, taking the signs of the angles into consideration. A good way to avoid errors is to change the sign of the angle of the denominator and then add it to the angle of the numerator.

EXAMPLE 19-16:
Given the circuit of Fig. 19-18, solve for total current.

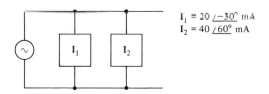

Figure 19-18

Solution:
1. $I_T = I_1 + I_2$
2. Change the currents from polar to rectangular form.

$$I_1 = 20\underline{/-30°} \text{ mA}$$
$$= 20 \cos -30° + j20 \sin -30°$$
$$= 17.32 - j10 \text{ mA}$$

$$I_2 = 40\underline{/60°} \text{ mA}$$
$$= 40 \cos 60° + j40 \sin 60°$$
$$= 20 + j34.64 \text{ mA}$$

3. Take the algebraic sum of the phasors.

$$\begin{aligned} I_1 &= 17.32 - j10 \\ +I_2 &= 20 + j43.64 \\ \hline I_T &= 37.32 + j24.64 \end{aligned}$$

4. Convert the rectangular form of I_T into polar form.

(a) $\theta = \tan^{-1} \dfrac{24.64}{37.32}$

$= 33.43°$

(b) $I_T = \dfrac{37.32}{\cos 33.43°} = \dfrac{24.64}{\sin 33.43°}$

= 44.72

$\mathbf{I}_T = 44.72\underline{/33.43°}$ mA

EXAMPLE 19-17:
Given the circuit of Fig. 19-19, solve for applied voltage:

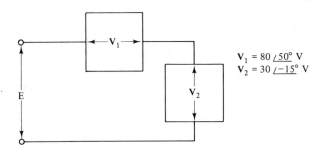

Figure 19-19

Solution:
1. The applied voltage equals the sum of the voltage drops.
2. Change the voltage drops into rectangular form.

$$V_1 = 80 \cos 50° + j80 \sin 50°$$
$$= 51.42 + j61.28$$

$$V_2 = 30 \cos -15° + j30 \sin -15°$$
$$= 28.98 - j7.76$$

3. Add the voltages.

$$\begin{array}{r} V_1 = 51.42 + j61.28 \\ + V_2 = 28.98 - j7.76 \\ \hline E = 80.4 + j53.52 \text{ V} \end{array}$$

4. Change E to polar form.

(a) $\theta = \tan^{-1} \dfrac{53.52}{80.4}$

= 33.65°

(b) $E = \dfrac{80.4}{\cos 33.65°} = \dfrac{53.52}{\sin 33.65°}$

= 96.58 V

$\mathbf{E} = 96.58\underline{/33.65°}$ V

EXAMPLE 19-18:
Take the ratio of the following phasors.
(a) $8\underline{/40} \div 2\underline{/-10°}$
(b) $25\underline{/30°} \div 12.5\underline{/70°}$
(c) $30 - j40 \div 10\underline{/20°}$

Solution:

(a) $\dfrac{8/40°}{2/-10°} = 4/50°$

(b) $\dfrac{25/30°}{12.5/70°} = 2/-40°$

(c) Change $30 - j40$ to polar form:

$$30 - j40 = 50/-53.1°$$

$$\dfrac{50/-53.1°}{10/20°} = 5/-73.1°$$

EXAMPLE 19-19:
Take the product of the following phasors.
(a) $40/10° \times 30/120°$
(b) $35/30° \times 10/-45°$
(c) $30/65° \times 12/-40°$
(d) $(9 - j12) \times (4 + j3)$

Solution:
(a) $(40 \times 30)/10 + 120° = 1200/130°$
(b) $(35 \times 10)/30 - 45° = 350/-15°$
(c) $(30 \times 12)/65 - 40° = 360/25°$
(d) Change the rectangular phasors into polar form.

$$9 - j12 = 15/-53.1°$$
$$4 + j3 = 5/36.9°$$
$$(15 \times 5)/-53.1 + 36.9° = 75/-16.2°$$

SUMMARY

1. The rate of change of a sine wave is maximum at 0°, 180°, and 360°. The rate of change is zero at 90° and 270°. The rate of change curve is a cosine wave.
2. With pure resistance circuits, Ohm's law calculations are the same as with dc circuits. However, power calculations require that we work with effective values.
3. Reactance is a frequency-dependent form of opposition to current. It is caused by capacitance and inductance when operated in ac circuits.
4. Inductive reactance increases linearly with frequency and inductance. $X_L = 2\pi f L$ Ω.
5. The current in a *purely* inductive circuit lags voltage by 90°.
6. Capacitive reactance is inversely proportional to frequency and capacitance. $X_C = 1/2\pi f C$ Ω.
7. In a pure capacitance circuit, current leads voltage by 90°.
8. All vectors represent magnitude and direction.
9. A polar vector represents the hypotenuse (magnitude) and an included angle of a right triangle.

10. A rectangular vector represents the other two sides of a right triangle. These are the side adjacent and the side opposite the included angle of a polar vector.
11. Phasors cannot be added or subtracted unless they are in rectangular form.
12. Multiplication of phasors is most convenient when they are in polar form.

PROBLEMS

19-1. A sine wave of voltage with a peak value of 21.2 V is applied across a resistance of 120 Ω. What is the peak value of the current?

19-2. What is the effective value of the current if a sine wave of voltage with a peak value of 141.4 V is applied across 1 kΩ of resistance?

19-3. In a pure resistance circuit, $R = 8.2$ kΩ and $I_{max} = 8.5$ mA. Solve for V_{max} and V_{rms}.

19-4. In a pure resistance circuit, $R = 2.2$ kΩ and $v = 25.4 \sin \theta$ V. Solve for i at 20° of the cycle.

19-5. Given the data of Prob. 19.4, solve for the average power dissipated in R.

19-6. The current in a 1-kΩ resistor is given as $i = 35.4 \sin (\theta + 20°)$ mA. (a) Solve for for the average power dissipated in R. (b) Solve for v at $\theta = 25°$.

19-7. Given the circuit of Fig. 19-20, solve for the average power dissipation in each resistor, given that $e = 68 \sin 377t$ V.

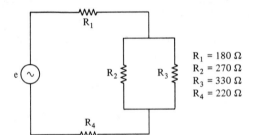

$R_1 = 180$ Ω
$R_2 = 270$ Ω
$R_3 = 330$ Ω
$R_4 = 220$ Ω

Figure 19-20

19-8. Given the data of Prob. 19-7, solve for i_3. Write the periodic function of the current.

19-9. Given the data of Prob. 19-7, what is the frequency of the source?

19-10. Given the data of Prob. 19-7, solve for the voltage drop across R_4 when $t = 5$ ms.

19-11. A current is given as $i = 120 \sin 94.2 \times 10^3 t$ mA. (a) What is the frequency of the current? (b) What is the current at $t = 15$ μs? (c) What is the current at $t = 66.7$ μs?

19-12. The current of Prob. 19-11 flows in a 15-Ω resistor. What is the power dissipated in the resistor?

19-13. Change the following effective voltages to peak-to-peak values:

28 V	48 V
15 V	8 V
120 V	600 V

19-14. Change the following voltages to rms values:
17 V_{p-p} 62 V_{p-p}
28 V_{max} 339 V_{p-p}
99 V_{p-p} 40 V_{max}

19-15. A coil with an inductance of 50 mH is used at each of the listed frequencies. Calculate X_L for each frequency: 15 Hz, 200 Hz, 3 kHz, 10 kHz, 80 kHz, 225 kHz, 500 kHz, 1 MHz.

19-16. Given that the frequency is 300 Hz, solve for the reactance of the following inductances: 10 µH, 50 mH, 300 µH, 250 mH, 0.6 mH, 1 H, 10 mH, 2.5 H.

19-17. Given the sketch of Fig. 19-21, solve for the voltage drop across the inductor for each of the following currents: (a) $i = 22.5 \sin 2512t$ mA; (b) $i = 930 \sin 62.8 \times 10^3 t$ µA.

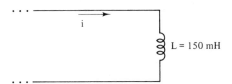

Figure 19-21

19-18. Given the circuit of Fig. 19-22, solve for i when $t = 8$ µs.

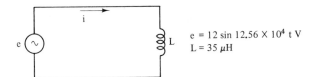

$e = 12 \sin 12.56 \times 10^4 t$ V
$L = 35$ µH

Figure 19-22

19-19. A 0.01-µF capacitor is used at each of the listed frequencies. Solve for X_C at each frequency: 10 Hz, 100 Hz, 500 Hz, 4 kHz, 50 kHz, 250 kHz, 500 kHz, 2 MHz.

19-20. Calculate the capacitive reactance of each listed capacitor at $f = 1$ kHz: 100 pF, 2 nF, 50 nF, 0.15 µF, 0.8 µF, 1.2 µF, 10 µF, 200 µF.

19-21. The voltage drop across a 1-nF capacitor is given as $v_C = 50 \sin 38.3 \times 10^3 t$ V.
(a) Solve for the effective value of the current.
(b) Solve for current at $t = 61$ µs.

19-22. Use a scientific calculator to solve for the following quantities.
sin 15°   cos −12°   tan 80°
sin 40°   cos 45°   tan 45°
sin −42.6°   cos −18.5°   tan 5°
sin 162.3°   cos 2.5°   tan −45°

19-23. Use a scientific calculator to solve for each angle given the function:
sin α = 0.25   tan γ = −1.7
cos β = 0.61   cos x = −0.51
tan θ = 1.2   sin y = −0.8

19-24. Sketch each of the following on a set of coordinate axes.
15/30°   12/110°
20/−45°   25/−170°
150/60°   10/15°

Chap. 19   Problems   377

19-25. Convert each of the polar quantities of Prob. 19-24 to rectangular form.

19-26. Sketch each of the following:

$15 - j12$ $25 - j25$
$22 + j30$ $10 + j30$
$45 + j80$ $-40 - j15$

19-27. Convert each of the rectangular quantities of Prob. 19-26 into polar form.

19-28. Solve each of the following:

$3\underline{/30°} \times 4\underline{/45°} \times 2\underline{/-15°}$
$15\underline{/20°} \div 75\underline{/-40°}$
$300\underline{/0°} \div 20\underline{/40°}$
$200\underline{/50°} \div 50\underline{/-70°}$
$120\underline{/70°} \times 3\underline{/-20°}$

19-29. Solve each of the following:

$(4 + j6) + (8 - j2)$
$(120 - j80) + (30 - j70)$
$(15 - j14) - (5 - j30)$

19-30. Solve each of the following:

$18\underline{/40°} + 5\underline{/60°}$
$200\underline{/-30°} - 150\underline{/0°}$
$(40 + j18)(22 - j20)$

19-31. A circuit has the following voltage drops:

$$\mathbf{V}_1 = 25 + j40$$
$$\mathbf{V}_2 = 10 + j35$$
$$\mathbf{V}_3 = 40 + j0$$

Solve for total voltage. State the result in polar form.

19-32. Given that $\mathbf{I} = 40\underline{/-30°}$ and $X_L = 250\underline{/90°}$ Ω, solve for $\mathbf{V}_L$.

19-33. Given the voltage across a capacitor is $\mathbf{V}_C = 25\underline{/-15°}$ and $X_C = 400\underline{/-90°}$, solve for $\mathbf{I}_C$.

19-34. The branch currents of a parallel circuit are given. Solve for total current. State $\mathbf{I}_T$ in polar form.

$$\mathbf{I}_1 = 150 - j300 \text{ mA}$$
$$\mathbf{I}_2 = 120 + j150 \text{ mA}$$
$$\mathbf{I}_3 = 80 - j100 \text{ mA}$$

19-35. The voltage drop across an inductor is given as $20\underline{/70°}$ V. The current is specified as $4\underline{/-20°}$ mA. Solve for X_L.

19-36. In Prob. 19-35, the operating frequency is 120 kHz. Solve for the inductance of the coil.

20

Series AC Circuits

20-1. IMPEDANCE OF A SERIES *RL* CIRCUIT

Impedance is the general expression for opposition to current flow in ac circuits. Remember that resistance opposes current and dissipates power, while reactance offers opposition to current that varies with frequency. Also, reactance does not dissipate power.

We use impedance to represent any form of opposition to current, whether resistance, reactance, or combinations of resistance and reactance. The symbol for impedance is Z, and the unit of opposition is the ohm. Impedance (sketched in Fig. 20-1) usually represents combinations of resistance and reactance and is therefore a vector quantity.

$$\mathbf{Z} = R \pm jX \qquad (20\text{-}1)$$

We apply Ohm's law to ac circuits. Impedance is used in place of total resistance when making total circuit calculations. However, we must follow the rules of phasor mathematics outlined in Chapter 19.

$$\mathbf{I} = \frac{\mathbf{E}}{\mathbf{Z}} \qquad (20\text{-}2)$$

$$\mathbf{E} = \mathbf{IZ} \qquad (20\text{-}3)$$

$$\mathbf{Z} = \frac{\mathbf{E}}{\mathbf{I}} \qquad (20\text{-}4)$$

Recall that inductance causes current to lag voltage by 90°. We handle this phase shift in the math by assigning $+j$ to X_L. This means that the

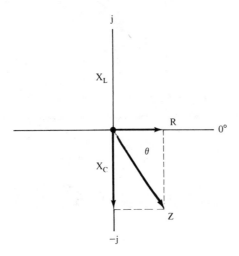

Figure 20-1 The complex plane is used to sketch impedance diagrams.

impedance of a series *RL* circuit has a positive phase angle between 0° and 90°. Positive phase angles for impedance result in lagging currents and leading voltages when used with the Ohm's law formulas.

EXAMPLE 20-1:
Solve for the impedance of the circuit of Fig. 20-2 when the frequency of the applied voltage is 2 kHz. Sketch the impedance on the coordinate axes. (Fig. 20-3)

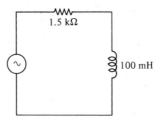

Figure 20-2

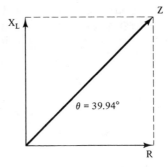

Figure 20-3

Solution:

$$\mathbf{Z} = R + jX_L$$

$$\begin{aligned}X_L &= 2\pi fL \\ &= 6.28 \times 2 \times 10^3 \times 100 \times 10^{-3} \\ &= 1256 \ \Omega\end{aligned}$$

$$\mathbf{Z} = 1500 + j1256$$

$$\theta = \tan^{-1} \frac{1256}{1500} = 39.94°$$

$$|\mathbf{Z}| = \frac{R}{\cos\theta} = \frac{X_L}{\sin\theta} = 1956 \ \Omega$$

$$\mathbf{Z} = 1956 \ \underline{/39.94°} \ \Omega$$

EXAMPLE 20-2:

What is the impedance of the circuit of Example 20-1 if the frequency of the applied voltage is 10 kHz? Sketch the impedance diagram.

Solution:

$$\begin{aligned}X_L &= 2\pi fL \\ &= 6.28 \times 10 \times 10^3 \times 100 \times 10^{-3} \\ &= 6280 \ \Omega\end{aligned}$$

$$\begin{aligned}\mathbf{Z} &= R + jX_L \\ &= 1500 + j6280\end{aligned}$$

$$\theta = \tan^{-1} \frac{6280}{1500} = 76.57°$$

$$|\mathbf{Z}| = \frac{6280}{\sin\theta} = 6457 \ \Omega$$

$$\mathbf{Z} = 6457 \ \underline{/76.57°} \ \Omega$$

The impedance diagram is shown in Fig. 20-4.

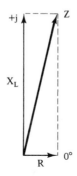

Figure 20-4

Sec. 20-1. Impedance of a Series *RL* Circuit

EXAMPLE 20-3:
Given the circuit of Example 20-1, (a) at what frequency is the phase angle of the impedance 45°? (b) What is the impedance when the operating frequency is 200 Hz?

Solution:
(a) The tangent is the ratio of the reactance to the resistance. We solve the formula for X_L. Once we know X_L, we solve the reactance formula for frequency.

$$\frac{X_L}{R} = \tan \theta$$

$$\begin{aligned} X_L &= \tan \theta \times R \\ &= \tan 45° \times 1500 = 1 \times 1500 \\ &= 1500 \ \Omega \end{aligned}$$

$$X_L = 2\pi f L$$

Therefore,

$$\begin{aligned} f &= \frac{X_L}{2\pi L} \\ &= \frac{1500}{6.28 \times 100 \times 10^{-3}} \\ &= 2.389 \text{ kHz} \end{aligned}$$

(b) $X_L = 6.28 \times 200 \times 100 \times 10^{-3}$
    $= 125.6 \ \Omega$

$$\begin{aligned} \mathbf{Z} &= R + jX_L \\ &= 1500 + j125.6 \end{aligned}$$

$$\begin{aligned} \theta &= \tan^{-1} \frac{125.6}{1500} \\ &= 4.79° \end{aligned}$$

$$|Z| = \frac{R}{\cos \theta} = 1505$$

$$\mathbf{Z} = 1505 \ \underline{/4.79°} \ \Omega$$

When we examine the results of these examples, we note that, when inductive reactance and resistance are equal, the angle of the impedance is 45°. In such a circuit, the impedance is 1.414 times the resistance; current lags source voltage by 45°. The voltage drop across the inductance, as always, *leads current by 90°.*

When inductive reactance is much greater than resistance, the phase angle increases and the impedance is very nearly the same as the inductive reactance. When the inductive reactance is 10 or more times greater than the resistance ($X_L \geq 10R$), we assume that the impedance equals the inductive reactance and that the phase angle is 90°.

The third example showed the effect upon impedance when resistance is very much greater than inductive reactance. When resistance is at least 10 times greater than inductive reactance ($R \geq 10X_L$), the impedance is approximately equal to the resistance, and θ approaches 0°. In this case, the circuit appears to be purely resistive for all practical purposes.

20-2. IMPEDANCE OF A SERIES *RC* CIRCUIT

In a capacitive circuit, current leads voltage. When we use current as the reference, voltage across a capacitor lags current by 90°.

$$\mathbf{Z} = R - jX_C \qquad (20\text{-}5)$$

The phase angles of impedance for capacitive circuits range between 0° and −90°. When the magnitude of the capacitive reactance is equal to the resistance, the phase angle is −45°. For smaller angles, resistance is greater than capacitive reactance. Clearly, at angles greater than −45° the circuit has a greater amount of reactance than resistance. We treat the circuit as purely resistive or purely capacitive when there is at least a 10-to-1 relationship between resistance and reactance.

$$R \geq 10X_C \ \ldots \ R - jX_C \sim R$$
$$X_C \geq 10R \ \ldots \ R - jX_C \simeq -jX_C$$

EXAMPLE 20-4:
The operating frequency of the circuit of Fig. 20-5 is 500 Hz. Solve for the impedance.

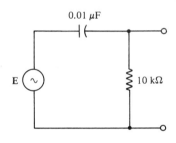

Figure 20-5

Solution:

$$X_C = \frac{1}{2\pi fC}$$
$$= \frac{1}{6.28 \times 5 \times 10^2 \times 0.01 \times 10^{-6}}$$
$$= 3.185 \times 10^4$$
$$= 31.85 \text{ k}\Omega$$

$$\mathbf{Z} = R - jX_C$$
$$= 10 \text{ k}\Omega - j31.85 \text{ k}\Omega$$

$$\theta = \tan^{-1} \frac{-31.85 \text{ k}\Omega}{10 \text{ k}\Omega}$$
$$= -72.57°$$

$$|Z| = \frac{R}{\cos\theta} = \frac{10{,}000}{\cos -72.57°}$$
$$= 33.38 \text{ k}\Omega$$

$$\mathbf{Z} = 33.38 \underline{/-72.57°} \text{ k}\Omega$$

EXAMPLE 20-5:
Solve for the impedance of the circuit of Example 20-4 when (a) frequency = 1.5 kHz, and (b) when frequency = 20 kHz.

Solution:
(a) At 1500 Hz:

$$X_C = \frac{1}{6.28 \times 1.5 \times 10^3 \times 0.01 \times 10^{-6}}$$
$$= 10{,}616 \, \Omega \cong 10.6 \text{ k}\Omega$$

$$\theta = \tan^{-1} \frac{-10.6 \text{ k}\Omega}{10 \text{ k}\Omega}$$
$$= -46.67°$$

$$|Z| = \frac{R}{\cos\theta} = \frac{10{,}000}{\cos -46.67°}$$
$$= 14.573$$

$$\mathbf{Z} = 14.6 \underline{/-46.67°} \text{ k}\Omega$$

(b) At 20 kHz:

$$X_C = \frac{1}{6.28 \times 20 \times 10^3 \times 0.01 \times 10^{-6}}$$
$$= 796 \, \Omega$$

$$\theta = \tan^{-1} \frac{-796}{10{,}000}$$
$$= -4.55°$$

$$|Z| = \frac{R}{\cos\theta} = \frac{10{,}000}{\cos -4.55°}$$
$$= 10{,}032$$

$$\mathbf{Z} = 10.032 \underline{/-4.55°} \text{ k}\Omega$$

20-3. VOLTAGE AND CURRENT RELATIONSHIPS

The rules of series ac circuits are the same as for series dc circuits, except that we must account for phase shift caused by reactance. As a result, our calculations, except for purely resistive circuits, require that we work with phasors.

1. *Current*: Current is the same throughout in any series circuit. Current is a phasor whose angle depends upon impedance and voltage phasors. Working with the total circuit, we have

$$I = \frac{E}{Z_T} \tag{20-6}$$

When working with voltage drops,

$$I = \frac{V_1}{Z_1} = \frac{V_2}{Z_2} = \frac{V_n}{Z_n} \tag{20-7}$$

2. *Impedance*: The total impedance of a series circuit is the phasor sum of the impedances of the circuit.

$$Z_T = Z_1 + Z_2 + \cdots + Z_n \tag{20-8}$$

We may also use Ohm's law to find impedances:

$$Z_T = \frac{E}{I} \tag{20-9}$$

Similarly,

$$Z_1 = \frac{V_1}{I} : Z_2 = \frac{V_2}{I} : Z_n = \frac{V_n}{I} \tag{20-10}$$

3. *Voltage*: The phasor sum of the voltage drops is equal to the applied voltage.

$$E = V_1 + V_2 + \cdots + V_n \tag{20-11}$$

Working with Ohm's law, we have

$$E = IZ_T \tag{20-12}$$

$$V_1 = IZ_1 : V_2 = IZ_2 : V_n = IZ_n \tag{20-13}$$

Practical considerations of impedance require that we be aware that circuits frequently contain R, L, and C. Impedance is the phasor sum of resistance and the *net* reactance.

$$Z = R + j(X_L - X_C) \tag{20-14}$$

The j term is negative when capacitive reactance is greater than inductive reactance.

EXAMPLE 20-6:
A series circuit consists of two impedances. Solve for the total impedance of the circuit, given that

$$Z_1 = 400 \underline{/30°} \: \Omega$$
$$Z_2 = 250 \underline{/40°} \: \Omega$$
$$E = 30 \underline{/0°} \: V$$

Solution:

$$Z_T = Z_1 + Z_2$$

$$Z_1 = 400 \cos 30° + j400 \sin 30°$$
$$= 346 + j200$$
$$Z_2 = 250 \cos 40° + j250 \sin 40°$$
$$= 192 + j161$$
$$Z_T = (346 + 192) + j(200 + 161)$$
$$= 538 + j361 \ \Omega$$
$$\theta = \tan^{-1} \frac{361}{538} = 33.9°$$
$$|Z| = \frac{538}{\cos 33.9°} = 648$$
$$Z_T = 648\underline{/33.9°} \ \Omega$$

A sketch of Z_1, Z_2, and Z_T are shown in Fig. 20-6.

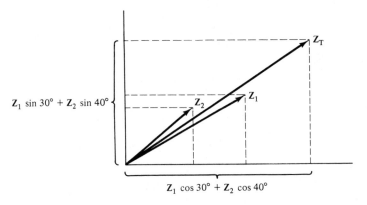

Figure 20-6

EXAMPLE 20-7:

Given the circuit of Example 20-6, (a) solve for current; (b) solve for each voltage drop; (c) demonstrate that the phasor sum of the voltage drops is equal to the applied voltage.

Solution:

(a) $\quad I = \dfrac{E}{Z_T}$
$$= \frac{30\underline{/0°}}{648\underline{/33.9°}}$$
$$= 0.0463\underline{/-33.9°} \ A$$

(b) $\quad V_1 = IZ_1$
$$= 0.0463\underline{/-33.9°} \times 400\underline{/30°}$$
$$= 18.52\underline{/-3.9°} \ V$$

$\quad V_2 = IZ_2$
$$= 0.0463\underline{/-33.9°} \times 250\underline{/40°}$$
$$= 11.58\underline{/6.1°} \ V$$

(c) $\mathbf{E} = \mathbf{V}_1 + \mathbf{V}_2$

$V_1 = 18.52 \cos -3.9° + j18.52 \sin -3.9°$
$= 18.48 - j1.26$ V

$V_2 = 11.58 \cos 6.1° + j11.58 \sin 6.1°$
$= 11.51 + j1.23$ V

$\mathbf{E} = \mathbf{V}_1 + \mathbf{V}_2 = (18.52 + 11.51) + (j1.23 - j1.26)$

$\mathbf{E} = 29.99 - j0.03$

$\mathbf{E} \approx 30\underline{/0°}$ V

Phasor diagrams of the computations:
1. $\mathbf{E}, \mathbf{I}, \mathbf{V}_1, \mathbf{V}_2$ (Fig. 20-7)

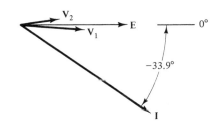

Figure 20-7

2. $\mathbf{E} = \mathbf{V}_1 + \mathbf{V}_2$ (Fig. 20-8)

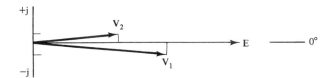

Figure 20-8

EXAMPLE 20-8:
Given a series *RLC* circuit with $R = 1.2$ kΩ, $X_L = 2.8$ kΩ, and $X_C = 3.9$ kΩ. The applied voltage is $50\underline{/0°}$ V. (a) Draw a phasor diagram of the impedance. (b) Solve for current. (c) Solve for each voltage drop. (d) Draw a phasor diagram of the current and voltages. (e) Demonstrate that the phasor sum of the voltage drops equals applied voltage.

Solution:
(a) See Fig. 20-9.

$\mathbf{Z}_T = R + j(X_L - X_C)$
$= 1200 + j(2800 - 3900)$
$= 1200 - j1100$ Ω

$\theta = \tan^{-1} \dfrac{-1100}{1200}$
$= -42.51°$

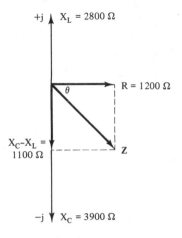

Figure 20-9

$$|Z_T| = \frac{1200}{\cos -42.51°} = 1628$$

$$\mathbf{Z}_T = 1628\underline{/-42.51°}\ \Omega$$

(b) $\mathbf{I} = \dfrac{\mathbf{E}}{\mathbf{Z}_T}$

$$= \frac{50\underline{/0°}}{1628\underline{/-42.51°}}$$

$$= 30.7\underline{/42.51°}\ \text{mA}$$

(c) $\mathbf{V}_R = \mathbf{I}R$
$= 30.7\underline{/42.51°}\ \text{mA} \times 1.2\ \text{k}\Omega$

$= 36.84\underline{/42.51°}\ \text{V}$

$\mathbf{V}_L = \mathbf{I}(jX_L)$
$= 30.7\underline{/42.51°}\ \text{mA} \times 2.8\underline{/90°}\ \text{k}\Omega$
$= 85.96\underline{/132.51°}\ \text{V}$

$\mathbf{V}_C = \mathbf{I}(-jX_C)$
$= 30.7\underline{/42.51°}\ \text{mA} \times 3.9\underline{/-90°}\ \text{k}\Omega$
$= 119.73\underline{/-47.49°}\ \text{V}$

(d) See Fig. 20-10.
(e) $\mathbf{E} = \mathbf{V}_R + \mathbf{V}_L + \mathbf{V}_C$

$\mathbf{V}_R = 36.84 \cos 42.51° + j36.84 \sin 42.51°$
$= 27.16 + j24.89$

$\mathbf{V}_L = 85.96 \cos 132.51° + j85.96 \sin 132.51°$
$= -58.08 + j63.37$

$\mathbf{V}_C = 119.73 \cos -47.49° + j119.73 \sin -47.49°$
$= 80.9 - j88.26$

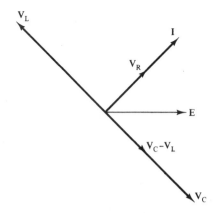

Figure 20-10

$$E = \begin{array}{r} 27.16 + j24.89 \\ + -58.08 + j63.37 \\ + \underline{80.90 - j88.26} \\ 49.98 \mid j0 \end{array}$$

$E \simeq 50\underline{/0°}$ V

20-4. ANALYZING SERIES AC CIRCUITS

In this section we continue with solutions to series ac circuits. We add some examples that review calculator usage.

We begin with applications of the voltage-division rule.

VDR for use with ac circuits: The voltage across an impedance in a series circuit is the product of the applied voltage and the ratio of that impedance to the total impedance.

$$V_x = E \frac{Z_x}{Z_T} \qquad (20\text{-}15)$$

We may rearrange the VDR formula to find other quantities:

$$Z_T = \frac{EZ_x}{V_x} \qquad (20\text{-}16)$$

$$E = \frac{V_x Z_T}{Z_x} \qquad (20\text{-}17)$$

EXAMPLE 20-9:
Given the circuit of Fig. 20-11, solve for V_2 by use of the VDR.

Sec. 20-4.   Analyzing Series ac Circuits

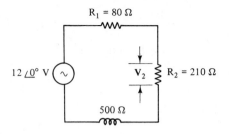

Figure 20-11

Solution:

1. $\mathbf{Z}_T = 80 + 210 + j500$
   $= 290 + j500$

   $\theta = \tan^{-1} \dfrac{500}{290}$

   $500 \boxed{\div} 290 \boxed{=} \boxed{\text{inv}} \boxed{\text{tan}}$

   $\theta = 59.89°$

   $|\mathbf{Z}_T| = \dfrac{500}{\sin 59.89°}$

   $500 \boxed{\div} 59.89 \boxed{\sin} \boxed{=} 577.99\ldots$

   $\mathbf{Z}_T = 578\underline{/59.89°}\ \Omega$

2. $\mathbf{Z}_2 = R_2 = 210\underline{/0°}\ \Omega$

3. $\mathbf{V}_2 = \mathbf{E}\dfrac{\mathbf{Z}_2}{\mathbf{Z}_T}$

   $= 12\underline{/30°} \times \dfrac{210\underline{/0°}}{578\underline{/59.89°}}$

   $= \dfrac{12 \times 210}{578}\underline{/30 - 59.89°}$

   $= 4.36\underline{/-29.89°}\ \text{V}$

EXAMPLE 20-10:

The total impedance of a series circuit is $380\underline{/-40°}\ \Omega$. We measure 15 V across a 100-Ω resistor. What is the magnitude of the applied voltage? (We can compute magnitude only, because we do not know the phase angle of the voltage across the resistor.)

Solution:

$$\mathbf{E} = \dfrac{\mathbf{V}_x \mathbf{Z}_T}{\mathbf{Z}_x}$$

$$= \dfrac{15 \times 380}{100}$$

$$|\mathbf{E}| = 57\ \text{V}$$

Clearly, we may make use of Ohm's law in the analysis of circuits. The following examples show typical series circuit problems and their solutions.

EXAMPLE 20-11:

The voltage drop across the capacitor in the circuit of Fig. 20-12 is 179 V_{rms}. Solve for total impedance by Ohm's law.

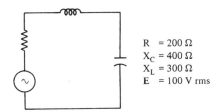

R = 200 Ω
X_C = 400 Ω
X_L = 300 Ω
E = 100 V rms

Figure 20-12

Solution:

1. $|I| = \dfrac{V_C}{X_C} = \dfrac{179}{400}$
   = 0.477 A_{rms}

2. $|Z_T| = \dfrac{E}{|I|} = \dfrac{100}{0.447}$
   = 223.6 Ω

Check:

$$Z_T = R + j(X_L - X_C)$$
$$= 200 + j(300 - 400)$$
$$= 200 - j100$$

$$\theta = \tan^{-1}\dfrac{-100}{200}$$

100 [+/-] [÷] 200 [=] [inv] [tan]

$\theta = -26.57°$

$$|Z_T| = \dfrac{R}{\cos\theta} = \dfrac{X}{\sin\theta}$$
$$= \dfrac{100}{\sin -26.57°}$$

100 [÷] 26.57 [+/-] [sin] =

= 223.6 Ω

EXAMPLE 20-12:

Given the circuit of Fig. 20-13, solve for **E**.

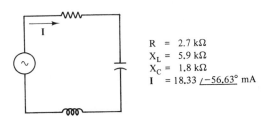

R = 2.7 kΩ
X_L = 5.9 kΩ
X_C = 1.8 kΩ
I = 18.33 /−56.63° mA

Figure 20-13

Solution:

$$E = IZ_T$$

$$\begin{aligned}Z_T &= R + j(X_L - X_C) \\ &= 2700 + j(5900 - 1800) \\ &= 2700 + j4100\end{aligned}$$

$$\theta = \tan^{-1} \frac{4100}{2700}$$

$\boxed{4100} \boxed{\div} \boxed{2700} \boxed{=} \boxed{\text{inv}} \boxed{\text{tan}}$

$$\theta = 56.63°$$

$$\begin{aligned}|Z_T| &= \frac{X}{\sin \theta} \\ &= \frac{4100}{\sin 56.63°}\end{aligned}$$

$\boxed{4100} \boxed{\div} \boxed{56.63} \boxed{\sin} \boxed{=} = 4909\ \Omega$

$$Z_T = 4909 \underline{/56.63°}\ \Omega$$

$$\begin{aligned}E &= I \times Z_T \\ &= (18.33 \times 10^{-3} \underline{/-56.63°})(4.909 \times 10^3 \underline{/56.63°}) \\ &= 89.98 \underline{/-56.63 + 56.63°} \\ &= 90 \underline{/0°}\ V\end{aligned}$$

20-5. THE MEANING OF Q

We have learned that inductors and capacitors are energy-storage devices. In our previous discussions we treated these devices as purely inductive or purely capacitive. We now consider how effective these devices are as energy-storage parts of a circuit.

We have a means of describing the measure of ability of a reactive device to store energy as opposed to the dissipation of energy in the device. We use the letter Q as a figure of merit. High Q means little energy dissipation with respect to energy storage, while low Q means that energy dissipation is nearly as great as energy storage.

$$Q = \frac{\text{energy stored}}{\text{energy dissipated}}$$

We apply the concept of Q to individual coils and capacitors and also to entire circuits. It is an important quantity when used in a special category of ac circuits, resonant circuits, to be studied in Chapter 23. Its importance in these circuits will be developed. We shall find its meanings extend well beyond our definition for Q in this section.

20-6. THE Q OF AN INDUCTOR

It can be shown that energy storage in an inductor is proportional to inductive reactance, while energy dissipation is proportional to the resistance of the coil.

$$\text{energy stored} = I^2 X_L$$
$$\text{energy dissipated} = I^2 R_o$$

Taking the ratio of the expressions, we have the Q of a coil.

$$Q = \frac{X_L}{R_o} = \frac{\omega L}{R_o} \qquad (20\text{-}18)$$

The resistance, R_o, represents the dissipative losses in the coil. The value of R_o tends to be constant at low frequencies, and increases with frequency at high frequencies. It is affected by radiation losses into surrounding conductive materials, hysteresis losses in magnetic materials, and skin effect. We find that coils designed to operate over some range of high frequencies have fairly constant Q over the range. Values of Q for coils may be as low as 5 to as high as 100.

20-7. THE Q OF A CAPACITOR

Capacitors, except for electrolytic types, have extremely low losses and therefore very high Q. We find Q in the same manner as for inductors.

$$Q = \frac{X_C}{R_o} = \frac{1}{\omega R_o C} \qquad (20\text{-}19)$$

In this instance, R_o represents the effective *series* resistance of a capacitor. This quantity usually approaches zero.

Because Q is extremely high in capacitors, we treat them as pure capacitance. We shall find that calculations for Q in circuits will involve inductive reactance and resistance; rarely do we concern ourselves with Q for capacitors.

As an example, consider that mica capacitors have values of Q ranging from 1400 to 10,000 when operating at high frequencies. Clearly, at lower frequencies Q is very much higher.

20-8. EFFECTIVE Q

We have stated that the concept of circuit Q is extremely important in resonant circuits. We discuss it in this chapter in order to reinforce a few ideas.

Q is defined as the ratio of energy stored to energy dissipated. In a circuit consisting of an inductor and a capacitor, nearly all energy dissipation takes place in the resistance, R_o, of the inductor. We find then that the Q of such a circuit is the same as the Q of the coil. We refer to circuit Q as Q' and the Q of the coil as Q_o.

We can increase the amount of energy dissipation in a circuit by placing a resistance in series with the remainder of the circuit. The added resistance reduces the Q of the circuit. In those instances where the resistance of the coil is still appreciable, we cannot yet make any accurate calculations for Q'. We shall learn how to do so in Chapter 23.

Where the added resistance is much greater than the resistance of the coil, Q' is found to be frequency dependent. As long as the added resistance remains much greater than the resistance of the coil, Q increases as frequency increases. Recall that, with the Q of a coil, we agreed that it changed slightly with frequency because its resistance also changed. In the case of effective Q, the resistance of the circuit is added and need not vary with frequency. Effective Q is found by the formula

$$Q' = \frac{2\pi f L}{R_o + R} \qquad (20\text{-}20)$$

where R_o is the resistance of the coil and R is added resistance.

EXAMPLE 20-13:
The intrinsic Q (Q_o) of a 100-μH inductor is 80 when operated in the 400-kHz range. What is the resistance, R_o, of the coil?

Solution:

$$Q_o = \frac{\omega L}{R_o}$$

Therefore,

$$R_o = \frac{\omega L}{Q_o}$$
$$= \frac{6.28 \times 0.4 \times 10^6 \times 100 \times 10^{-6}}{80}$$
$$= 3.14 \ \Omega$$

EXAMPLE 20-14:
Given the circuit of Fig. 20-14, solve for the effective Q. The inductor is the same as for Example 20-13.

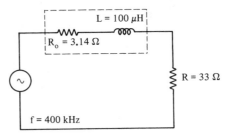

Figure 20-14

Solution:

$$Q' = \frac{X_L}{R_o + R}$$
$$= \frac{6.28 \times 0.4 \times 10^6 \times 100 \times 10^{-6}}{3.14 + 33.3}$$
$$= 6.89$$

EXAMPLE 20-15:

An inductor has a Q_o of 150 at 3 MHz. If the inductance is 75 μH, what is the resistance, R_o? What value of resistance must be added to the coil to cause Q' to be 50?

Solution:

$$R_o = \frac{X_L}{Q_o} = \frac{6.28 \times 3 \times 10^6 \times 75 \times 10^{-6}}{150}$$
$$= 9.42 \ \Omega$$

$$R + R_o = \frac{X_L}{Q'} = \frac{6.28 \times 3 \times 10^6 \times 75 \times 10^{-6}}{50}$$
$$= 28.26 \ \Omega$$

$$R = 28.26 - 9.42 = 18.84 \ \Omega$$

SUMMARY

1. Impedance is a general expression for opposition to current in an ac circuit. It can represent any form from pure resistance to pure reactance. $\mathbf{Z} = R \pm jX$.
2. We use Ohm's law in ac circuits by substituting $\mathbf{Z}$ for R_T and following the rules of vector mathematics.
3. The impedance of an inductive circuit has a leading angle. The tangent of the angle is X_L/R in a series circuit.
4. Capacitive circuits have impedances with lagging angles. The tangent of the angle is X_C/R in a series circuit.
5. The general rules of series circuits are the same in ac as in dc. We must, however, take vector quantities into account.
6. Because ac circuits can contain energy-storage devices, we are interested in the ratio of energy stored to energy dissipated. This ratio is called Q.
7. The Q_o of a coil is $\omega L/R_o$.
8. The Q_o of a capacitor is $1/\omega R_o C$.
9. Effective Q defines Q for a circuit.
10. We reduce effective Q by adding resistance to a circuit.

PROBLEMS

20-1. Solve for the missing quantity:

$E = 20\underline{/15°}$ V $Z_T = 150\underline{/-40°}$ Ω $I_T =$
$I_T = 30\underline{/25°}$ mA $Z_T = 400\underline{/-25°}$ Ω $E =$
$E = 120\underline{/0°}$ V $I_T = 7.5\underline{/-40°}$ A $Z_T =$
$E = 48\underline{/-33°}$ mV $Z_T = 2.7\underline{/0°}$ kΩ $I_T =$
$E = 15\underline{/0°}$ V $I_T = 5\underline{/-60°}$ mA $Z_T =$

20-2. In the circuit of Fig. 20-15, $X_L = 400$ Ω and $R = 800$ Ω. Solve for **Z**.

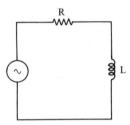

Figure 20-15

20-3. In the circuit of Fig. 20-15, $X_L = 400$ Ω and $R = 100$ Ω. Solve for **Z**.

20-4. In the circuit of Fig. 20-15, operating frequency is 2 kHz, inductance is 100 mH, and resistance is 100 Ω. Solve for the impedance.

20-5. Given the data of Prob. 20-3, solve for each voltage drop, given that $\mathbf{E} = 40\underline{/0°}$ V. Demonstrate that the phasor sum of the voltage drops is equal to the applied voltage.

20-6. Solve for the impedance of the circuit of Fig. 20-16 when $X_C = 1200$ Ω and $R = 400$ Ω.

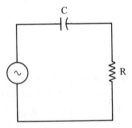

Figure 20-16

20-7. What is the impedance of the circuit of Fig. 20-16 when $X_C = 300$ Ω and $R = 1500$ Ω?

20-8. Given the data of Prob. 20-6, solve for **I** given that $\mathbf{E} = 30\underline{/0°}$ V.

20-9. Given the data of Prob. 20-7, solve for **E**, given that $\mathbf{I} = 150\underline{/11.31°}$ mA.

20-10. In the circuit of Fig. 20-16, $X_C = 2500$ Ω and $R = 700$ Ω. Given that $\mathbf{E} = 28\underline{/0°}$ V, solve for $\mathbf{V}_R$.

20-11. In a certain two-element series circuit, $\mathbf{E} = 18\underline{/20°}$ V and $\mathbf{V}_1 = 9\underline{/-30°}$ V. Solve for V_2.

20-12. In the circuit of Fig. 20-17, $X_L = 3$ kΩ and $R = 4.7$ kΩ. The voltage drop across R is given. Solve for **E**.

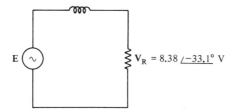

Figure 20-17

20-13. Given the data of Prob. 20-12, solve for the operating frequency if $L = 175 \; \mu H$.

20-14. Given the circuit of Fig. 20-18, with $R = 1 \; k\Omega$, $X_C = 2.5 \; k\Omega$, and $X_L = 4.5 \; k\Omega$, solve for $V_R, V_L,$ and V_C.

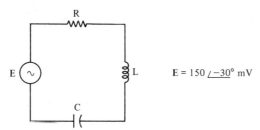

Figure 20-18

20-15. In the circuit of Fig. 20-18, $X_L = 15 \; k\Omega$, $X_C = 16 \; k\Omega$, and $R = 1 \; k\Omega$. Solve for V_L.

20-16. In the circuit of Fig. 20-18, if $R = 4 \; k\Omega$ and $\mathbf{Z} = 7.21/{-56.3°} \; k\Omega$, what is the net reactance of the circuit?

20-17. In the circuit of Fig. 20-18, solve for $\mathbf{Z}$ if $R = 3.3 \; k\Omega$ and $V_R = 99/{-48.7°}$ mV.

20-18. Given the data of Prob. 20-17, if $X_C = 10 \; k\Omega$, what is X_L?

20-19. Apply the voltage-division rule (VDR) to a series RLC circuit and solve for V_R, given that $\mathbf{E} = 180/0°$ V, $R = 1500 \; \Omega$, $X_C = 2000 \; \Omega$, and $X_L = 4000 \; \Omega$.

20-20. The intrinsic Q of a 15-μH coil when operating at 10 MHz is 90. Solve for R_o, the resistance of the coil.

20-21. The resistance of a 50-mH coil is 30 Ω. What is the intrinsic Q of the coil at 5 kHz?

20-22. A certain coil with an inductance of 20 mH and a Q_o of 40 is to be used in a circuit so that the effective Q is 15. If the operating frequency is 20 kHz, how much resistance must be added to obtain the desired Q'?

20-23. The Q_o of a 100-pF capacitor, operating at 20 MHz, is 900. Solve for the equivalent series resistance of the capacitor.

20-24. In the circuit of Fig. 20-19, solve for R, given that $Q' = 12$ and $f = 10.7$ MHz.

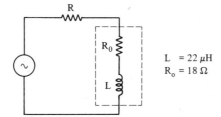

Figure 20-19

Chap. 20 Problems

21

Analysis of AC Circuits

21-1. RULES OF PARALLEL CIRCUITS

In parallel ac circuits there are some rules that are similar to those for dc circuits.

1. The voltage across each branch of the circuit is the same.
2. The total current drawn from a source is the *phasor* sum of the branch currents.

However, total current, unlike that for dc circuits, is not always greater than any of the branch currents. We shall find this demonstrated by some examples in the next section. *When branches have the same type of reactance, total current is greater than any branch current.*

3. The total impedance (Z_{eq}) is not always less than the impedance of any of the branches.

We need only refer to our discussion on total current. Certainly, if we have a circuit where total current is less than branch current, then total impedance must be greater than branch impedance. For those parallel circuits where each of the branches contains the same type of reactance, total impedance is less than any branch impedance.

21-2. TWO-BRANCH PARALLEL CIRCUITS

We make use of the general formula for a two branch parallel circuit.

$$Z_{eq} = \frac{Z_1 Z_2}{Z_1 + Z_2} \qquad (21\text{-}1)$$

When we solve for Z_{eq}, we should keep Z_1 and Z_2 in polar form in the numerator in order to make multiplication easier. Simply multiply magnitudes and add angles algebraically. In the denominator, we must change Z_1 and Z_2 into rectangular form in order to add them. The resultant sum should then be changed into polar form in order to divide. Simply divide the numerator by the denominator and subtract the angles algebraically.

When both branches have the same kind of reactance, then Z_{eq} will be an impedance that is smaller than either branch impedance and of the same kind of reactance.

When one branch is capacitive and the other inductive, Z_{eq} may be larger than a branch impedance. It will represent a reactance of the same type as the branch with the larger current.

The following examples illustrate these general concepts.

EXAMPLE 21-1:

Solve for the total impedance of the circuit of Fig. 21-1, given that $Z_1 = 20\underline{/40°}$ Ω and $Z_2 = 25\underline{/30°}$ Ω.

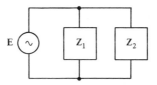

Figure 21-1

Solution:

$$Z_{eq} = \frac{Z_1 \times Z_2}{Z_1 + Z_2}$$

1. $Z_1 = 20\underline{/40°} = 20 \cos 40° + j20 \sin 40°$
   $= 15.32 + j12.86$

   $Z_2 = 25\underline{/30°} = 25 \cos 30° + j25 \sin 30°$
   $= 21.65 + j12.5$

   $Z_1 + Z_2 = 36.97 + j25.36$

   $\theta = \tan^{-1} \frac{25.36}{36.97}$
   $= 34.45°$

   $|Z_1 + Z_2| = \frac{36.97}{\cos 34.45°}$

   $Z_1 + Z_2 = 44.83\underline{/34.45°}$ Ω

2. $Z_1 \times Z_2 = 20\underline{/40°} \times 25\underline{/30°}$
   $= 20 \times 25\underline{/40° + 30°}$
   $= 500\underline{/70°}$

3. $Z_{eq} = \dfrac{500\underline{/70°}}{44.83\underline{/34.45°}}$
   $= 11.15\underline{/35.55°}\ \Omega$

EXAMPLE 21-2:
Assume a supply voltage of $10\underline{/0°}$ V for the circuit of Example 21-1. Solve for the total current and demonstrate that the ratio of **E** to I_T is equal to the Z_{eq} found in Example 21-1.

Solution:

$$I_T = I_1 + I_2$$

$$I_1 = \frac{E}{Z_1} = \frac{10\underline{/0°}}{20\underline{/40°}}$$
$$= 0.5\underline{/-40°}\ A$$
$$= 0.5 \cos -40° + j0.5 \sin -40°$$
$$= 0.383 - j0.321\ A$$

$$I_2 = \frac{E}{Z_2} = \frac{10\underline{/0°}}{25\underline{/30°}}$$
$$= 0.4\underline{/-30°}\ A$$
$$= 0.4 \cos -30° + j0.4 \sin -30°$$
$$= 0.346 - j0.2$$

$$I_1 + I_2 = 0.729 - j0.521$$

$$\alpha = \tan^{-1} \frac{-0.521}{0.729}$$
$$= -35.55°$$

$$|I_T| = \frac{0.729}{\cos -35.55°}$$
$$= 0.896$$

$$I_T = 0.896\underline{/-35.55°}\ A$$

$$Z_{eq} = \frac{E}{I_T} = \frac{10\underline{/0°}}{0.896\underline{/-35.55°}} = 11.16\underline{/35.55°}\ \Omega$$

EXAMPLE 21-3:
In the circuit of Fig. 21-2, $Z_1 = 100\underline{/45°}$ and $Z_2 = 150\underline{/-50°}\ \Omega$. (a) Solve for total impedance. (b) Assume that $E = 50\underline{/0°}$ V; solve for I_T.

Solution:

(a) 1. $Z_1 = 100\underline{/45°}$
        $= 100 \cos 45° + j100 \sin 45°$

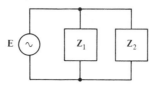

Figure 21-2

$$Z_2 = 150\underline{/-50°}$$
$$= 150 \cos -50° + j150 \sin -50°$$

$$\begin{aligned}Z_1 &= 70.7 + jj70.7\\ Z_2 &= 96.4 - j114.9\\ \hline Z_1 + Z_2 &= 167.1 - j44.2\end{aligned}$$

$$\theta = \tan^{-1} \frac{-44.2}{167.1}$$
$$= -14.82°$$

$$|Z_1 + Z_2| = \frac{44.2}{\sin -14.82°}$$
$$= 172.8$$

$$Z_1 + Z_2 = 172.8\underline{/-14.82°}\ \Omega$$

2. $Z_1 \times Z_2 = 100 \times 150\underline{/45 - 50°}$
$$= 15{,}000\underline{/-5°}$$

3. $Z_{eq} = \dfrac{Z_1 \times Z_2}{Z_1 + Z_2}$
$$= \frac{15000\underline{/-5°}}{172.8\underline{/-14.82°}}$$
$$= 86.8\underline{/9.82°}\ \Omega$$

(b) $I_T = I_1 + I_2$

$$I_1 = \frac{E}{Z_1} = \frac{50\underline{/0°}}{100\underline{/45°}}$$
$$= 0.5\underline{/-45°}\ A$$

$$I_2 = \frac{E}{Z_2} = \frac{50\underline{/0°}}{150\underline{/-50°}}$$
$$= 0.333\underline{/50°}\ A$$

$I_1 = 0.5 \cos -45° + j0.5 \sin -45°$

$I_2 = 0.333 \cos 50° + j0.333 \sin 50°$

$$\begin{aligned}I_1 &= 0.3535 - j0.3535\ A\\ I_2 &= \underline{0.214 + j0.2551\ A}\\ I_T &= 0.5675 - j0.0984\ A\end{aligned}$$

$$\alpha = \tan^{-1} \frac{-0.0984}{0.5675}$$
$$= -9.84°$$

$$|I_T| = \frac{0.5675}{\cos -9.84°}$$
$$= 0.576$$
$$\mathbf{I}_T = 0.576\underline{/-9.84°} \text{ A}$$

Figure 21-3 is a phasor diagram of the currents.

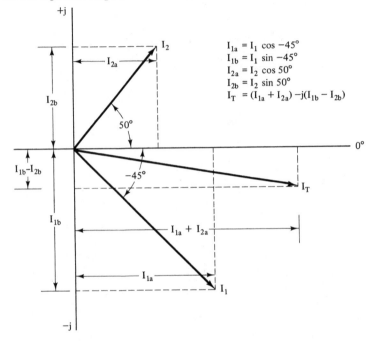

Figure 21-3 The currents of Example 21-3.

Check:
$$\mathbf{I}_T = \frac{\mathbf{E}}{\mathbf{Z}_{eq}} = \frac{50\underline{/0°}}{86.8\underline{/9.82°}}$$
$$= 0.576\underline{/-9.82°} \text{ A}$$

EXAMPLE 21-4:
In the circuit of Fig. 21-4, $\mathbf{Z}_1 = 100\underline{/70°}$ and $\mathbf{Z}_2 = 150\underline{/-60°}$ Ω: (a) Solve for total impedance. (b) Assume that $\mathbf{E} = 50\underline{/0°}$ V; solve for I_T.

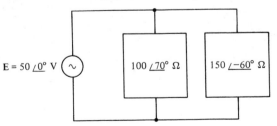

Figure 21-4

Solution:

(a) $\quad Z_1 = 100 \cos 70° + j100 \sin 70°$

$\quad\quad Z_2 = 150 \cos -60° + j150 \sin -60°$

$$\begin{aligned}Z_1 &= 34.2 + j93.97 \ \Omega \\ Z_2 &= 75 \ - j129.9 \ \Omega \\ Z_1 + Z_2 &= 109.2 - j35.93 \ \Omega \\ &= 114.96 \underline{/-18.21°} \ \Omega\end{aligned}$$

$$\begin{aligned}Z_1 \times Z_2 &= (100 \times 150)\underline{/70 - 60°} \\ &= 15{,}000\underline{/10°}\end{aligned}$$

$$\begin{aligned}Z_{eq} &= \frac{Z_1 \times Z_2}{Z_1 + Z_2} \\ &= \frac{15{,}000\underline{/10°}}{114.96\underline{/-18.21°}} \\ &= 130.5\underline{/28.21°} \ \Omega\end{aligned}$$

(b) I_T may be found by use of Ohm's law. We shall solve for I_T by finding each branch of the current and then take the sum of the currents.

$$I_T = I_1 + I_2$$

$$\begin{aligned}I_1 &= \frac{E}{Z_1} = \frac{50\underline{/0°}}{100\underline{/70°}} \\ &= 0.5\underline{/-70°} \ A \\ &= 0.5 \cos -70° + j0.5 \sin -70°\end{aligned}$$

$$\begin{aligned}I_2 &= \frac{E}{Z_2} = \frac{50\underline{/0°}}{150\underline{/-60°}} \\ &= 0.333\underline{/60°} \ A \\ &= 0.333 \cos 60° - j0.333 \sin 60°\end{aligned}$$

$$\begin{aligned}I_1 &= 0.171 \ - j0.4698 \ A \\ I_2 &= 0.1665 + j0.2884 \ A \\ I_1 + I_2 &= 0.3375 - j0.1814 \ A \\ I_T &= 0.383\underline{/-28.26°} \ A\end{aligned}$$

Check:

$$Z_{eq} = \frac{E}{I_T} = \frac{50\underline{/0°}}{0.383\underline{/-28.26°}}$$
$$= 130.5\underline{/28.26°} \ \Omega$$

The slight difference between phase angles is the result of rounding off.

The preceding examples enable us to clarify our comments on equivalent impedance when the branches have different types of reactance. Note that in Example 21-3 the total impedance was less than the impedance of either branch. This is due to the moderate angles of the branch impedances, indicating that the resistance in each branch was important and comparable to reactance. In Example 21-4, we used the same values of branch impedance but increased the phase angles. In this case, reactance in each branch was much greater than resistance. This resulted in a greater amount of phase shift to the branch currents. However, the reactive branch currents oppose each other in the common lines, causing total current to be less than one of the branch currents. Therefore, the total impedance was greater than that particular branch impedance. We shall see truly significant effects relating to cancellation of reactive currents in our studies of resonant circuits.

In any event, the total impedance does tend to take on the properties of the smallest branch impedance. Simply stated, it is total current that determines equivalent impedance. Very often, total current is nearly the same as the current in the smallest branch impedance.

21-3. MULTIBRANCH PARALLEL CIRCUITS

The general formula for the total impedance of a multibranch parallel circuit has the same general form as for dc circuits.

$$Z_{eq} = \frac{1}{1/Z_1 + 1/Z_2 + \cdots + 1/Z_n} \qquad (21\text{-}2)$$

Each reciprocal of a branch impedance must be in rectangular form for us to be able to take the sum of the reciprocals. This sum is then changed to polar form in order to take the reciprocal of the sum, which results in Z_{eq}.

We may use an Ohm's law approach to multibranch ac circuits. We assume any convenient voltage; then solve for each branch current. The phasor sum of the currents is equal to the total current. The ratio of the assumed voltage to the total current is the total impedance, Z_{eq}. We shall demonstrate both methods in the next example.

EXAMPLE 21-5:
Solve the circuit of Fig. 21-5 for total impedance. $Z_1 = 200\underline{/-40°}$ Ω, $Z_2 = 350\underline{/35°}$ Ω, and $Z_3 = 300\underline{/70°}$ Ω.

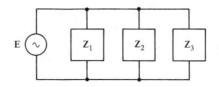

Figure 21-5

Solution:

$$Z_{eq} = \frac{1}{1/Z_1 + 1/Z_2 + 1/Z_3}$$

$$\frac{1}{Z_1} = \frac{1}{200/\underline{-40°}} = 5 \times 10^{-3}/\underline{40°}$$
$$= 5 \times 10^{-3} \cos 40° + j5 \times 10^{-3} \sin 40°$$
$$= 3.83 \times 10^{-3} + j3.214 \times 10^{-3}$$

$$\frac{1}{Z_2} = \frac{1}{350/\underline{35°}} = 2.86 \times 10^{-3}/\underline{-35°}$$
$$= 2.342 \times 10^{-3} - j1.64 \times 10^{-3}$$

$$\frac{1}{Z_3} = \frac{1}{300/\underline{70°}} = 3.33 \times 10^{-3}/\underline{-70°}$$
$$= 1.14 \times 10^{-3} - j3.13 \times 10^{-3}$$

$$\frac{1}{Z_1} + \frac{1}{Z_2} + \frac{1}{Z_3} = 7.312 \times 10^{-3} - j1.556 \times 10^{-3}$$

$$\theta = \tan^{-1} \frac{-1.556}{7.312}$$
$$= -12.01°$$

$$\left|\Sigma\left(\frac{1}{Z}\right)\right| = \text{magnitude of the sum of reciprocals}$$
$$= 7.312 \times 10^{-3}/\cos 12.01°$$

$$\Sigma\left(\frac{1}{Z}\right) = 7.476 \times 10^{-3}/\underline{-12.01°}$$

$$Z_{eq} = \frac{1}{\Sigma(1/Z)} = \frac{1}{7.476 \times 10^{-3}/\underline{-12.01°}}$$
$$= 133.8/\underline{12.01°} \; \Omega$$

An alternate method of solution can be used by assuming any convenient voltage. By choosing $\mathbf{E} = 350/\underline{0°}$ V, all branch currents are convenient values to work with.

$$\mathbf{I}_1 = \frac{\mathbf{E}}{\mathbf{Z}_1} = \frac{350/\underline{0°}}{200/\underline{-40°}} = 1.75/\underline{40°} \text{ A}$$

$$\mathbf{I}_2 = \frac{\mathbf{E}}{\mathbf{Z}_2} = \frac{350/\underline{0°}}{350/\underline{35°}} = 1.0/\underline{-35°} \text{ A}$$

$$\mathbf{I}_3 = \frac{\mathbf{E}}{\mathbf{Z}_3} = \frac{350/\underline{0°}}{300/\underline{70°}} = 1.167/\underline{-70°} \text{ A}$$

In rectangular form,

$$\mathbf{I}_1 = 1.75 \cos 40° + j1.75 \sin 40°$$
$$\mathbf{I}_2 = 1.0 \cos -35° - j1.0 \sin -35°$$
$$\mathbf{I}_3 = 1.167 \cos -70° - j1.167 \sin -35°$$

$$\mathbf{I}_1 = 1.341 + j1.125 \text{ A}$$
$$\mathbf{I}_2 = 0.819 - j0.573 \text{ A}$$
$$\mathbf{I}_3 = 0.399 - j1.097 \text{ A}$$
$$\mathbf{I}_T = 2.559 - j0.545 \text{ A}$$

$$\alpha = \tan^{-1}\frac{-0.545}{2.559}$$
$$= -12.02°$$
$$|I_T| = \frac{-0.545}{\sin -12.02°}$$
$$= 2.617$$
$$\mathbf{I}_T = 2.617\underline{/-12.02°}$$
$$\mathbf{Z}_{eq} = \frac{\mathbf{E}}{\mathbf{I}_T} = \frac{350\underline{/0°}}{2.617\underline{/-12.02°}}$$
$$= 133.7\underline{/12.02°}\ \Omega$$

21-4. SUSCEPTANCE AND ADMITTANCE

Conductance is a measure of the ease with which current may flow in a *resistive* circuit. Conductance and resistance are reciprocals of one another.

In a similar manner, *susceptance* and reactance are related as shown in Fig. 21-6. Susceptance, whose symbol is B, is a measure of the ease with which current may flow in a reactance. The unit of susceptance is the siemen (S). We have two forms of susceptance:

1. *Inductive susceptance:*

$$B_L = \frac{1}{jX_L} = -j\frac{1}{2\pi fL} \tag{21-3}$$

2. *Capacitive susceptance:*

$$B_C = \frac{1}{-jX_C} = j2\pi fC \tag{21-4}$$

Admittance is a general term relating to the degree with which any circuit permits current flow. The symbol is $\mathbf{Y}$ and the unit is the siemen.

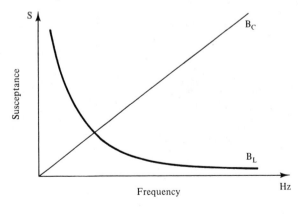

Figure 21-6 The variation of susceptance with frequency. Inductive susceptance (B_L) decreases as frequency increases. Note that this change is not linear. Capacitive susceptance (B_C) is a linear function of frequency. The unit of susceptance is the sieman (S). Be aware that many engineers and technicians still use the old unit, the mho (Ω).

$$Y = \frac{1}{Z} \tag{21-5}$$

We can solve for the total impedance of a parallel circuit by solving for admittance. To find Z_{eq}, we take the reciprocal of the admittance.

$$Y_T = \frac{1}{Z_1} + \frac{1}{Z_2} + \cdots + \frac{1}{Z_n}$$

$$Z_{eq} = \frac{1}{Y_T} \tag{21-6}$$

A special application of admittance is found in two-branch parallel circuits, when one branch is entirely resistive and the other entirely reactive.

$$Y = \frac{1}{R} \pm j\frac{1}{X}$$

$$Y = G \pm jB \tag{21-7}$$

21-5. PARALLEL *RC* CIRCUITS

In this section we deal with resistance in parallel with capacitance. While we may find total impedance by use of the product-over-sum formula (21-1), we have much more knowledge to gain by working with the admittance of the circuit.

$$Z_{eq} = \frac{1}{Y}$$

$$Z_{eq}\underline{/\theta} = \frac{1}{G + jB_C}\underline{/\theta} \tag{21-8}$$

where $G = 1/R$ and $B_C = 2\pi fC$.

The phase angle of the equivalent impedance is the *negative* of the angle of the denominator. Let α be the angle of the denominator. Then

$$\theta = -\alpha$$

Given that the angle of the denominator is found by

$$\alpha = \tan^{-1}\frac{B_C}{G} = \frac{2\pi fC}{1/R}$$

$$\alpha = \tan^{-1} \omega RC \tag{21-9}$$

$$\theta = -\alpha = -\tan^{-1} \omega RC \tag{21-10}$$

we may also find θ by

$$\theta = -\alpha = -\tan^{-1}\frac{R}{X_C} \tag{21-11}$$

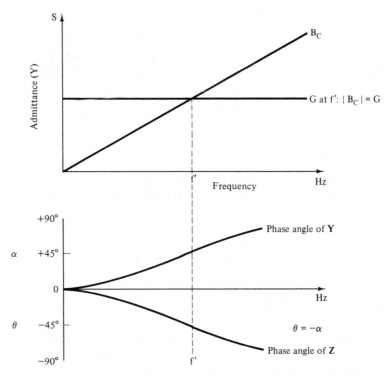

Figure 21-7 The admittance of a parallel RC circuit increases as frequency increases because of the increasing amount of capacitance susceptance.

The admittance of a parallel *RC* circuit increases with frequency (Fig. 21-7). This means that impedance *decreases* as frequency increases. However, the phase shift increases with frequency, approaching 90° as a limit. The impedance becomes purely capacitive at some high frequency. This simply means that the capacitive reactance, *in comparison to resistance*, has become so low that for all practical purposes total current and the current in the capacitive branch are equal.

It can be extremely instructive to change the parallel circuit to a series equivalent circuit. This simply involves taking the rectangular form of Z_{eq}. It becomes most instructive for the reader to learn that a *large parallel resistance is equivalent to a small amount of series resistance*, while a small parallel resistance is equivalent to a large amount of series resistance.

EXAMPLE 21-6:
A capacitive reactance of 1.2 kΩ is in parallel with a 10-kΩ resistor. (a) What is the phase angle of the impedance of the circuit? (b) What value of series resistance may be used in place of the 10-kΩ parallel resistor in order to cause the same amount of phase shift?

Solution:

(a) $\theta = -\tan^{-1} \dfrac{R_p}{X_C}$ (parallel)

$$= -\tan^{-1}\frac{10000}{1200}$$
$$= -83.16°$$

(b) $\tan\theta = \dfrac{X_C}{R_s}$ (series)

Therefore,
$$R_s = \frac{X_C}{\tan\theta} = \frac{1200}{\tan 83.16°}$$
$$= 144\ \Omega$$

In terms of *phase shift*, both circuits produce the same result; current leads voltage by 83.16°. The series circuit is *not* equivalent, in all respects, to the parallel circuit. It would be if we simply used $\mathbf{Z}_{eq}\cos\theta$ for R and $\mathbf{Z}_{eq}\sin\theta$ for X_C.

EXAMPLE 21-7:

Given the circuit of Fig. 21-8, calculate the admittance at 500 Hz, 2 kHz, 5 kHz, and 10 kHz.

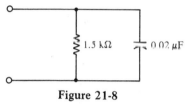

Figure 21-8

Solution:
At 500 Hz:
$$B_C = 2\pi fC$$
$$= 6.28 \times 5 \times 10^2 \times 2 \times 10^{-8}$$
$$= 62.8 \times 10^{-6} = 0.0628 \times 10^{-3}\ S$$
$$G = \frac{1}{R} = \frac{1}{1500} \times 10^3$$
$$= 0.667 \times 10^{-3}\ S$$
$$\mathbf{Y} = 0.667 + j0.0628\ mS$$
$$\alpha = \tan^{-1}\frac{0.0628}{0.667}$$
$$= 5.38°$$
$$|Y| = \frac{0.667}{\cos 5.38°}$$
$$= 0.67$$
$$\mathbf{Y} = 0.67\underline{/5.38°}\ mS$$

At 2 kHz:
$$B_C = 6.28 \times 2 \times 10^3 \times 2 \times 10^{-8}$$
$$= 25.12 \times 10^{-5} = 0.2512\ mS$$

Sec. 21-5.   Parallel *RC* Circuits

$$G = 0.667 \text{ mS}$$
$$\mathbf{Y} = 0.667 + j0.2512 \text{ mS}$$
$$\alpha = \tan^{-1} \frac{0.2512}{0.667}$$
$$= 20.6°$$
$$|Y| = \frac{0.2512}{\sin 20.6°}$$
$$= 0.714$$
$$\mathbf{Y} = 0.714 \underline{/20.6°} \text{ mS}$$

At 5 kHz:
$$B_C = 6.28 \times 5 \times 10^3 \times 2 \times 10^{-8}$$
$$= 62.8 \times 10^{-5} = 0.628 \text{ mS}$$
$$G = 0.667 \text{ mS}$$
$$\mathbf{Y} = 0.667 + j0.628 \text{ mS}$$
$$\alpha = \tan^{-1} \frac{0.628}{0.667}$$
$$= 43.3°$$
$$|Y| = \frac{0.667}{\cos 43.3°}$$
$$= 0.916$$
$$\mathbf{Y} = 0.916 \underline{/43.3°} \text{ mS}$$

At 10 kHz:
$$B_C = 6.28 \times 10 \times 10^3 \times 2 \times 10^{-8}$$
$$= 125.6 \times 10^{-5} = 1.256 \text{ mS}$$
$$G = 0.667 \text{ mS}$$
$$\mathbf{Y} = 0.667 + j1.256 \text{ mS}$$
$$\alpha = \tan^{-1} \frac{1.256}{0.667}$$
$$= 62.03°$$
$$|Y| = \frac{0.667}{\cos 62.03°}$$
$$= 1.422$$
$$\mathbf{Y} = 1.422 \underline{/62.03°} \text{ mS}$$

We now turn our attention to the series equivalent of a parallel RC circuit. The concept of a series equivalent was developed when we studied series–parallel dc circuits. We find similar needs for series equivalent values in ac circuits.

EXAMPLE 21-8:

Solve for the series equivalent impedance of the parallel *RC* circuit of Fig. 21-9. The operating frequency is 2 kHz.

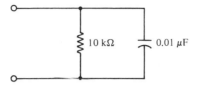

Figure 21-9

Solution:

$$B_C = 2\pi fC = 6.28 \times 2 \times 10^3 \times 0.01 \times 10^{-6}$$
$$= 0.1256 \times 10^{-3} = 0.1256 \text{ mS}$$

$$G = \frac{1}{R} = \frac{1}{10 \times 10^3} = 0.1 \text{ mS}$$

$$\mathbf{Y} = 0.1 + j0.1256 \text{ mS}$$

$$\alpha = \tan^{-1} \frac{0.1256}{0.1}$$
$$= 51.47°$$

$$|Y| = \frac{0.1256}{\sin 51.47°}$$
$$= 0.1605$$

$$\mathbf{Y} = 0.1605 \underline{/51.47°} \text{ mS}$$

$$\mathbf{Z}_{eq} = \frac{1}{\mathbf{Y}} = \frac{1}{0.1605 \times 10^{-3} \underline{/51.47°}}$$
$$= 6.23 \underline{/-51.47°} \text{ k}\Omega$$

To find the series equivalent,

$$R = 6.23 \times 10^3 \times \cos -51.47°$$
$$= 3.88 \text{ k}\Omega$$

$$X_C = 6.23 \times 10^3 \times \sin -51.47°$$
$$= 4.874 \text{ k}\Omega$$

To find the series capacitance,

$$C = \frac{1}{2\pi f X_C}$$
$$= \frac{1}{6.28 \times 2 \times 10^3 \times 4.874 \times 10^3}$$
$$= 0.016 \text{ }\mu\text{F}$$

The series equivalent circuit is sketched in Fig. 21-10.

$R_S = 3.88$ kΩ $C_S = 0.016$ μF   **Figure 21-10**

21-6. PARALLEL *RL* CIRCUITS

The parallel combination of a capacitor and resistor is very common, particularly in electronic circuits. A resistor and an inductor in parallel are much less widely used. However, we can gain valuable knowledge and experience in circuit analysis by studying this type of parallel circuit.

The admittance of the circuit is found.

$$\mathbf{Y} = G - jB_L \tag{21-12}$$

where

$$G = \frac{1}{R}$$

$$B_L = \frac{1}{X_L} = \frac{1}{2\pi fL}$$

$$\mathbf{Z}_{eq} = \frac{1}{\mathbf{Y}} = \frac{1}{G - jB_L} \tag{21-13}$$

As frequency increases, B_L decreases. We find that the admittance approaches G as a limit. This means that admittance decreases with frequency, which results in *increased* impedance for the parallel circuit (Fig. 21-11). The phase angle of **Y**, α, varies from $-90°$ at those frequencies

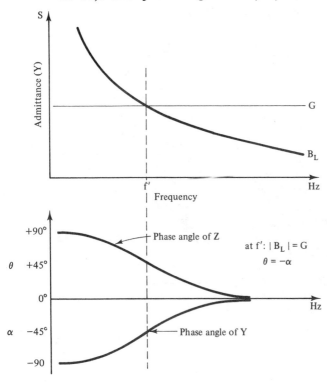

Figure 21-11 The admittance of a parallel RL circuit decreases as frequency increases.

where B_L is very much greater than G, to approximately 0° at those frequencies where B_L is very much less than G.

The phase angle of the admittance (α) is the angle whose tangent is the ratio of B_L to G.

$$\alpha = \tan^{-1} \frac{-B_L}{G} \qquad (21\text{-}14)$$

If we use resistance and inductive reactance, we find α as

$$\alpha = \tan^{-1} -\frac{R}{X_L} \qquad (21\text{-}15)$$

The phase angle of the impedance is simply the negative of the phase angle of the admittance.

$$\theta = -\alpha$$

The phase angle of the impedance (θ) varies from +90° at very low frequencies toward 0° at that range of frequencies where X_L is very much greater than R.

We use the same methods as with any parallel circuit to find series equivalent impedance.

EXAMPLE 21-9:

The circuit of Fig. 21-12 is used over a range of frequencies. (a) Calculate the susceptance of the coil at each of the listed frequencies. (b) Calculate the impedance at each frequency.

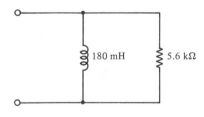

Figure 21-12

Solution:
(a) At 100 Hz:

$$B_L = \frac{1}{6.28 \times 100 \times 0.18}$$
$$= 8.85 \text{ mS}$$

At 1 kHz:

$$B_L = \frac{1}{6.28 \times 1000 \times 0.18}$$
$$= 0.885 \text{ mS}$$

Sec. 21-6. Parallel *RC* Circuits

At 5 kHz:
$$B_L = \frac{1}{6.28 \times 5 \times 10^3 \times 0.18}$$
$$= 0.177 \text{ mS}$$

At 10 kHz:
$$B_L = \frac{1}{6.28 \times 10 \times 10^3 \times 0.18}$$
$$= 0.0885 \text{ mS}$$

At 50 kHz:
$$B_L = \frac{1}{6.28 \times 50 \times 10^3 \times 0.18}$$
$$= 0.0177 \text{ mS}$$

(b) $Z = \dfrac{1}{G - jB_L} = \dfrac{1}{Y}$

$G = \dfrac{1}{R} = \dfrac{1}{5600}$
$= 0.179 \text{ mS}$

At 100 Hz:
$$Y = 0.179 - j8.85 \text{ mS}$$
$$= 8.852\underline{/-88.8°} \text{ mS}$$
$$Z_{eq} = \frac{1}{8.852 \times 10^{-3}\underline{/-88.8°}}$$
$$= 113\underline{/88.8°} \text{ Ω}$$

At 1 kHz:
$$Y = 0.179 - j0.885 \text{ mS}$$
$$= 0.903\underline{/-78.6°} \text{ mS}$$
$$Z_{eq} = \frac{1}{0.903 \times 10^{-3}\underline{/-78.6°}}$$
$$= 1107\underline{/78.6°} \text{ Ω}$$

At 5 kHz:
$$Y = 0.179 - j0.177 \text{ mS}$$
$$= 0.252\underline{/-44.7°} \text{ mS}$$
$$Z_{eq} = \frac{1}{0.252 \times 10^{-3}\underline{/-44.7°}}$$
$$= 3.97\underline{/44.7°} \text{ kΩ}$$

At 10 kHz:
$$Y = 0.179 - j0.0885 \text{ mS}$$
$$= 0.2\underline{/-26.3°} \text{ mS}$$

$$Z_{eq} = \frac{1}{0.2 \times 10^{-3}/-26.3°}$$
$$= 5/26.3° \text{ k}\Omega$$

At 50 kHz:

$$Y = 0.179 - j0.0177 \text{ mS}$$
$$= 0.18/-5.65° \text{ mS}$$

$$Z_{eq} = \frac{1}{0.18 \times 10^{-3}/-5.65°}$$
$$= 5.56/5.65° \text{ k}\Omega$$

The reader should note that Z_{eq} at 100 Hz was nearly the same as the inductive reactance of the coil. Similarly, at 50 kHz, Z_{eq} was approximately equal to the resistance.

21-7. PARALLEL *RLC* CIRCUITS

This circuit arrangement is presented now in order to complete our analysis of pure branch parallel circuits. The practical importance of this circuit will be seen in our studies of resonant circuits.

The admittance of the circuit is readily found.

$$Y = G + j(B_C - B_L) \qquad (21\text{-}16)$$

For those frequencies where $B_L > B_C$, the circuit appears inductive. The admittance has a lagging phase angle. Parallel *RLC* circuits normally appear inductive at low frequencies. Clearly, as frequency is increased we reach a frequency where $B_C > B_L$, which makes the circuit capacitive, and the admittance has a leading phase angle.

It is important to recall that in a parallel circuit the largest branch current determines the nature of the circuit. At low frequencies, $|X_L| < |X_C|$; therefore $|I_L| > |I_C|$. At high frequencies, $|X_C| < |X_L|$ and $|I_C| > |I_L|$.

Equivalent impedance is readily found from the admittance.

$$Z_{eq} = \frac{1}{Y} = \frac{1}{G + j(B_C - B_L)} \qquad (21\text{-}17)$$

We shall reserve example problems of this type of circuit for our studies of resonant circuits.

21-8. CIRCUIT ANALYSIS

In this section we use the principles of parallel ac circuits in some practical circuit applications. We shall also review scientific calculator procedures.

We begin with an unusual problem. We wish to find the *parallel equivalent* of a series circuit. This is particularly useful in working with high Q_0

coils, where it is useful to represent the resistance of the coil by some parallel resistance.

We start with the impedance of the series circuit in *polar* form. We refer to this impedance as $\mathbf{Z}_s$.

1. Take the reciprocal of $\mathbf{Z}_s$ to obtain $\mathbf{Y}_s$. If the parallel circuit is to be equivalent to the series circuit, it must have the same admittance.

$$\mathbf{Y}_s = \mathbf{Y}_p$$

2. Change $\mathbf{Y}_p$ to rectangular form.

$$G = \mathbf{Y}_p \cos \alpha$$
$$B = \mathbf{Y}_p \sin \alpha$$

3. Take the reciprocals of G and B to obtain the parallel components of the equivalent circuit.

$$R = \frac{1}{G}$$
$$X = \frac{1}{B}$$

EXAMPLE 21-10:
Given the circuit of Fig. 21-13, at the operating frequency the reactance of the coil is 1500 Ω and total series resistance is 50 Ω. Solve for the parallel equivalent of the series circuit.

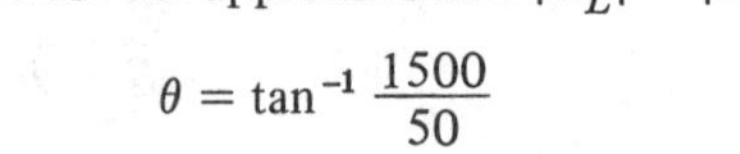

$$X_{L_S} = 1.5\ k\Omega \qquad R_S = 50\ \Omega$$

Figure 21-13

Solution:

$$\mathbf{Z}_s = 50 + j1500\ \Omega$$

Because $|X_L| = 30R$, we use the approximation $|X_L| = |Z|$. Solve for θ.

$$\theta = \tan^{-1} \frac{1500}{50}$$
$$= 88.1°$$

Calculator operations:

1500 ÷ 50 = inv tan 88.0908 . . .

$$\mathbf{Z}_p = \mathbf{Z}_s = 1500 \underline{/88.1°} \qquad \text{(approximation)}$$

$$\mathbf{Y}_p = \frac{1}{\mathbf{Z}_p} = \frac{1}{1500\underline{/88.1°}}$$
$$= 0.667\underline{/-88.1°}\ \text{mS}$$

$$G_p = 0.667 \times 10^{-3} \times \cos -88.1°$$
$$= 0.0221\ \text{mS}$$

$$B_L = 0.667 \times 10^{-3} \times \sin -88.1°$$
$$= 0.6666 \text{ mS}$$

$$R_p = \frac{1}{G_p} = \frac{1}{0.0221 \times 10^{-3}}$$
$$= 45.25 \text{ k}\Omega$$

$$X_{Lp} = \frac{1}{B_L} = \frac{1}{0.6666 \times 10^{-3}}$$
$$= 1500 \text{ }\Omega$$

Note that in the series circuit X_L was 30 times larger than R. In the parallel equivalent circuit we find that R_p is 30 times greater than X_L. In this particular example, X_L did not change because of the very large ratio of X_L to R in the series circuit.

In a circuit where X_L and R are comparable values, both R_p and X_{Lp} would be appreciably different from the series values. In a practical sense, we would have no reason for replacing this kind of series circuit with a parallel equivalent circuit.

Another area of great interest is in the use of capacitors to effectively short out a resistor so far as ac currents are concerned. This is called *bypassing*. Bypassing is required in many electronic circuits where a resistor is used in a dc voltage-divider application; yet for ac signals we require that negligible amounts of signal voltage be developed across the resistor. In this application the bypass capacitor is selected so that its reactance is very small in comparison to the *proper* resistance (Fig. 21-14).

$$X_C \leq 0.1R$$

The selection of R depends upon the resistance values in the circuit. As shown in Fig. 21-15, R must be the lesser resistance. It may be the resistor that is bypassed, or it may be that resistance that is in series with the RC combination. We simply select the smaller amount of resistance for R in our calculation.

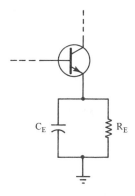

Figure 21-14 An example of bypassing. The circuit is part of a transistor amplifier circuit.

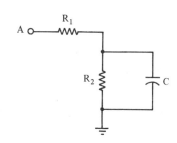

Figure 21-15

Sec. 21-7. Parallel *RLC* Circuits

$$C \geqslant \frac{10}{2\pi fR} \qquad (21\text{-}18)$$

EXAMPLE 21-11:
In the circuit of Fig. 21-15, R_2 is part of a dc voltage-divider system. (a) Given that $R_1 = 12$ kΩ and $R_2 = 2.2$ kΩ, solve for the bypass capacitor needed if the lowest signal frequency is 50 Hz, (b) Change R_1 to 1.2 kΩ and repeat the problem.

Solution:
(a) $R_2 < R_1$; therefore,

$$C \geqslant \frac{10}{2\pi fR_2}$$
$$\geqslant \frac{10}{6.28 \times 50 \times 2.2 \times 10^3}$$
$$\geqslant 14.5 \ \mu F$$

In this instance $X_C \leqslant 220 \ \Omega$ at 50 Hz. With a 50-Hz voltage applied at terminal A, the total impedance is $12\underline{/0°}$ kΩ ($\mathbf{Z} = 12{,}000 - j220$). This results in a very small voltage across the capacitor at the signal frequency. For all higher-frequency signals, the bypass is even more effective. Using the VDR,

$$V_C = V_{in} \frac{220}{12{,}000} = 0.0183(V_{in})$$

(b) $R_1 < R_2$; therefore,

$$C \geqslant \frac{10}{2\pi fR_1}$$
$$\geqslant \frac{10}{6.28 \times 50 \times 1.2 \times 10^3}$$
$$\geqslant 26.5 \ \mu F$$

In this instance $X_C \leqslant 120 \ \Omega$ at 50 Hz. With a signal of this frequency applied at A, total impedance is approximately $1200\underline{/0°}$ ($\mathbf{Z} = 1200 - j120$). By use of the VDR,

$$V_C = V_{in} \frac{120}{1200} = 0.1(V_{in})$$

Note that in this case the bypass is less effective than in part (a). This is because the series resistance is only 10 times the capacitive reactance. Of course, at higher signal frequencies, the reactance decreases, and even more effective bypassing takes place.

The following examples illustrate the application of the principles of parallel circuits in the solution of problems.

EXAMPLE 21-12:
Given the circuit of Fig. 21-16, calculate total impedance, each branch current, and total current. Sketch a phasor diagram of currents and voltage.

$$\mathbf{E} = 25\underline{/0°} \text{ V at 100 kHz}$$

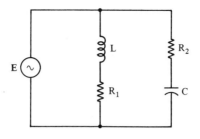

Figure 21-16

$$R_1 = 150\ \Omega, \quad R_2 = 300\ \Omega$$
$$L = 318\ \mu H, \quad C = 0.01\ \mu F$$

Solution:

1. Solve for each reactance.

$$X_L = 2\pi f L$$
= 6.28 [EE] 5 [×] 318 [EE] 6 [+/−] [=]
= 200 Ω

$$X_C = \frac{1}{2\pi f C}$$
= 6.28 [EE] 5 [×] 0.01 [EE] 6 [+/−] [=] [1/x]
= 159 Ω

2. Solve for Z_1 and Z_2.

$\mathbf{Z_1} = 150 + j200$
$\quad \theta$: 200 [÷] 150 [=] [inv] [tan] ... 53.13°
$\quad |Z_1|$: 150 ÷ 53.13 [cos] [=] ... 249.99

$\mathbf{Z_1} = 250\underline{/53.13°}\ \Omega$

$\mathbf{Z_2} = 300 - j159$
$\quad \theta$: 159 [+/−] [÷] 300 [=] [inv] [tan] ... −27.92°
$\quad |Z_2|$: 159 [+/−] [÷] 27.92 [+/−] [sin] [=] 339.57 ...

$\mathbf{Z_2} = 340\underline{/-27.92°}\ \Omega$

3. Solve for each branch current.

$$\mathbf{I_1} = \frac{\mathbf{E}}{\mathbf{Z_1}} = \frac{25\underline{/0°}}{250\underline{/53.13°}}$$
$\quad = 0.1\underline{/-53.13°}$ A

Polar → rectangular:
0.1 [×] 53.13 [+/−] [cos] [=] ... 0.06
0.1 [×] 53.13 [+/−] [sin] [=] ... −0.08

$\mathbf{I_1} = 0.06 - j0.08$ A

Sec. 21-7.   Parallel *RLC* Circuits

$$I_2 = \frac{E}{Z_2} = \frac{25/0°}{340/-27.92°}$$
$$= 0.0735/27.92° \text{ A}$$

Polar → rectangular:

0.0735 [X] 27.92 [cos] [=] ... 0.065

0.0735 [X] 27.92 [sin] [=] ... 0.0344

$I_2 = 0.065 + j0.0344$ A

4. The total current is the sum of the branch currents.

$$\begin{aligned} I_T = I_1 + I_2 \\ = 0.06 - j0.08 \\ \underline{0.065 + j0.0344} \\ I_T = 0.125 - j0.0456 \end{aligned}$$

Rectangular → polar:

θ: 0.0456 [+/−] [÷] 0.125 [=] [inv] [tan] ... −20.04°

$|I_T|$: 0.125 [÷] 20.04 [+/−] [cos] [=] ... 0.13305

$I_T = 0.133/-20°$ A

5. $Z_T = \dfrac{E}{I_T} = \dfrac{25/0°}{0.133/-20°}$
$= 188/20°$ Ω

6. The phasor diagram of currents and applied voltage is given in Fig. 21-17.

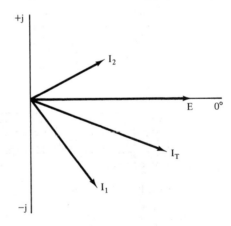

Figure 21-17

EXAMPLE 21-13:

Total current in the circuit of Fig. 21-18 is $150/-30°$ mA. The applied voltage of $30/0°$ V is at a frequency of 200 kHz. Solve for the impedance, Z_2.

Solution:
We solve for the current I_2 by subtracting I_1 from I_T. Once I_2 is known, we use Ohm's law to find Z_2.

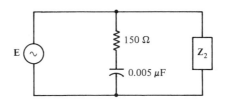

Figure 21-18

1. Change I_T from polar to rectangular.

   150 [X] 30 [+/−] [cos] [=] 129.9 ...
   150 [X] 30 [+/−] [sin] [=] −75 ...
   $I_T = 130 - j75$

2. Solve for I_1.
   (a) Solve for X_C.

   $X_C = 6.28$ [X] 2 [EE] 5 [X] 0.005
   [EE] 6 [+/−] [=] [1/x] ... 159.23 ...

   (b) Solve for Z_1.

   $Z_1 = 150 - j159$
   θ: 159 [+/−] [÷] 150 [=] [inv] [tan] −46.668 ...
   $|Z_1|$: 159 [÷] 46.7 [+/−] [cos] [=] ... 218.71 ...
   $Z_1 = 219\underline{/-46.7°}\ \Omega$

   (c) $I_1 = \dfrac{E}{Z_1} = \dfrac{30\underline{/0°}}{219\underline{/-46.7°}}$
   $= 137\underline{/46.7°}$ mA
   $= 94 + j100$

3. Solve for I_2.

   $I_2 = I_T - I_1$
   $ = 150 - j75$ mA
   $ -(94 + j100)$ mA
   $ = 56 - j175$ mA
   $I_2 = 184\underline{/-72.26°}$ mA

4. Solve for Z_2.

   $Z_2 = \dfrac{E}{I_2} = \dfrac{30\underline{/0°}}{184\underline{/-72.46°}}$
   $= 163\underline{/72.26°}\ \Omega$

5. Change Z_2 to rectangular form.

   $Z_2 = 50 + j155\ \Omega$

Sec. 21-7. Parallel *RLC* Circuits

6. The solution shows that this branch contains 50 Ω of resistance and 155 Ω of inductive reactance. We now find the inductance required for $X_L = 155$ Ω.

$$L = \frac{X_L}{2\pi f} = \frac{155}{6.28 \times 2 \times 10^5}$$
$$= 123 \; \mu H$$

21-9. THE Q OF A PARALLEL CIRCUIT

We discuss briefly some additional concepts on circuit Q. The definition of Q that we developed in series circuits is equally accurate for parallel circuits. Q is always the ratio of energy stored to energy dissipated per cycle.

Parallel *RC* Circuit

$$Q = \frac{R_p}{X_C} = 2\pi f C R_p$$
$$Q = \omega R_p C \tag{21-19}$$

Parallel *RL* Circuit

$$Q = \frac{R_p}{X_L} = \frac{R_p}{2\pi f L}$$
$$Q = \frac{R_p}{\omega L} \tag{21-20}$$

Note that in a series circuit we found Q as the ratio of reactance to resistance. In a parallel circuit we find that Q is the ratio of resistance to reactance.

Series Circuit

$$Q = \frac{X}{R_s}$$

Parallel Circuit

$$Q = \frac{R_p}{X}$$

We should note that these ratios equal the *magnitude of the tangent of the phase angle of the impedance*. We calculate the Q of any circuit whose impedance is known by

$$Q = |\tan \theta| \tag{21-21}$$

21-10. SERIES–PARALLEL CIRCUITS

The solution of problems in series–parallel ac circuits is done in the same way that we solved series–parallel dc circuits.

1. Reduce all parallel circuits to series equivalent impedances. We calculate these impedances in rectangular and polar forms, for we shall work with both forms.
2. The total impedance of the circuit is the phasor sum of all the impedances. This sum must be obtained by using the rectangular form of the impedances. The final result is then changed from rectangular to polar form.
3. Total current is found by Ohm's law.

$$I_T = \frac{E}{Z_T}$$

4. Each voltage drop is calculated using Ohm's law.

$$V_1 = I_T Z_1$$
$$V_2 = I_T Z_2$$
$$V_n = I_T Z_n$$

5. Individual branch currents are found by dividing parallel circuit voltage drops by branch impedances.

SUMMARY

1. The rules of parallel circuits are the same as for dc circuits except for effects caused by phasor quantities and reactances.
2. The total impedance of a parallel circuit is defined as *equivalent* impedance (Z_{eq}).
3. For a two-branch parallel circuit, Z_{eq} is found by the product-over-sum formula.
4. The formula for Z_{eq} of a many-branched parallel circuit is of the same general form as for pure resistance circuits.
5. Susceptance is the reciprocal of reactance. The unit is in siemens (S).
6. Capacitive susceptance (B_C) increases linearly with frequency and capacitance.
7. Inductive susceptance (B_L) is inversely proportional to frequency and inductance.
8. Admittance (**Y**) is the reciprocal of impedance.
9. The phase angle of an impedance is the negative of the phase angle of its admittance.
10. Capacitors are very widely used in electronic circuits as coupling and bypass devices.
11. The tangent of the phase angle of an impedance is equal to the Q of the circuit.
12. We analyze series–parallel ac circuits as in dc, but with all of the necessary mathematics relating to phasor quantities.

PROBLEMS

21-1. A two-branch parallel circuit consists of $Z_1 = 4250\underline{/-48°}$ Ω and $Z_2 = 1590\underline{/-60°}$ Ω. Solve for Z_{eq}.

21-2. A two-branch parallel circuit has $Z_1 = 4200\underline{/15°}$ Ω and $Z_2 = 3000\underline{/30°}$ Ω. Solve for Z_{eq}.

21-3. Given the circuit of Fig. 21-19, (a) solve for Z_{eq}; (b) solve for each branch current and total current, given that $E = 36\underline{/0°}$ V; (c) sketch the phasor diagram of currents, using E as the reference.

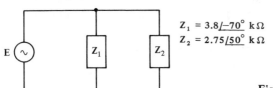

$Z_1 = 3.8\underline{/-70°}$ kΩ
$Z_2 = 2.75\underline{/50°}$ kΩ

Figure 21-19

21-4. In the circuit of Fig. 21-19, solve for E and I_2, given that $I_1 = 15.8\underline{/70°}$ mA.

21-5. The equivalent impedance of a two-branched parallel circuit is $682\underline{/40.2°}$ Ω. Given that $Z_1 = 1.1\underline{/45°}$ kΩ and $I_T = 29.3\underline{/-25.2°}$ mA, (a) solve for the voltage drop across the circuit; (b) solve for Z_2.

21-6. The circuit of Figure 21-20 is operated over a range of frequencies. Solve for Z_{eq} at each of the following frequencies: 15 Hz, 40 Hz, 500 Hz, 8 kHz, 50 kHz.

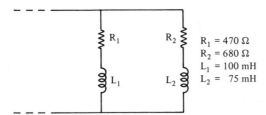

$R_1 = 470$ Ω
$R_2 = 680$ Ω
$L_1 = 100$ mH
$L_2 = 75$ mH

Figure 21-20

21-7. Calculate the impedance of the circuit of Fig. 21-21 at each of the listed frequencies: 15 Hz, 75 Hz, 1 kHz, 12 kHz, 100 kHz.

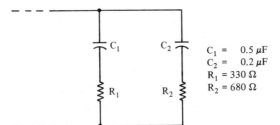

$C_1 = 0.5$ μF
$C_2 = 0.2$ μF
$R_1 = 330$ Ω
$R_2 = 680$ Ω

Figure 21-21

21-8. In the circuit of Fig. 21-22, solve for each branch current and total current.

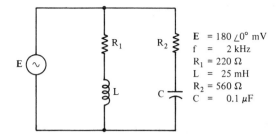

E = 180 ∠0° mV
f = 2 kHz
R_1 = 220 Ω
L = 25 mH
R_2 = 560 Ω
C = 0.1 μF

Figure 21-22

21-9. In the circuit of Fig. 21-23, $Z_1 = 150 - j300$, $Z_2 = 75 + j400$, and $Z_3 = 150 + j150$ Ω. Solve for Z_{eq}.

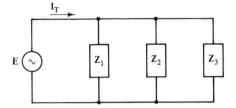

Figure 21-23

21-10. Given the data of Prob. 21.9, solve for E if $I_T = 89.6 \underline{/-30.9°}$ mA.

21-11. Solve for the susceptance of a 75-μH coil at each of the listed frequencies: 1 kHz, 15 kHz, 200 kHz, 3 MHz.

21-12. Solve for the susceptance of a 5-μF capacitor at each of the listed frequencies: 30 Hz, 400 Hz, 5 kHz, 75 kHz.

21-13. A 15-kΩ resistor is in parallel with a capacitive reactance of 9.2 kΩ. Solve for total admittance.

21-14. The admittance of a two-branch parallel circuit is $807\underline{/-34.3°}$ μS. (a) Solve for the parallel circuit. (b) Solve for the series equivalent of the parallel circuit.

21-15. Given the circuit of Fig. 21-24, solve for Y and circuit Q, given that frequency is 75 kHz.

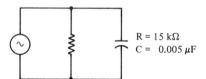

R = 15 kΩ
C = 0.005 μF

Figure 21-24

21-16. Given the data of Prob. 21.15, solve for the series equivalent impedance.

21-17. In the circuit of Fig. 21-24, what is the operating frequency if Q = 50?

21-18. In the circuit of Fig. 21-25, solve for circuit Q, given that operating frequency is 100 kHz.

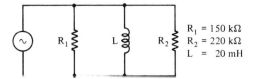

R_1 = 150 kΩ
R_2 = 220 kΩ
L = 20 mH

Figure 21-25

21-19. Solve for the admittance of each branch of the circuit of Fig. 21-26.

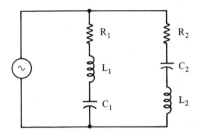

Figure 21-26

21-20. Given the data of Prob. 21-19, solve for **E** given that $I_1 = 30\underline{/63°}$.mA

21-21. Given the circuit of Fig. 21-26, if the frequency is changed to 1 MHz, solve for the total admittance.

21-22. In the circuit of Fig. 21-27, point *a* must be at signal ground for frequencies from 30 Hz to 20 kHz. Given that $R_1 = 100\ \Omega$ and $R_2 = 400\ \Omega$, solve for C.

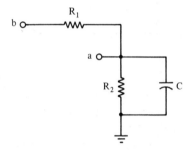

Figure 21-27

21-23. Given the data of Prob. 21-22, what is the approximate impedance from point *b* to ground over the range of frequencies?

21-24. In the circuit of Fig. 21-27, $R_1 = 470\ \Omega$ and $R_2 = 56\ \Omega$. Repeat Prob. 21-22.

21-25. Calculate the total impedance of the circuit of Fig. 21-28.

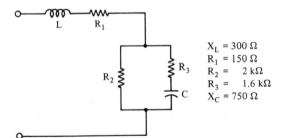

$X_L = 300\ \Omega$
$R_1 = 150\ \Omega$
$R_2 = 2\ k\Omega$
$R_3 = 1.6\ k\Omega$
$X_C = 750\ \Omega$

Figure 21-28

22

Power in AC Circuits

22-1. INTRODUCTION

Power is the rate at which work is done. We have learned that resistance is the only circuit element that can dissipate electrical power. We have also learned that inductance and capacitance have energy-storage properties. They do not dissipate power. Instead, capacitance and inductance exchange energy with a source. Energy taken from a source is returned to the source. Whatever losses there are in capacitors and inductors are due to the small amounts of resistance associated with these devices.

In alternating-current circuits we are concerned with the effects of reactance upon the amount of load current supplied by a source. Only some portion of load current develops power. However, *all* the load current must be supplied by a source. The current supplied by the source produces losses in the source and in all the connecting wires and power lines (Fig. 22-1). These

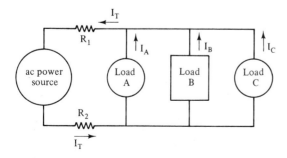

Figure 22-1 *All* sources of electrical energy have internal resistance. In addition, the lines between loads and source have resistance.

427

losses are unimportant in a small circuit. But consider the power losses in an electrical utility system as a result of reactive loads. These losses can be enormous. We therefore have to study *three* forms of "power," only one being "true" power; the others are simply means of rating systems and components in ac circuits, in terms of loading and dissipation.

22-2. APPARENT POWER

Apparent power is simply the product of source voltage and current, as shown in Fig. 22-2. It is the amount of power that would be dissipated if the load were purely resistive. The unit of apparent power is the *volt-ampere* (VA) and, very frequently, the kilovolt-ampere (kVA). In this text we shall use P_A as the symbol for apparent power.

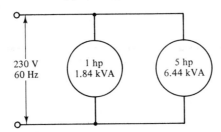

Figure 22-2 Apparent power is simply the product of source current and voltage. It is also the sum of the kVA ratings of each load (assuming that the loads are operated at their rated values).

We make use of the apparent power rating of ac equipment as a means of determining current drawn by the equipment.

$$P_A = V_{rms} I_{rms} \qquad (22\text{-}1)$$

$$I_{rms} = \frac{P_A}{V_{rms}} \qquad (22\text{-}2)$$

EXAMPLE 22-1:
An ac motor is rated at 11.5 kVA at a line voltage of 230 V, 60 Hz. What is the full-load current drawn by the motor?

Solution:

$$I_{rms} = \frac{P_A}{V_{rms}} = \frac{kVA \times 10^3}{V_{rms}}$$
$$= \frac{11{,}500}{230}$$
$$= 50 \text{ A}$$

EXAMPLE 22-2:
The rated power output of the motor of Example 22-1 is 10 hp. (a) Calculate the amount of power output in watts. (b) Calculate the current required to develop 10 hp in a purely resistive load.

Solution:

(a) 1 hp = 746 W

   10 hp = 7460 W

(b) $P = V_{rms} I_{rms}$

$$I_{rms} = \frac{P}{V} = \frac{7460}{230}$$
$$= 32.43 \text{ A}$$

Note that the motor actually draws 50 A under full load. However, only 32.43 A is used to develop useful power. Motors are highly inductive devices. These two examples illustrate one of the differences between apparent power and true power.

22-3. TRUE POWER

True power is often referred to as *average power*. It is the power dissipated by a circuit, the power that we are used to working with in dc circuits. The unit of power is the watt (W), and we shall continue to use the symbol P for power. We have several formulas for calculating power.

$$P = I_{rms}^2 R \tag{22-3}$$

$$P = \frac{V_{rms}^2}{R} \tag{22-4}$$

$$P = V_{rms} I_{rms} \cos \theta \tag{22-5}$$

where θ is the phase angle of the current with respect to the voltage.

We shall explain the meaning of the factor, $\cos \theta$, in the next section. This quantity is called the *power factor* of the circuit.

22-4. POWER FACTOR

The power factor enables us to relate apparent power to true power. The power factor of a circuit is the ratio of power to volt-amperes.

$$\text{Power factor} = \frac{\text{power}}{\text{volt-amperes}} \tag{22-6a}$$

$$= \frac{V_{rms} I_{rms} \cos \theta}{V_{rms} I_{rms}}$$

$$= \cos \theta \tag{22-6b}$$

We shall use the following examples to demonstrate the relationship between apparent power and true power.

EXAMPLE 22-3:

The impedance of a circuit is $150/30°$ Ω. The applied voltage is $230/0°$ V. (a) Calculate the current supplied by the source. (b) Calculate the portion of the current that is in phase with line voltage. *This is the component of current that will dissipate power.* (c) Calculate the portion of current that lags line voltage by 90°. *It is this portion of current that is used for energy exchange between source and load.* We refer to this current and line voltage product as reactive volt-amperes or reactive power. The unit is the var or kvar.

Solution:

(a) $\quad I = \dfrac{E}{Z}$

$$I_{rms} = \dfrac{230/0°}{150/30°}$$
$$= 1.533/\underline{-30°} \text{ A}$$

(b) The current in phase with line voltage is the product of line current and the cosine of the phase angle.

$$I' = I_{rms} \cos \theta$$
$$= 1.533 \times \cos -30°$$
$$= 1.33 \text{ A}$$

(c) The current that lags line voltage by 90° is the product of line current and the sine of the phase angle.

$$I'' = I_{rms} \sin \theta$$
$$= 1.533 \times \sin -30°$$
$$= -j0.767 \text{ A}$$

Voltage and current relations are sketched in Fig. 22-3.

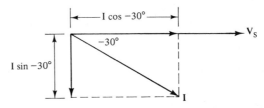

Figure 22-3 Voltage and current relations in the circuit of Example 22-3.

EXAMPLE 22-4:

Given the results of Example 22-3, calculate (a) apparent power, (b) power, and (c) reactive power.

Solution:

(a) apparent power $= E_{rms} I_{rms}$
$$= 230 \times 1.533$$
$$= 352.6 \text{ VA}$$

(b) power $= E_{rms} I_{rms} \cos \theta$
$$= 230 \times 1.533 \times \cos 30°$$
$$= 305.3 \text{ W}$$

(c) reactive power = $E_{rms}I_{rms} \sin \theta$
= 230 × 1.533 × sin 30°
= 176.3 vars

EXAMPLE 22-5:
The circuit of Example 22-3 is drawn in Fig. 22-4. Demonstrate that $P = I_{rms}^2 R$ equals the true power calculated in Example 22-4.

$$P = 1.533^2 \times 130$$
$$= 305.5 \text{ W}$$

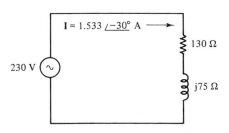

Figure 22-4

We can also demonstrate that apparent power is current squared times impedance, and that reactive power is current squared times reactance.

$$P_A = I_{rms}^2 Z \quad \text{(VA)} \tag{22-7}$$

$$P_{reac} = I_{rms}^2 X_L \quad \text{(vars)} \tag{22-8}$$

Figure 22-5 illustrates true, apparent, and reactive power relationships.

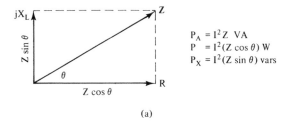

(a)

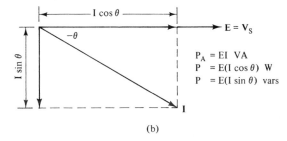

(b)

Figure 22-5 The power triangles.

Sec. 22-4. Power Factor

We have used a series circuit as the model for our discussions. All our work is equally accurate for parallel circuits. All formulas apply. We must recall, however, that the phase angle of the impedance of a parallel circuit is the angle whose tangent is R_p/X_p.

22-5. POWER FACTOR CORRECTION

Loads with leading or lagging power factors cause the waste of electrical power in sources and connecting lines. We can minimize this waste of power by placing reactances across a load so that the reactive currents of the load and compensating reactance cancel each other.

Typically, industrial electrical loads are inductive, causing a lagging power factor. There are commercially available capacitors especially designed to compensate for the inductive portion of the load. The actual current *in the load* is not affected by the compensating capacitor. However, the current in the source and in the connecting lines is *reduced*, which results in lowered line and source power losses. This results in higher efficiency.

Suppose that the inductive portion of current in a load is 27 A. We select a compensating capacitor whose reactance is such that the current it draws is 27 A. These currents are 180° out of phase and cancel each other in the lines. As far as the source and the lines are concerned, the load is resistive.

The selection of a capacitor is based upon our discussion of currents.

1. The power factor ($\cos \theta$) is known.
2. Let $I_{L(X)}$ = the reactive portion of the current drawn by the load.

$$I_{L(X)} = I_L \sin \theta$$

3. Select a capacitor so that the current drawn by it is equal to the magnitude of the current, $I_{L(X)}$.

$$|I_C| = |I_{L(X)}|$$

$$I_C = \frac{V_L}{X_C}$$

$$C = \frac{I_{L(X)}}{2\pi f V_L}$$

$$= \frac{I_L \sin \theta}{2\pi f V_L} \qquad (22\text{-}9)$$

EXAMPLE 22-6:
A single-phase system, operating at 230 V, 60 Hz, works into a load of 69 kVA, with a lagging power factor of 0.75. (a) Calculate line current. (b) Calculate the portion of line current (I_{LX}) used to supply reactive power (kvars). (c) Calculate the portion of line current (I_{LW}) used to supply power (kW).

Solution:

(a) $I_L = \dfrac{kVA}{V_L} \times 10^3$

$= \dfrac{69 \times 10^3}{230}$

$= 300$ A

(b) $I_{L(X)} = I_L \sin \theta$

$\theta = \cos^{-1} 0.75$
$= 41.41°$

$I_{L(X)} = 300 \times \sin 41.41°$
$= 198$ A

(c) $I_{L(W)} = I_L \cos 41.41°$
$= I_L \times 0.75$
$= 300 \times 0.75$
$= 225$ A

EXAMPLE 22-7:

Given the circuit of Example 22-6, (a) calculate the capacitance needed to make the power factor of the circuit 1 (unity power factor); (b) demonstrate that the line current is equal to 225 A.

Solution:

(a) $C = \dfrac{I_L \sin \theta}{2\pi f V_L}$

$= \dfrac{300 \times \sin 41.41°}{6.28 \times 60 \times 230}$

$= 2285 \; \mu F$

This rather large amount of capacitance would be supplied by a bank of capacitors. Despite the initial cost, the power saved as a result of higher efficiency would justify the expense. Also, public utilities have the right to add premiums to the kilowatt-hour cost when the power factor is significantly less than 1.

(b) $|I_C| = \dfrac{V_L}{X_C}$

$= \dfrac{230}{1/(6.28 \times 60 \times 2.285 \times 10^{-3})}$

$= 198$ A

$I_L = I_{L(W)} + j(I_C - I_{L(X)})$
$= 225 + j(198 - 198)$ A
$= 225$ A

Our discussion of power factor correction has related to lagging power factor. This is because most industrial loads cause lagging power factors. Low-power loads, as in most electronic circuits, may cause leading power factors, but the losses caused by reactive power are so low as to be unimportant.

Sec. 22-5. Power Factor Correction

SUMMARY

1. In ac circuits the product of the effective value of source voltage and current does not always equal average power. This would be true only for pure resistance circuits.
2. *Apparent power* is the product of effective source voltage and current and is measured in volt-amperes.
3. *True power* is the power dissipated in resistance and is in watts.
4. The ratio of true power to apparent power is called the *power factor* (p.f.).
5. Power factor is equal to the cosine of the phase angle of current in a load.
6. The portion of the apparent power that is used to supply energy-storage devices is called reactive power (var).
7. To minimize losses in a power source, we attempt to eliminate reactive power. This is done by power factor correction methods.
8. Because industrial loads contain inductance, practical uncorrected power factors are lagging.
9. Lagging power factors are corrected by connecting capacitors across an inductive load.

PROBLEMS

22-1. An ac motor is rated at 13.8 kVA. When operated from a 230-V line, find the full-load current of the motor.

22-2. The input voltage to a transformer is 4.8 kV. If current is 120 A, what is the apparent power input?

22-3. The motor of Prob. 22-1 has a power factor of 0.73 lagging. What is the true power developed in the motor?

22-4. An impedance is given as $100\underline{/37°}$ Ω. What is the power factor of the impedance?

22-5. The power dissipation of a 440-V ac circuit is 6 kW. If line current is 18.7 A, find (a) the volt-amperes of the circuit, and (b) the power factor.

22-6. Given the circuit of Fig. 22-6: (a) Solve for the power factor. (b) If 220 V is applied, solve for the reactive volt-amperes (vars). (c) What is the true power dissipation? (d) What is the apparent power?

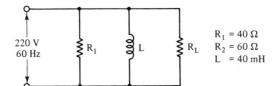

Figure 22-6

22-7. In the circuit of Fig. 22-7, current is 100 mA. (a) Solve for the true power. (b) Solve for the power factor. (c) Solve for the apparent power. (d) Solve for the line voltage, **E**.

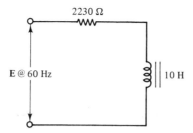

Figure 22-7

22-8. A 230 V/60 Hz motor is rated at 25 kVA, with a power factor of 0.78 lagging. Solve for the capacitance needed across the line to cause unity power factor.

22-9. A single-phase power line at a manufacturing plant is found to have a 0.6 lagging power factor when the total load is 100 kVA. Line voltage is 440 V/60 Hz. What is the amount of capacitance needed to bring the power factor to approximately 1?

22-10. Given the data of Prob. 22-9: (a) What is the reactive power (kvars) prior to power factor correction? (b) What is the line current prior to power factor correction? (c) What is the line current after power factor correction?

23

Resonant Circuits

23-1. INTRODUCTION

Resonance can occur in any physical system, provided that the system has opposite types of energy-storage devices. Whenever the energy stored in these devices is exchanged between them so that added energy is not taken from a source, a *resonant* condition exists.

We have all experienced this condition when driving a car with an unbalanced front tire. At some specific speed, the vibrations of the unbalanced wheel are at a rate equal to the natural resonant frequency of the front end suspension. Energy stored in the springs and mass of the system is exchanged between them, which results in a large amount of vibration and movement of the steering wheel that is certainly evident to the driver!

In this chapter we examine the properties and some applications of resonance in electric circuits. Resonance in electric circuits requires that reactive quantities must cancel each other. In a series RLC circuit, this requires that reactive voltage drops cancel. In a parallel RLC circuit, it is reactive currents that cancel.

23-2. SERIES RESONANCE

The impedance of a series circuit is the phasor sum of resistance and reactance. In a series RLC circuit, there is *one* frequency where the inductive and capacitive reactances are equal. At this frequency, called the *resonant frequency*, the reactances cancel, resulting in zero net reactance (Fig. 23-1).

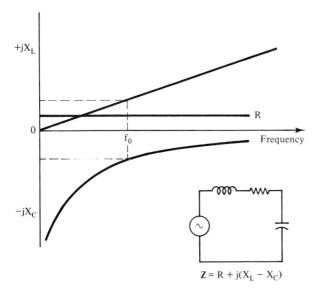

Figure 23-1 Series RLC circuit. The curve of capacitive reactance is drawn below the horizontal axis, simply to remind that X_C causes $-90°$ of lag for voltage drops.

At the resonant frequency (f_0), $|X_L| = |X_C|$.

$$2\pi f_0 L = \frac{1}{2\pi f_0 C}$$

Solving for f_0,

$$f_0 = \frac{1}{2\pi\sqrt{LC}} \qquad (23\text{-}1)$$

and

$$\omega_0 = \frac{1}{\sqrt{LC}} \qquad (23\text{-}2)$$

Frequently, we must determine the inductance or capacitance needed to cause resonance. If inductance and resonant frequency are known, we wish to determine the capacitance needed to cause resonance, we use the following:

$$C = \frac{1}{\omega_0^2 L} \qquad (23\text{-}3)$$

If C and f_0 are known, and a solution for L is required,

$$L = \frac{1}{\omega_0^2 C} \qquad (23\text{-}4)$$

EXAMPLE 23-1:
Given the circuit of Fig. 23-2, solve for the resonant frequency for each set of L and C, if (a) $C = 47$ pF, $L = 22$ μH, and (b) $C = 0.01$ μF, $L = 40$ mH.

Sec. 23-2. Series Resonance

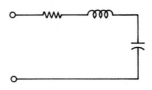

Figure 23-2

Solution: $f_0 = 1/2\pi\sqrt{LC}$.

(a) $f_0 = \dfrac{1}{6.28\sqrt{22 \times 10^{-6} \times 47 \times 10^{-12}}}$
$= 4.952 \times 10^6$
$= 4.952$ MHz

(b) $f_0 = \dfrac{1}{6.28\sqrt{40 \times 10^{-3} \times 0.01 \times 10^{-6}}}$
$= 7.962 \times 10^3$
$= 7.962$ kHz

EXAMPLE 23-2:

A series circuit must resonate at 9.3 MHz. The capacitance is 56 pF. What value of inductance is required?

Solution:

$$L = \dfrac{1}{\omega_0^2 C}$$

$\omega_0 = 6.28 \times 9.3 \times 10^6$
$= 58.4 \times 10^6$

$\omega_0^2 = 3411 \times 10^{12}$

$L = \dfrac{1}{3.411 \times 10^{15} \times 56 \times 10^{-12}}$
$= \dfrac{1}{191 \times 10^3}$
$= 5.235\ \mu H$

EXAMPLE 23-3:

A series resonant circuit must tune a range of frequencies from 4.1 to 5.8 MHz. The coil to be used in the circuit has an inductance of 25 μH. Specify the maximum and minimum values of capacitance.

Solution:

1. C_{min} is required at the highest frequency.

$C_{min} = \dfrac{1}{\omega_0^2 L}$
$= \dfrac{1}{(6.28 \times 5.8 \times 10^6)^2 \times 25 \times 10^{-6}}$
$= \dfrac{1}{3.317} \times 10^{10}$
$= 30.15$ pF

2. C_{max} is required at the lowest operating frequency.

$$C_{max} = \frac{1}{\omega_0^2 L}$$
$$= \frac{1}{(6.28 \times 4.1 \times 10^6)^2 \times 25 \times 10^{-6}}$$
$$= \frac{1}{6.58} \times 10^{10}$$
$$= 60.33 \text{ pF}$$

Note that the frequency range is 1.414 to 1. The capacitance range is 2 to 1. This is because the resonant frequency of a circuit varies inversely with the square root of the LC product. The square root of 2 is 1.414.

Impedance Properties

The impedance of a series RLC circuit is *minimum* at resonance. At the resonant frequency the impedance is exactly equal to the total resistance of the circuit.

$$\text{At } f_0: \quad Z = R_T$$

At frequencies lower than f_0, capacitive reactance is greater than inductive reactance. Therefore, the total impedance of the circuit is the phasor sum of resistance and *net* reactance. Clearly, the farther the operating frequency is below resonant frequency, the greater the amount of net reactance and the greater the total impedance. Also, the phase angle of the impedance, which is lagging, increases. Simply put, series RLC circuits are capacitive for frequencies less than resonant frequency.

At frequencies greater than f_0, inductive reactance is greater than capacitive reactance. The impedance is then the phasor sum of resistance and net inductive reactance. With increasing frequency, above resonance, the more inductive the circuit becomes. The leading phase angle of the impedance increases, as does the total impedance. We summarize the frequency-related effects as follows:

$$f < f_0: \quad \mathbf{Z}_T = R - j(X_C - X_L)\underline{/-\theta}$$
$$\theta = \tan^{-1} \frac{X_C - X_L}{R}$$
$$f = f_0: \quad \mathbf{Z}_T = R\underline{/0°}$$
$$f > f_0: \quad \mathbf{Z}_T = R + j(X_L - X_C)\underline{/\theta}$$
$$\theta = \tan^{-1} \frac{X_L - X_C}{R}$$

Figures 23-3 and 23-4 illustrate the impedance properties of a series RLC circuit.

The variation of impedance above and below resonant frequency is not completely symmetrical, but depends upon the Q of the circuit. The higher

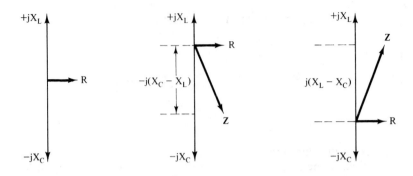

Figure 23-3 The series RLC circuit at resonance, below resonance, and above resonance. (a) $f = f_0$ (b) $f < f_0$ (c) $f > f_0$.

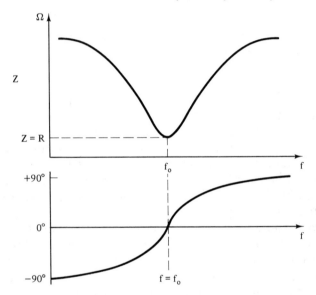

Figure 23-4 Impedance and phase angle variation with frequency, of a series RLC circuit.

the Q is, the greater the amount of symmetry. We shall review Q as it relates to resonant circuits in this chapter.

EXAMPLE 23-4:
In the circuit of Fig. 23-5, (a) solve for f_0, (b) solve for $\mathbf{Z}_T$ at f_0, (c) solve for $\mathbf{Z}_T$ at 4 kHz below resonance, and (d) solve for $\mathbf{Z}_T$ at 4 kHz above resonance.

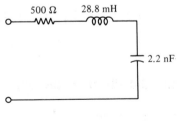

Figure 23-5

Solution:

(a) $f_0 = \dfrac{1}{2\pi\sqrt{LC}}$

$= \dfrac{1}{6.28\sqrt{28.8 \times 10^{-3} \times 2.2 \times 10^{-9}}}$

$= 20 \text{ kHz}$

(b) At $f = f_0$, $Z_T = R$:
$$Z_T = 500\underline{/0°} \; \Omega$$

(c) At $f = 16$ kHz:
$$X_C = \dfrac{1}{6.28 \times 16 \times 10^3 \times 2.2 \times 10^{-9}}$$
$$= \dfrac{1}{2.211} \times 10^{-4}$$
$$= 4524 \; \Omega$$

$X_L = 6.28 \times 16 \times 10^3 \times 28.8 \times 10^{-3}$
$\quad\; = 2894 \; \Omega$

$$Z_T = R - j(X_C - X_L)$$
$$= 500 - j(4524 - 2894)$$
$$= 500 - j1630$$

$$\theta = \tan^{-1}\dfrac{1630}{500}$$

$$\theta = -72.9°$$

$$|Z_T| = \dfrac{500}{\cos -72.9°} = 1700 \; \Omega$$

$$Z_T = 1700\underline{/-72.9°} \; \Omega$$

(d) At $f = 24$ kHz:
$$X_C = \dfrac{1}{6.28 \times 24 \times 10^3 \times 2.2 \times 10^{-9}}$$
$$= \dfrac{1}{3.316} \times 10^{-4}$$
$$= 3016 \; \Omega$$

$X_L = 6.28 \times 24 \times 10^3 \times 28.8 \times 10^{-3}$
$\quad\; = 4341 \; \Omega$

$$\mathbf{Z}_T = R + j(X_L - X_C)$$
$$= 500 + j(4341 - 3016)$$
$$= 500 + j1325$$

$$\theta = \tan^{-1}\dfrac{1325}{500}$$

$$\theta = 69.3°$$

$$|Z_T| = \frac{500}{\cos 69.3°} = 1415 \ \Omega$$

$$Z_T = 1415\underline{/69.3°} \ \Omega$$

Current

Given that impedance is minimum at the resonant frequency, then, clearly, current is maximum at resonance. At resonance current is in phase with source voltage because the impedance is purely resistance.

For frequencies below resonance, current decreases as the impedance increases. Current leads source voltage because the impedance is capacitive.

At frequencies above resonance, current lags the applied voltage because the impedance is inductive. Current decreases with increased frequency because the impedance of the circuit increases.

Voltage Drops

As with any series circuit, the phasor sum of the voltage drops is equal to the applied voltage.

$$\mathbf{E} = \mathbf{I}R + j\mathbf{I}X_L - j\mathbf{I}X_C$$

For $f = f_0$,

$$\mathbf{E} = \mathbf{I}R$$

For $f < f_0$,

$$\mathbf{E} = \mathbf{I}R - j\mathbf{I}(X_C - X_L)$$

For $f > f_0$,

$$\mathbf{E} = \mathbf{I}R + j\mathbf{I}(X_L - X_C)$$

We shall find that, at the resonant frequency, voltage drops across the reactances may be many times greater than source voltage.

23-3. CIRCUIT Q

We have previously discussed Q in terms of a quality factor applied to reactive circuits. We now consider Q in relation to resonance, in this instance, series resonant circuits.

When a circuit is operated at its resonant frequency, we find that the energy stored by capacitance is exactly equal to the energy stored by inductance. In fact, each quarter-cycle these properties *exchange* energy. During the quarter-cycle that inductance stores energy, capacitance returns energy to the circuit. On the next 90° of the cycle, inductance returns energy to the circuit as capacitance stores energy. If it were not for dissipation caused by resistance, no added energy would be required from an external source. We use Q as a measure of the ratio of energy stored to energy dissipated.

$$Q = \frac{\omega L}{R} \qquad (23\text{-}5)$$

and

$$Q = \frac{1/\omega C}{R}$$
$$= \frac{1}{\omega RC} \qquad (23\text{-}6)$$

In both of the formulas for Q, R is the *total* resistance of the circuit. This quantity includes the resistance of the coil.

Q Rise of Voltage

At resonance, the current drawn by a series circuit is limited only by the resistance of the circuit. The current is the same throughout the circuit. *We find that this maximum current produces voltage drops across the reactances that are Q times source voltage.* At resonance,

$$\mathbf{I} = \frac{E}{R}$$

$$\mathbf{V}_L = jX_L \frac{E}{R}$$
$$= j\mathbf{E}\frac{\omega L}{R}$$
$$\mathbf{V}_L = j\mathbf{E}Q \qquad (23\text{-}7)$$

$$\mathbf{V}_C = -jX_C \frac{E}{R}$$
$$= -j\mathbf{E}\frac{1}{\omega RC}$$
$$\mathbf{V}_C = -j\mathbf{E}Q \qquad (23\text{-}8)$$

It is possible for these reactive drops to be in the thousands of volts. One must consider insulation and safety requirements in such cases. For instance, if we have a circuit with a Q of 150 and an applied voltage of 30 V, the voltage across each reactance at resonance is 4500 V.

$$\mathbf{V}_C = -\mathbf{V}_L = 30 \times 150 = 4500 \text{ V}$$

EXAMPLE 23-5:
Given the circuit of Fig. 23-6, solve for current and each voltage drop at $f = f_0$.

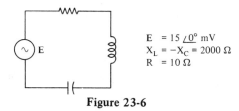

Figure 23-6

Solution:

$$Z_T = R = 10\underline{/0°}\ \Omega$$

$$I = \frac{E}{R} = \frac{15 \times 10^{-3}\underline{/0°}}{10\underline{/0°}}$$
$$= 1.5\underline{/0°}\ \text{mA}$$

$$V_R = IR$$
$$= (1.5 \times 10^{-3})\underline{/0°} \times 10\underline{/0°}$$
$$= 15\underline{/0°}\ \text{mV}$$

$$V_L = I(jX_L)$$
$$= (1.5 \times 10^{-3})\underline{/0°} \times (2 \times 10^3)\underline{/90°}$$
$$= 3\underline{/90°}\ \text{V}$$

$$V_C = I(-jX_C)$$
$$= (1.5 \times 10^{-3})\underline{/0°} \times (2 \times 10^3)\underline{/90°}$$
$$= 3\underline{/-90°}\ \text{V}$$

Note that if we solve for Q and then multiply the applied voltage by Q, we find the magnitude of the voltage drop across each reactance.

$$Q = \frac{X_L}{R} = \frac{X_C}{R} = \frac{2000}{10}$$
$$= 200$$

$$|V_L| = |V_C| = EQ = 15\ \text{mV} \times 200 = 3\ \text{V}$$

Bandwidth

We know that the current in a series RLC circuit is maximum at resonance and falls off at frequencies above and below resonance. *We define the bandwidth of any resonant circuit as the range of frequencies between half-power points.* In the series resonant circuit, current is 0.707 of the maximum at each half-power point. For this to occur, the impedance must be 1.414 times the impedance at resonance. *It is therefore 1.414R.* We define the frequency below f_0 for which $I = 0.707I_{max}$ as f_1, and the frequency above the resonant frequency at which current again falls to $0.707I_{max}$ as f_2, as shown in Fig. 23-7.

$$f = f_1: \quad Z_T = 1.414R = R - j(X_C - X_L)$$
$$f = f_2: \quad Z_T = 1.414R = R + j(X_L - X_C)$$

In order for Z to equal $1.414R$, the *net* reactance must be equal to R.

A solution for the bandwidth $(f_2 - f_1)$ results in the standard formula for bandwidth.

$$BW = f_2 - f_1 = \frac{f_0}{Q} \tag{23-9}$$

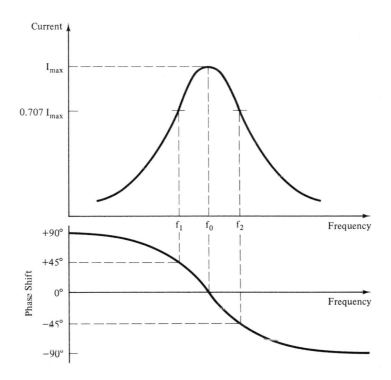

Figure 23-7 Current and phase shift variation with frequency in a series RLC circuit. At the frequencies f_1 and f_2 the net reactance equals the resistance of the circuit. This is the reason for the 45° phase angles. The frequencies f_1 and f_2 define the band width of a resonant circuit.

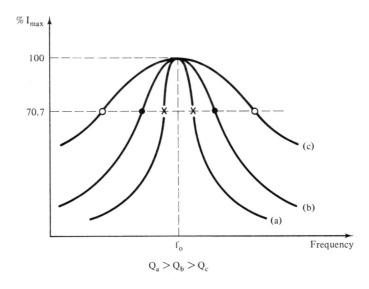

Figure 23-8 The effect of Q upon the bandwidth of a resonant circuit.

Sec. 23-3. Circuit Q 445

Bandwidth is simply one means of defining the *selectivity* of a resonant circuit. Selectivity is a measure of the ability of a circuit to reject frequencies above and below resonant frequency. Selectivity is proportional to Q, while bandwidth is inversely proportional to Q, as shown in Fig. 23-8. The highest value of Q for a circuit is the intrinsic Q (Q_0) of the coil itself. In many instances, this is too high a value of Q to allow the desired bandwidth. We add resistance as needed to obtain the required Q for the necessary bandwidth. The overall Q is usually referred to as effective Q (Q').

$$Q_0 = \frac{\omega_0 L}{R_0} \quad \text{and} \quad R_0 = \frac{\omega_0 L}{Q_0}$$

$$Q' = \frac{\omega_0 L}{R_T} \quad \text{and} \quad R_T = \frac{\omega_0 L}{Q'}$$

The added resistance is found as

$$\begin{aligned} R &= R_T - R_0 \\ &= \frac{\omega_0 L}{Q'} - \frac{\omega_0 L}{Q_0} \\ &= \omega_0 L \left(\frac{1}{Q'} - \frac{1}{Q_0} \right) \end{aligned} \qquad (23\text{-}10)$$

EXAMPLE 23-6:
A 140-μH coil, with $Q_0 = 75$, is used in a series resonant circuit. The resonant frequency is 2.4 MHz. (a) What is the bandwidth of the circuit? (b) Solve for f_1 and f_2.

Solution:

(a) $\text{BW} = \dfrac{f_0}{Q}$ (we shall use Q_0 on the assumption that $R_0 = R_T$)

$\text{BW} = \dfrac{2.4 \times 10^6}{75} = 32 \times 10^3$

$= 32 \text{ kHz}$

(b) $\text{BW} = f_2 - f_1$. In this example, we assume that the response curve is symmetrical about the resonant frequency because $Q > 10$.

$f_1 = f_0 - \dfrac{\text{BW}}{2}$

$= 2400 \text{ kHz} - \dfrac{32}{2} \text{ kHz}$

$= 2384 \text{ kHz}$

$= 2.384 \text{ MHz}$

$f_2 = f_0 + \dfrac{\text{BW}}{2}$

$= 2400 \text{ kHz} + 16 \text{ kHz}$

$= 2416 \text{ kHz}$

$= 2.416 \text{ MHz}$

EXAMPLE 23-7.
The circuit conditions of Example 23-6 are changed so that the required bandwidth is 100 kHz. What value of resistance must be placed in series in order to reduce the Q of the circuit?

Solution:
1. Solve for the effective Q.
$$Q' = \frac{f_0}{BW} = \frac{2.4 \times 10^6}{0.1 \times 10^6}$$
$$= 24$$

2. Solve for R.
$$R = \omega_0 L \left(\frac{1}{Q'} - \frac{1}{Q_0}\right)$$
$$= 6.28 \times 2.4 \times 10^6 \times 140 \times 10^{-6} \times \left(\frac{1}{24} - \frac{1}{75}\right)$$
$$= 59.79 \, \Omega \approx 60 \, \Omega$$

The keystroke sequence for the reciprocals is worth noting: When we reach the last part of the calculation, use the following steps:
(24 1/x - 75 1/x) and then = in order to obtain the solution.

Check:
$$R_T = 60 + \frac{\omega_0 L}{Q_0} = 60 + 28 = 88 \, \Omega$$
$$Q' = \frac{\omega_0 L}{R_T}$$
$$= \frac{6.28 \times 2.4 \times 10^6 \times 140 \times 10^{-6}}{88}$$
$$= 24$$

23-4. IDEAL PARALLEL RESONANT CIRCUIT

We begin our analysis of parallel resonance by considering a parallel circuit consisting of an ideal capacitor and inductor, as in Fig. 23-9. To simplify our analysis, we shall drive the parallel circuit from a current source.

The total current drawn by the circuit must be the phasor sum of the branch currents. Using susceptances and allowing $\mathbf{V}_L = \mathbf{V}_C = \mathbf{V}$ across the parallel circuit, we have

$$\mathbf{I}_T = \mathbf{I}_C + \mathbf{I}_L$$
$$= j\mathbf{V}B_C - j\mathbf{V}B_L$$
$$= (\mathbf{V})j(B_C - B_L)$$

At resonance, the reactive currents must equal each other so that stored energy is exchanged between capacitance and inductance as in series resonance. For the branch currents to be equal, the susceptances must be equal.

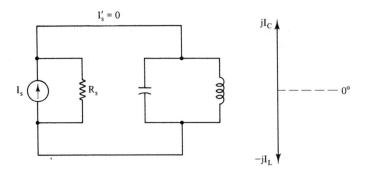

Figure 23-9 The ideal parallel resonant circuit.

$$f = f_0: \quad B_C = B_L \quad (23\text{-}10)$$

$$\frac{1}{2\pi f_0 C} = 2\pi f_0 L$$

Solving for f_0,

$$f_0 = \frac{1}{2\pi\sqrt{LC}} \quad (23\text{-}11)$$

At frequencies less than resonant frequency, the current in the inductive branch is greater than current in the capacitive branch, and the circuit appears as *pure* inductance.

At frequencies greater than resonant frequency, capacitive susceptance increases, resulting in greater current in the capacitive branch. The circuit appears as *pure* capacitance.

At resonance, the ideal circuit draws no current at all from the source. Instead, the branch currents circulate between the inductance and the capacitance. This current, often called *tank* current, is a circulating current; it is sketched in Fig. 23-10. Tank current is confined to the tank circuit and does not enter the supply lines. It is common practice to refer to *all* parallel resonant circuits as "tank circuits" because of their energy-storage capability.

The impedance of an ideal parallel resonant circuit is infinite, since zero current is drawn from a source (Fig. 23-11).

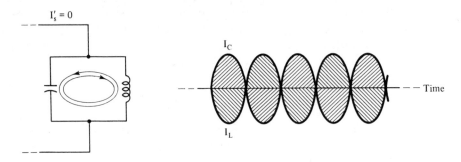

Figure 23-10 Circulating tank current.

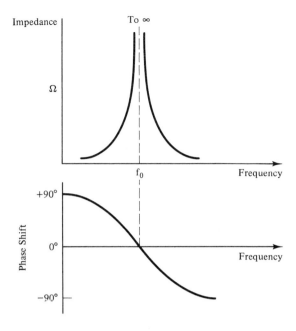

Figure 23-11 Impedance and phase characteristics for an ideal parallel resonant circuit. Because the circuit draws no current at resonance, it has infinite impedance at that frequency.

$$Z_{eq} = \frac{V}{I_T} = \infty$$

Clearly, the Q of a circuit made with ideal reactances is infinite. In actual practice, we do have devices, such as quartz crystals, with such high Q as to approach the properties of a resonant circuit with infinite Q. You will find these devices in your studies of electronics, especially in radio-frequency communications systems.

23-5. PRACTICAL PARALLEL RESONANT CIRCUITS

In a parallel circuit consisting of inductance and capacitance, we do have resistance. This resistance consists of the intrinsic resistance of the coil and the dissipative effects of any loads connected to the circuit.

The effect of resistance is to cause the phase shift of current in the inductive branch to be less than 90°. For the circuit to be resonant, there must be a component of current in the inductive branch that is exactly equal to the capacitive current. For this to be so, the inductive branch current must be slightly greater than the capacitive branch current. This will occur at a frequency that is lower than the frequency we would find by use of Eq. (23-1). For example, if the Q of the circuit is 5, meaning that $X_L = 5R$, the inductive branch current must be approximately 1.02 times the capacitive branch current. A practical parallel resonant circuit is shown in Fig. 23-12. Total current, as affected by circuit Q, is shown in Fig. 23-13.

A general solution for the parallel circuit can be made.

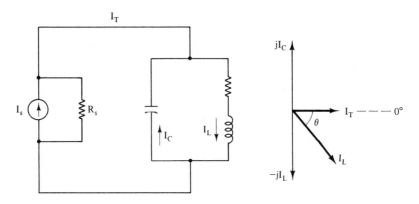

Figure 23-12 Practical parallel resonant circuit.

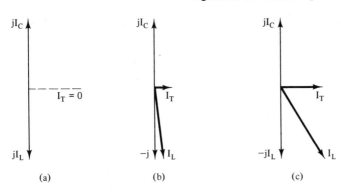

Figure 23-13 Total current drawn by a tank circuit at resonance is an inverse function of Q. The higher the Q, the lower the total current. (a) Infinite Q. At resonance, the impedance is infinite and total current equals zero. (b) High Q. Impedance is high and total current is small. (c) Low Q. Impedance is lower and total current is greater than in the other examples.

$$Z_{eq} = \frac{Z_1 Z_2}{Z_1 + Z_2}$$

where $Z_1 = -jX_C$
$Z_2 = R + jX_L$
This "reduces" to

$$Z_{eq} = \frac{RX_C^2 - j(R^2 X_C + X_L^2 X_C - X_L X_C^2)}{R^2 + (X_L - X_C)^2}$$

For the circuit to be resonant (purely resistive), the second term of the numerator must be equal to zero.

$$R^2 X_C + X_L^2 X_C - X_L X_C^2 = 0$$

$$\frac{R^2}{2\pi f_0 C} + \frac{(2\pi f_0 L)^2}{2\pi f_0 C} - \frac{2\pi f_0 L}{(2\pi f_0 C)^2} = 0$$

This equation is then solved for f_0 so that we have a formula for the resonant frequency of a practical parallel RLC circuit.

$$f_0 = \frac{1}{2\pi L}\sqrt{\frac{L}{C} - R^2} \qquad (23\text{-}12)$$

Note that the effect of R can be so great as to *prevent* resonance. Whenever $R^2 \geqslant L/C$, the circuit cannot resonate. This is one reason why a high L to C ratio is desirable in parallel resonant circuits.

In most cases, resistance will be very much less than L/C, so R^2 can be ignored. Given that $L/C \gg R^2$, the formula for resonant frequency reduces to the same formula as for series resonance.

$$f_0 = \frac{1}{2\pi L}\sqrt{\frac{L}{C}}$$
$$= \frac{1}{2\pi\sqrt{LC}} \quad (23\text{-}1)$$

EXAMPLE 23-8:
Given the circuit of Fig. 23-14, (a) solve for the resonant frequency of the circuit with $R = 0$; (b) solve for f_0 with $R = 200\ \Omega$; (c) solve for f_0 with $R = 1\ k\Omega$; (d) solve for f_0 with $R = 10\ \Omega$.

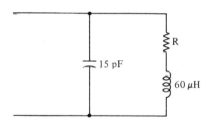

Figure 23-14

Solution:

(a) $f_0 = \dfrac{1}{2\pi\sqrt{LC}}$

$= \dfrac{1}{6.28\sqrt{60 \times 10^{-6} \times 15 \times 10^{-12}}}$

$= \dfrac{1}{6.28 \times 30 \times 10^{-9}}$

$= 5.308$ MHz

(b) $f_0 = \dfrac{1}{2\pi L}\sqrt{\dfrac{L}{C} - R^2}$

$= \dfrac{1}{6.28 \times 60 \times 10^{-6}}\sqrt{\dfrac{60 \times 10^{-6}}{15 \times 10^{-12}} - 200^2}$

$= 2.654 \times 10^3 \times \sqrt{4{,}000{,}000 - 40{,}000}$

$= 2.654 \times 10^3 \times 1.99 \times 10^3$

$= 5.281$ MHz

(c) $f_0 = 2.654 \times 10^3 \times \sqrt{4{,}000{,}000 - 1000^2}$

$= 2.654 \times 10^3 \times 1.732 \times 10^3$

$= 4.597$ MHz

Sec. 23-5. Practical Parallel Resonant Circuits

(d) $f_0 = 2.654 \times 10^3 \times \sqrt{4{,}000{,}000 - 10^2}$
    $= 2.654 \times 10^3 \times 1999.975$
    $= 2.654 \times 10^3 \times 2 \times 10^3$
    $= 5.308$ MHz

Impedance

If we use Eq. (23-1) in the solution of the impedance of a practical parallel resonant circuit, the reactances are considered to be equal. The impedance equation is then simplified.

$$Z_{eq} = \frac{Z_1 Z_2}{Z_1 + Z_2}$$
$$= \frac{X_L X_C - jRX_C}{R}$$

Using X_L in place of X_C, since they are equal,

$$Z_{eq} = \frac{X_L X_L}{R} - jX_L$$

Provided that $Q \geqslant 10$, the first term is 10 times greater than the second term of the equation, and we drop the second term.

$$Z_{eq} = \frac{X_L^2}{R} \tag{23-13}$$

Given that $Q = X_L/R$,

$$Z_{eq} = X_L Q \tag{23-14}$$

At resonance, the impedance is high and purely resistive.

$$Z_{eq} = R_p = \omega_0 L Q_0 \tag{23-15}$$

where Q_0 is the intrinsic Q of the coil.

When the circuit is under load, as in Fig. 23-15, we refer to the Q as effective with the symbol Q'. At $f = f_0$,

$$Z_{eq} = R_{eq} = \frac{R_p R_L}{R_p + R_L}$$

Assuming that $Q' \geqslant 10$,

$$Z_{eq} = \omega_0 L Q' \tag{23-16}$$

and

$$Q' = \frac{Z_{eq}}{\omega_0 L} \tag{23-17}$$

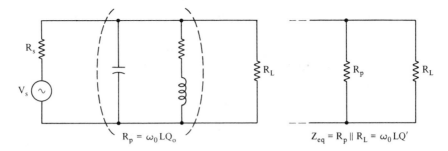

Figure 23-15 At the resonant frequency, the effect of loading a tank circuit can be seen by considering two resistors in parallel. R_p represents the parallel circuit *without* load.

EXAMPLE 23-9:
A 50-μH coil with an intrinsic resistance of 28 Ω is used in a parallel resonant circuit. The resonant frequency is 4.5 MHz. What is the impedance of the circuit at the resonant frequency?

Solution:
The impedance is purely resistive and equal to $\omega_0 L Q_0$, provided that $Q_0 \geqslant 10$.

1. $Q_0 = \dfrac{\omega_0 L}{R_0}$
   $= \dfrac{6.28 \times 4.5 \times 10^6 \times 50 \times 10^{-6}}{28}$
   $= 50.46$

2. $Z_{eq} = R_p = \omega_0 L Q_0$
   $= 6.28 \times 4.5 \times 10^6 \times 50 \times 10^{-6} \times 50.46$
   $= 71.3 \text{ k}\Omega$

EXAMPLE 23-10:
Given the tank circuit of Example 23-9, solve for the equivalent impedance at the resonant frequency and the effective Q when a 47-kΩ resistor is connected across the tuned circuit.

Solution:

1. $Z_{eq} = \dfrac{R_p R_L}{R_p + R_L}$
   $= \dfrac{71.3 \times 10^3 \times 47 \times 10^3}{71.3 \times 10^3 + 47 \times 10^3}$
   $= 28.33 \text{ k}\Omega$

2. $Q' = \dfrac{Z_{eq}}{\omega_0 L}$
   $= \dfrac{28.33 \times 10^3}{6.28 \times 4.5 \times 10^6 \times 50 \times 10^{-6}}$
   $= 20$

23-6. EFFECTIVE Q AND BANDWIDTH

Bandwidth, as always, is measured between half-power points. We find the frequencies for which the voltage drop across the parallel circuit is 0.707 of the voltage at resonance. The solution for f_2 (upper freqeuncy) and for f_1 (lower frequency) is the same as for series resonant circuits.

$$\text{BW} = f_2 - f_1 = \frac{f_0}{Q'} \tag{23-18}$$

The sharpness of the selectivity characteristic is a direct function of Q, as in the series circuit.

EXAMPLE 23-11:

A parallel circuit is resonant at 10.7 MHz. The effective Q of the circuit is 53.5. (a) What is the bandwidth of the circuit? (b) Specify f_1 and f_2.

Solution:

(a) $\text{BW} = \dfrac{f_0}{Q'}$

$= \dfrac{10.7 \times 10^6}{53.5}$

$= 200 \text{ kHz}$

(b) $f_1 = f_0 - \dfrac{\text{BW}}{2}$

$= 10.7 \text{ MHz} - 0.1 \text{ MHz}$

$= 10.6 \text{ MHz}$

$f_2 = f_0 + \dfrac{\text{BW}}{2}$

$= 10.7 \text{ MHz} + 0.1 \text{ MHz}$

$= 10.8 \text{ MHz}$

EXAMPLE 23-12:

Given the circuit of Fig. 23-16, (a) calculate f_0; (b) calculate the bandwidth.

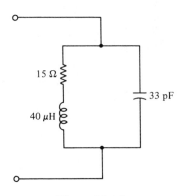

Figure 23-16

Solution:

(a) $L/C = \dfrac{40 \times 10^{-6}}{33 \times 10^{-12}}$
$= 1.21 \times 10^6$

L/C is very much greater than 15^2, therefore, we use the usual formula for f_0.

$$f_0 = \dfrac{1}{2\pi\sqrt{LC}}$$
$$= \dfrac{1}{6.28\sqrt{40 \times 10^{-6} \times 33 \times 10^{-12}}}$$
$$= \dfrac{1}{6.28 \times 3.6 \times 10^{-8}}$$
$$= 4.383 \text{ MHz}$$

(b) $Q_0 = \dfrac{\omega_0 L}{R_0}$
$= \dfrac{6.28 \times 4.383 \times 10^6 \times 40 \times 10^{-6}}{15}$
$= 73.4$

$$\text{BW} = \dfrac{f_0}{Q_0}$$
$$= \dfrac{4.383 \times 10^6}{73.4}$$
$$= 59.7 \text{ kHz}$$

EXAMPLE 23-13:
What value of resistance must be shunted across the tank circuit in order for the bandwidth of the circuit of Example 23-12 to be 210 kHz?

Solution:
1. Solve for Q'.

$$Q' = \dfrac{f_0}{\text{BW}}$$
$$= \dfrac{4.383 \times 10^6}{210 \times 10^3}$$
$$= 20.87$$

2. Solve for the equivalent impedance of the loaded tank circuit.

$Z_{eq} = \omega_0 L Q'$
$= 6.28 \times 4.383 \times 10^6 \times 40 \times 10^{-6} \times 20.87$
$= 22.98 \text{ k}\Omega$

3. Solve for the equivalent impedance of the unloaded tank circuit.

$R_p = \omega_0 L Q_0$
$= 6.28 \times 4.383 \times 10^6 \times 40 \times 10^{-6} \times 73.4$
$= 80.8 \text{ k}\Omega$

4. The impedance found in step 2 is the parallel combination of R_L and R_p. We use the product-over-difference formula for finding an unknown paral-

lel resistance when equivalent resistance and one branch resistance are known.

$$R_L = \frac{(\omega_0 L Q_0)(\omega_0 L Q')}{\omega_0 L Q_0 - \omega_0 L Q'}$$
$$= \frac{80.8 \times 10^3 \times 22.98 \times 10^3}{80.8 \times 10^3 - 22.98 \times 10^3}$$
$$= 32.11 \text{ k}\Omega$$

Circulating Tank Current

At resonance, the circulating tank current is much greater than any source current drawn by the parallel circuit.

The equivalent impedance of the circuit is Q times greater than the reactance of each branch. The current drawn by the circuit is therefore less than branch current by a factor of Q. Each branch current is the same as the circulating tank current (I_{cir}).

$$I_{cir} = I_L = I_C = QI_s \qquad (23\text{-}19)$$

where I_s is source current.

EXAMPLE 23-14:
Given the circuit of Fig. 23-17, solve for line current (I_s') and circulating tank current.

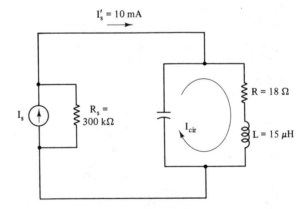

Figure 23-17

Solution:
1. Solve for R_p.

$$R_p = Z_{eq} = \frac{X_L^2}{R_0}$$
$$= \frac{(6.28 \times 30 \times 10^6 \times 15 \times 10^{-6})^2}{18}$$
$$= 444 \text{ k}\Omega$$

2. By use of the CDR solve for I'_s.

$$I'_s = \frac{R_i}{R_p + R_i} \times I_{gen}$$
$$= \frac{300 \times 10^3}{744 \times 10^3} \times 10 \text{ mA}$$
$$= 4.03\underline{/0°} \text{ mA}$$

3. Solve for the voltage drop across the tank circuit.

$$V_p = I'_s R_p$$
$$= 4.03 \times 10^{-3} \times 444 \times 10^3$$
$$= 1790\underline{/0°} \text{ V}$$

4. The current in the inductive branch is equal to the current in the capacitive branch and is equal to the circulating current.

$$I_{cir} = \frac{V_p}{\omega_0 L}$$
$$= \frac{1790}{6.28 \times 30 \times 10^6 \times 15 \times 10^{-6}}$$
$$= 633 \text{ mA}$$

EXAMPLE 23-15:

Demonstrate that the circulating tank current of Example 23-14 is Q times line current.

Solution:
1. Solve for Q_0.

$$Q_0 = \frac{\omega_0 L}{R_0}$$
$$= \frac{6.28 \times 30 \times 10^6 \times 15 \times 10^{-6}}{18}$$
$$= 157$$

2. $I_{cir} = Q I_s$
$= 157 \times 4.03 \text{ mA}$
$= 633 \text{ mA}$

23-7. APPLICATIONS OF RESONANT CIRCUITS

Resonant circuits have a great variety of applications where frequency selectivity is needed, particularly at higher frequencies. They are used as loads in radio-frequency amplifier circuits, as part of coupling networks, and as frequency-selective filters. Whenever you tune a radio receiver or change channels in a TV set, you are using resonant circuits to select the desired station and reject others. The output of an amplifier depends on several quantities, including load impedance. At its resonant frequency, the parallel circuit offers maximum impedance and therefore maximum output by the amplifier.

Resonant circuits are frequently used as *band-pass* and *band-reject* (band-stop) filters. An ideal band-pass filter passes all signal voltages at frequencies within the passband without loss and rejects signals at all other frequencies. Clearly, the properties of practical filter circuits differ from these ideal characteristics. By using multiple sections (cascading) and additional tuned circuits to steepen the selectivity curve, we can approach the properties of the ideal band-pass filter. Such complex circuits are beyond the scope of this text; we shall limit our discussions to basic circuits.

Band-pass Filters

The filter functions as a voltage divider; we use its changing impedance with frequency characteristic to cause maximum output over a narrow range of frequencies (Fig. 23-18).

When the load on the output of the filter is a large amount of resistance, we use a parallel resonant circuit across the load terminals. A series resis-

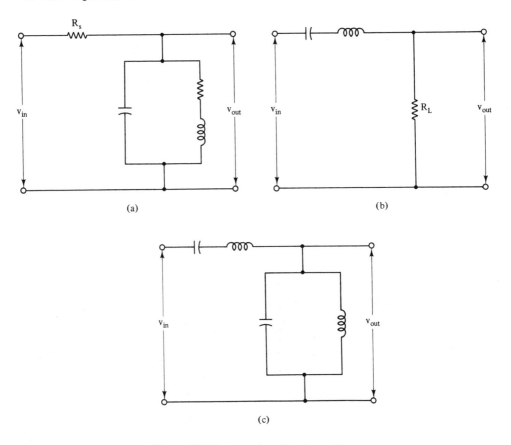

Figure 23-18 Examples of band-pass filters.

tance is connected between the source and load terminals. This resistance is selected so that it causes appreciable reduction in output except when the tank circuit offers high impedance. It is usually selected as equal to R_p, the impedance of the tank at resonance.

The output voltage is found by the voltage-divider rule.

$$v_{out} = v_{in} \frac{Z_{eq} \parallel R_L}{R_s + Z_{eq} \parallel R_L}$$

For frequencies outside of the passband, the impedance of the tuned circuit falls, resulting in greatly reduced output. For these frequencies, the equivalent impedance of the tuned circuit is approximately the same as the parallel combination of tuned circuit and load, because the tuned circuit impedance is so much less than the load resistance.

EXAMPLE 23-16:
In the band-pass filter circuit of Fig. 23-19, the load resistance is the input resistance of a MOSFET amplifier and is equal to 2 MΩ. The center frequency of the passband is 4.5 MHz, and the bandwidth is 200 kHz. (a) Calculate the output at 4.5 MHz. (b) Calculate the output at 4.4 and 4.6 MHz. (c) Calculate the output at 4 MHz.

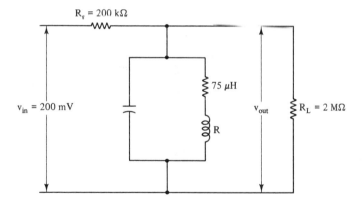

Figure 23-19

Solution:
(a) Calculate Q':

$$Q' = \frac{f_0}{BW}$$
$$= \frac{4.5 \times 10^6}{0.2 \times 10^6}$$
$$= 22.5$$

Calculate Z_{eq}:

$$Z_{eq} = \omega_0 L Q'$$
$$= 6.28 \times 4.5 \times 10^6 \times 75 \times 10^{-6} \times 22.5$$
$$= 47.7 \text{ k}\Omega$$

By use of the VDR:

$$v_o = v_{in} \frac{Z_{eq}}{R_s + Z_{eq}}$$
$$= 0.2 \times \frac{47.7 \text{ k}}{247.7 \text{ k}}$$
$$= 38.5 \text{ mV}$$

(b) At f_1 and f_2, $Z_{eq} = 0.707 R_p$.

$$Z_{eq} = 0.707 \times 47.7 \text{ k}\Omega$$
$$= 33.7 \underline{/\pm 45°} \text{ k}\Omega$$
$$= 23.8 \text{ k}\Omega \pm j23.8 \text{ k}\Omega$$

Therefore, $R_s + Z_{eq}$ is a phasor sum.

$$\begin{aligned} R_s + Z_{eq} = {} & 200 \text{ k}\Omega + j0 \\ & \underline{23.8 \text{ k}\Omega + j23.8 \text{ k}\Omega} \\ = {} & 223.8 \text{ k}\Omega \pm j23.8 \text{ k}\Omega \\ \simeq {} & 224 \text{ k}\Omega \end{aligned}$$

By use of the VDR:

$$v_o = 0.2 \times \frac{33.7 \text{ k}\Omega}{224 \text{ k}\Omega}$$
$$= 30.1 \text{ mV}$$

We should note some interesting items. First, the output at f_1 (4.4 MHz) and f_2 (4.6 MHz) is slightly greater than 0.707 of the output at 4.5 MHz. The effect is caused by the series resistance R_s. We can select a larger value of R_s. This will make the output response curve correspond to the shape of the frequency response curve of the resonant circuit. Second, to do so is to greatly reduce output voltage. Finally, as in most filter applications, it is the rejection *outside* the passband that is most important. One final note: we would not want to reduce R_s to some value considerably less than R_p. This would completely eliminate the effect of the filter!

When the output load resistance is a relatively low amount, we use a series resonant circuit rather than a parallel resonant circuit. The resonant circuit is connected between the input and output terminals, as in Fig. 23-19. At the resonant frequency, the impedance of the series resonant circuit is very low, equal to the resistance of the coil. Virtually all the applied voltage appears across the load resistance. For other frequencies, the impedance of the tuned circuit rises, which results in reduced output.

EXAMPLE 23-17:
Given the circuit of Fig. 23-20, demonstrate that $v_{out} = v_{in}$ at $f = f_0$, and that $v_{out} = 0.707 v_{in}$ at $f = f_1$ and f_2.

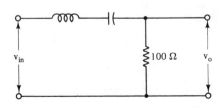

Figure 23-20

Solution:
1. At $f = f_0$,
$$Z = R$$
and
$$v_{out} = v_{in}$$

2. At $f = f_1$ and f_2,
$$Z = 1.414R$$
$$v_{out} = v_{in} \times \frac{R}{1.414R}$$
$$= v_{in} \times \frac{1}{1.414}$$
$$= 0.707 v_{in}$$

Band-reject Filters

In this case we wish to greatly reduce output over a narrow range of frequencies. Once again, the filter system works as a voltage divider. We use either series or parallel tuned circuits, or combinations of both types (Fig. 23-21).

If we use a series resonant circuit, it is connected *across* the output terminals. A fixed amount of resistance (R_s) is connected between the input and output terminals. At the resonant frequency, the impedance of the tuned circuit is very low and purely resistive. Because this amount of resistance is

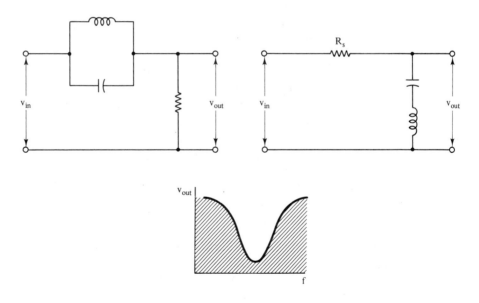

Figure 23-21 Examples of band-reject filters.

Sec. 23-7. Applications of Resonant Circuits

a small fraction of R_s, most of the input voltage is dropped across R_s. The degree of reduction depends upon the ratio of R_0 to R_s.

$$v_{out} = v_{in} \frac{R_0}{R_s + R_0}$$

For frequencies considerably off of resonant frequency, the impedance of the tuned circuit is purely reactive. The impedance can be considerably larger than R_s, which will result in greater output voltage than within the reject band.

When a parallel resonant circuit is used in a band-reject filter, it is connected in series between the input and output terminals. The load is connected across the output terminals. At resonance, the impedance of the tank circuit is very high and resistive. The output is much less than the input because of the voltage drop across the tank circuit. As with the series resonant circuit, we may demonstrate that the degree of rejection depends upon the intrinsic Q. The greater the Q_0, the lower is the output voltage across the load terminals. However, the higher the intrinsic Q of a resonant filter is, the narrower the range of frequencies affected. This is true for band-pass as well as for band-reject filters.

This brief discussion of frequency-selective filters is an attempt to demonstrate some of the applications of resonant circuits. Although we shall not cover other selective filters in this text, it must be pointed out that many types of frequency-selective filters do not make use of resonance.

SUMMARY

1. Resonance occurs when reactive quantities cancel.
2. There is only *one* frequency where reactive quantities cancel in a given circuit. This is called the resonant frequency (f_0) of the circuit.
3. In series resonance $|V_L| = |V_C|$ and therefore $|X_L| = |X_C|$.
4. The impedance of a series circuit at resonance is low and is pure resistance.
5. At resonance, the current in a series circuit is maximum.
6. Both inductance and capacitive reactances develop voltages at resonance that are equal but of opposite polarity. $V_L = -V_C$.
7. At resonance, the voltage across both forms of reactance is Q' times the source voltage.

$$V_L = Q'E, \quad V_C = -Q'E$$

8. The voltage across resistance at resonance equals source voltage in a series resonant circuit.
9. The bandwidth of a resonant circuit is the ratio of the resonant frequency to the circuit Q.
10. In a parallel resonant circuit it is the reactive currents that cancel. $|I_L| = |I_C|$.
11. In a parallel resonant circuit, with Q' less than 10, resistance placed in series with the inductance will cause the resonant frequency to decrease.
12. At resonance, the impedance of a parallel circuit is its highest and is pure resistance.
13. Parallel resonant circuits are often called tank circuits.

14. The circulating current in a parallel resonant circuit is Q times the source current entering the tank circuit.
15. Resonant circuits are widely used at radio frequencies. They are required for frequency-selective loads, as in tuning circuits, and are often used as selective filters.

PROBLEMS

23-1. In the circuit of Fig. 23-22, solve for the resonant frequency given that:
(a) $L = 250$ mH, $C = 3$ nF, $R = 100$ Ω
(b) $L = 10$ mH, $C = 500$ pF, $R = 80$ Ω
(c) $L = 15$ μH, $C = 50$ pF, $R = 3$ Ω
(d) $L = 300$ mH, $C = 0.01$ μF, $R = 120$ Ω
(e) $L = 150$ μH, $C = 300$ pF, $R = 25$ Ω

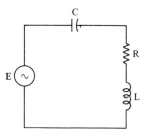

Figure 23-22

23-2. In the circuit of Fig. 23-22, $f_0 = 1.2$ MHz and $C = 300$ pF. Solve for L.
23-3. In the circuit of Fig. 23-22, $f_0 = 29$ MHz and $C = 15$ pF. Solve for L.
23-4. In the circuit of Fig. 23-22, $f_0 = 2.5$ MHz and $L = 40$ μH. Solve for C.
23-5. In the circuit of Fig. 23-22, $f_0 = 4.1$ kHz and $L = 150$ mH. Solve for C.
23-6. Given a series resonant circuit, with $L = 35$ μH, $R_0 = 6.5$ Ω, and $C = 75$ pF, solve for Q_0 at the resonant frequency.
23-7. Given that $R_0 = 8$ Ω, $L = 55$ μH, and $C = 100$ pF, solve for Q_0 at the resonant frequency of the series circuit.
23-8. Given the circuit of Fig. 23-23, (a) solve for the impedance at resonance; (b) at 10 kHz above resonance; and (c) at 10 kHz below resonance.

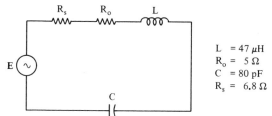

$L = 47$ μH
$R_0 = 5$ Ω
$C = 80$ pF
$R_s = 6.8$ Ω

Figure 23-23

23-9. In the circuit of Fig. 23-23, if $\mathbf{E} = 2.5\underline{/0°}$ mV, solve for $\mathbf{I}$ at resonance.
23-10. Given the data of Prob. 23-9, solve for $\mathbf{V}_c$ at resonance.
23-11. Given the data of Prob. 23-9, solve for the effective Q of the circuit.
23-12. What is the intrinsic Q of the coil at resonance in the circuit of Figure 23-23?

23-13. A series resonant circuit has a Q_0 of 110. The resonant frequency is 10.7 MHz. What is the bandwidth of the unloaded circuit?

23-14. Given the data of Prob. 23-13, what is the required effective Q for a bandwidth of 200 kHz?

23-15. Given the input voltage and Q', specify V_R, V_L, and V_C.
  (a) $E = 150/\underline{10°}$ mV, $Q' = 80$
  (b) $E = 1.2/\underline{15°}$ V $Q' = 150$
  (c) $E = 40/\underline{0°}$ V $Q' = 30$
  (d) $E = 25/\underline{0°}$ V $Q' = 15$
  (e) $E = 20/\underline{-90°}$ mV $Q' = 50$

23-16. Given a series resonant circuit with $L = 10$ mH, $R = 35$ Ω, and $f_0 = 20$ kHz, solve for R_s so that $Q' = 15.7$.

23-17. In the parallel circuit of Fig. 23-24, solve for the resonant frequency, given $L = 20\ \mu\text{H}$, $C = 110$ pF, and:
  (a) $R = 30$ Ω
  (b) $R = 75$ Ω
  (c) $R = 100$ Ω
  (d) $R = 300$ Ω
  (e) $R = 450$ Ω

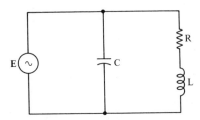

Figure 23-24

23-18. The circuit of Fig. 23-24 is resonant at 30 MHz, with $L = 2.8\ \mu\text{H}$ and $R_0 = 6$ Ω. Solve for the impedance at resonance and the bandwidth.

23-19. Given the data of Prob. 23-18, what value of R must be placed across the circuit to make the bandwidth 2 MHz?

23-20. The circuit of Fig. 23-24 is resonant at 16 MHz with $L = 9\ \mu\text{H}$, $R_0 = 12$ Ω, and $Q' = 80$. If $E = 1.2/\underline{0°}$ V, solve for line current.

23-21. Given the data of Prob. 23-20, solve for the circulating current.

23-22. Calculate the output voltage at resonance for the circuit of Fig. 23-25.

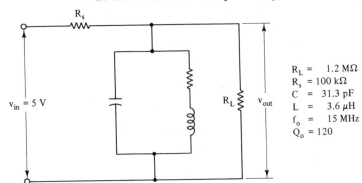

$R_L = 1.2$ MΩ
$R_s = 100$ kΩ
$C = 31.3$ pF
$L = 3.6\ \mu\text{H}$
$f_0 = 15$ MHz
$Q_0 = 120$

Figure 23-25

23-23. Given the data of Prob. 23-22, calculate the output voltage at 14 MHz.

23-24. Calculate the output voltage at resonance for the circuit of Fig. 23-26.

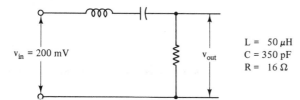

$v_{in} = 200$ mV

V_{out}

L = 50 μH
C = 350 pF
R = 16 Ω

Figure 23-26

23-25. Given the data of Prob. 23-24, solve for the bandwidth and the output at the f_1 and f_2 frequencies.

24

Transformers

24-1. TYPES OF TRANSFORMERS

Transformers allow ac voltage and current levels to be changed easily and efficiently. All transformers make use of the properties of electromagnetic induction and mutual inductance.

We classify transformers by application. Therefore, we have power transformers, audio transformers, pulse transformers, radio-frequency transformers, and so on. We find that the application itself defines the construction of the transformers. *Power* transformers are those that operate at line frequency. They may be as small as a few volt-amperes and as large as those used by electric utilities, rated in many hundreds of kilovolt-amperes. *Audio* transformers are similar in construction to power transformers, but must be designed to operate over a broad range of frequencies. *Radio-frequency* transformers are designed to be used at much higher frequencies. Typically, these are used in resonant circuits of RF amplifiers.

Audio and power transformers have high coefficients of coupling and very high efficiencies. In fact, in most calculations we assume these transformers to be "perfect" coupling devices. The error caused by this assumption is so small as to be unimportant. On the other hand, RF transformers have low-permeability flux paths and therefore low coefficients of coupling. Because of the "loose" coupling, RF transformers have considerable leakage loss.

24-2. TRANSFORMER CONSTRUCTION

All transformers consist of two or more coils wound upon a common core so that mutual inductance exists between the coils. In power and audio transformers, because the frequencies are low, the core is made of high-permeability material, always a ferromagnetic material, like transformer steel. The core is made of sheets (laminations) covered with an insulating varnish to insulate the laminations from one another. This is done to keep eddy current losses to a minimum. Eddy currents are simply induced currents in the core material, because the core is a conductor. The laminated design for the core increases the electrical resistance of the core without affecting the permeability of the flux path. Figure 24-1 shows several types of power and audio transformer constructions.

Cores for RF transformers are either air or powdered iron in a ceramic binder. Because of the high operating frequencies, losses would be excessive

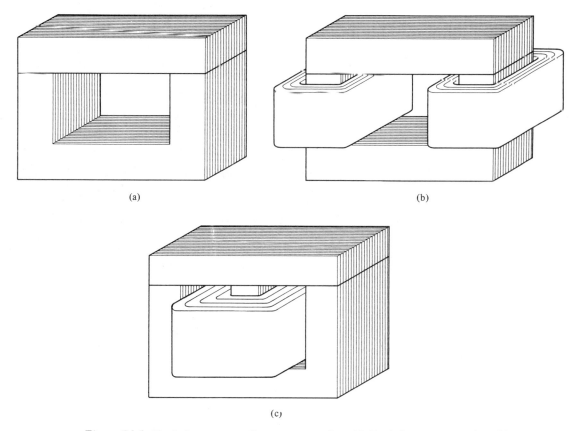

Figure 24-1 Typical power transformer construction. (a) Typical core construction. (b) Typical coil and core arrangement for what is called "core" type construction. (c) Shell type construction of an audio or power transformer. The primary and secondary windings are on the same leg of the core.

Sec. 24-2. Transformer Construction

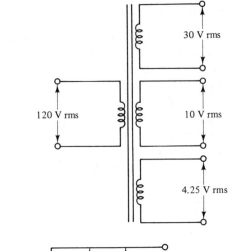

Figure 24-2 A small power transformer with three secondary windings. The separate secondaries allow the design of circuits with differing operating voltages but with a common power transformer.

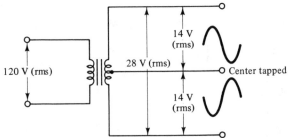

Figure 24-3 A center tapped (c.t.) power transformer. In this illustration it is the secondary winding with the center tap. This allows us to obtain two equal voltages with exactly opposite polarities. Also available, is the entire secondary winding voltage.

if ordinary metal cores were used. The powdered iron core has very low eddy current loss because the ceramic binder insulates the iron granules from one another. Examples of RF transformers are shown in the illustrations. When air is the core material, the coils are wound upon paper or plastic tubes.

It is standard practice to refer to the coil connected to the input as the *primary* winding. Any other coil is then a *secondary* winding. Transformers may have several secondaries, as in Fig. 24-2. Windings, primary and secondary, may have tap points. These allow us to modify voltage and current ratios. Of special importance is a center-tapped secondary, allowing two equal output voltages, but with opposite polarities (Fig. 24-3).

24-3. VOLTAGE RATIOS

For power and audio transformers, we assume an ideal transformer. We use Faraday's law to find induced voltages. The induced voltage is determined by the rate of change of flux linkages. If we define the primary winding as N_1, where N is the number of turns, the rate of change of the flux as the fraction $d\Phi/dt$, and k as the coefficient of coupling, then the primary induced voltage is

$$v_1 = kN_1 \frac{d\Phi}{dt}$$

In the same way, we find the induced voltage of a secondary as:

$$v_2 = kN_2 \frac{d\Phi}{dt}$$

where N_2 is the number of turns on the secondary. If we take the ratio of the voltages, we have

$$\frac{v_1}{v_2} = \frac{kN_1(d\Phi/dt)}{kN_2(d\Phi/dt)}$$

$$\frac{v_1}{v_2} = \frac{N_1}{N_2} \qquad (24\text{-}1)$$

The turns ratio is given the symbol a.

$$\frac{v_1}{v_2} = \frac{N_1}{N_2} = a \qquad (24\text{-}2)$$

A transformer with more primary turns than secondary turns is a *step-down* transformer. In this case a is greater than 1, and V_2 is a fraction of V_1. We may use instantaneous values, effective values, or phasors in our ratios. The voltage relationships are always controlled by the turns ratio.

$$V_2 = \frac{V_1}{a} \qquad (24\text{-}3)$$

A *step-up* transformer has more secondary turns than primary turns. By our definition, a is a fraction. Secondary voltage is greater than primary voltage. Some sources define a as the ratio of N_2/N_1. To minimize confusion, we shall write the ratio of turns as $N_1 : N_2$, so that $1 : 4$ is a step-up transformer with 4 turns on the secondary for each primary turn. In the same way, a transformer with 300 primary turns and 100 secondary turns is a $3 : 1$ step-down transformer. We shall identify these ratios simply as N. This variation in notation is for our convenience. When working with formulas, use a in its defined form.

EXAMPLE 24-1:

A power transformer has a secondary voltage of 440 V when 120 V is applied to the primary winding. (a) What is a? (b) If there are 90 turns on the primary, how many turns are there on the secondary?

Solution:

(a) $a = \dfrac{V_1}{V_2} = \dfrac{120}{440}$

$= 0.2727$

(b) Given that $a = N_1/N_2$, then

$$N_2 = \frac{N_1}{a}$$
$$= \frac{90}{0.2727}$$
$$= 330 \text{ turns}$$

Note that this is a 1 : 3.67 step-up transformer.

EXAMPLE 24-2:
We require an effective voltage of 12 V from a 120-V line. What is the turns ratio of the transformer?

Solution: By inspection of the voltage ratio, we find that the transformer is a 10 : 1 step-down unit.

$$a = \frac{V_1}{V_2}$$
$$= \frac{120}{12} = 10$$

24-4. CURRENT RATIOS

In all closely coupled transformers, we assume that input volt-amperes is equal to output volt-amperes.

$$V_1 I_1 = V_2 I_2$$

If we solve the equation for a, we have

$$a = \frac{V_1}{V_2} = \frac{I_2}{I_1} \qquad (24\text{-}4)$$

and

$$I_1 = \frac{I_2}{a} \qquad (24\text{-}5)$$

The current ratio is the reciprocal of the voltage ratio. A step-up transformer has a step-down ratio for current. A step-down transformer has a step-up ratio for current.

EXAMPLE 24-3:
A transformer with $a = 24$ is connected to a 120-V line. With a 10-Ω load connected across the secondary, primary current is 20.8 mA. Demonstrate that primary VA = secondary VA.

Solution:

1. $V_2 = \dfrac{V_1}{a}$

$$= \frac{120}{24}$$
$$= 5 \text{ V}$$

2. $I_2 = \dfrac{V_2}{R_L}$

$$= \frac{5}{10}$$
$$= 0.5 \text{ A}$$

3. $I_1 = \dfrac{I_2}{a}$

$$= \frac{0.5}{24}$$
$$= 20.8 \text{ mA}$$

4. $V_1 I_1 = 120 \times 0.0208$
$$= 2.496 \text{ VA} \approx 2.5 \text{ VA}$$

$V_2 I_2 = 5 \times 0.5$
$$= 2.5 \text{ VA}$$

EXAMPLE 24-4:
A 1:3.33 step-up transformer is connected to a load so that primary current is 12 A. What is secondary current?

Solution: *A step-up transformer steps down current.*

$$I_2 = \frac{12}{3.33} = 3.6 \text{ A}$$

If we examine Eq. (24-5), we might assume that without secondary current, primary current is zero. This is not exactly right! The primary current does depend upon secondary current. However, there is always a very small primary current, called *magnetizing current*, that flows under all conditions. In the absence of any secondary current, this current is just enough to maintain the flux. When a load draws secondary current, the primary current must increase to maintain the flux in spite of the opposition caused by the secondary current. Remember that the secondary current is supplied by an *induced* voltage. In accordance with Lenz's law, this current must establish a flux that opposes the original flux. This is why the primary current must change with changes in secondary current.

EXAMPLE 24-5:
Given the circuit of Fig. 24-4, find I_2 and I_1 for each of the listed load resistances: $R_L = 25 \, \Omega, 10 \, \Omega, 1 \, \Omega$.

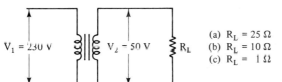

(a) $R_L = 25 \, \Omega$
(b) $R_L = 10 \, \Omega$
(c) $R_L = 1 \, \Omega$

Figure 24-4

Solution:

1. $a = \dfrac{V_1}{V_2}$

   $= \dfrac{230}{50}$

   $= 4.6$

2. $I_2 = \dfrac{V_2}{R_L} = \dfrac{50}{R_L}$

   $I_1 = \dfrac{I_2}{a} = \dfrac{I_2}{4.6}$

   For $R_L = 25\ \Omega$:

   $I_2 = \dfrac{50}{25} = 2\ \text{A}$

   $I_1 = \dfrac{2}{4.6} = 0.435\ \text{A}$

   For $R_L = 10\ \Omega$:

   $I_2 = \dfrac{50}{10} = 5\ \text{A}$

   $I_1 = \dfrac{5}{4.6} = 1.087\ \text{A}$

   For $R_L = 1\ \Omega$:

   $I_2 = \dfrac{50}{1} = 50\ \text{A}$

   $I_1 = \dfrac{50}{4.6} = 10.87\ \text{A}$

24-5. IMPEDANCE RATIOS

Impedance is another quantity that is "transformed" by transformers. In this case, we are able to demonstrate another way to illustrate the current change in the primary circuit as a result of load changes. We find that the secondary reflects impedance into the primary. This reflected impedance is in *parallel* with the primary winding.

Let Z_1 = primary impedance and $Z_2 = Z_L$ = secondary load impedance. Then

$$Z_1 = \dfrac{V_1}{I_1}$$

and

$$Z_2 = \dfrac{V_2}{I_2}$$

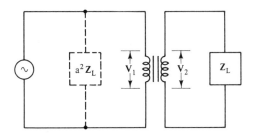

Figure 24-5 Impedance reflection in a closely coupled transformer. Recall that a is whole number for step-down transformers, while it is a fraction for step-up transformers. Therefore a step-down transformer reflects an impedance that is greater than load impedance, while a step-up transformer reflects an impedance that is less than the load impedance.

Taking the ratio of $Z_1 : Z_2$, we have

$$\frac{Z_1}{Z_2} = \frac{V_1}{V_2} \times \frac{I_2}{I_1} = a^2$$

Substituting Z_L for Z_2, and solving for Z_1,

$$Z_1 = a^2 Z_L \qquad (24\text{-}6)$$

where Z_1 is the *reflected* impedance from secondary into primary. Note that the characteristics of the reflected impedance are the same as those of the load impedance. Impedance reflection is illustrated in Fig. 24-5.

EXAMPLE 24-6:
A transformer with a 20:1 turns ratio is connected between an amplifier and its load. Calculate the reflected impedance if (a) $Z_L = R_L = 8\ \Omega$, (b) $Z_L = 6\underline{/30°}\ \Omega$, and (c) $Z_L = 5\underline{/-40°}\ \Omega$.

Solution:
1. Solve for a.

$$a = \frac{N_1}{N_2} = 20$$

2. $Z_1 = a^2 Z_L$.
   (a) $Z_1 = 20^2 \times 8\underline{/0°}$
      $= 3200\underline{/0°}\ \Omega$
   (b) $Z_1 = 20^2 \times 6\underline{/30°}$
      $= 2400\underline{/30°}\ \Omega$
   (c) $Z_1 = 20^2 \times 5\underline{/-40°}$
      $= 2000\underline{/-40°}\ \Omega$

Clearly, one use of transformers is in the "matching" or changing of impedance. This is a common application in low-frequency and audio power amplifier circuits. For example, assume that a loudspeaker is the actual load for an audio amplifier. Suppose that the impedance of the speaker is 8 Ω. If the amplifier is designed to work into a 2400-Ω load, we use a transformer so that the 8-Ω load is reflected into the primary as 2400 Ω. For these applications, we first solve for a.

$$a = \sqrt{\frac{Z_1}{Z_L}} \qquad (24\text{-}7)$$

1. $Z_1 > Z_L$: then $a > 1$ and we have a step-down transformer.
2. $Z_1 < Z_L$: then $a < 1$, and we have a step-up transformer.

EXAMPLE 24-7:
Given the circuit of Fig. 24-6, calculate the resistance reflected into the primary. $R_{in} = 26$ kΩ and $N_1 : N_2 = 1 : 4$.

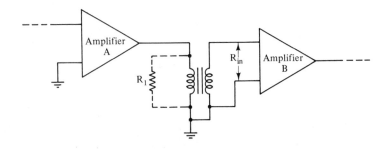

Figure 24-6

Solution:

$$a = \frac{N_1}{N_2} = \frac{1}{4}$$
$$= 0.25$$

$$R_1 = a^2 R_{in}$$
$$= 0.25^2 \times 26{,}000$$
$$= 1625 \text{ Ω}$$

EXAMPLE 24-8:
In the circuit of Fig. 24-7, specify the turns ratio of the transformer, given that the proper load impedance for the amplifier is 3200 Ω and the speaker impedance is 4 Ω.

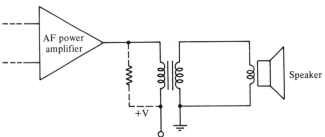

Figure 24-7

Solution:
1. Let $R_1 = R_L$ for the amplifier.
$$R_1 = 3200 \ \Omega$$
2. Solve for a.
$$a = \sqrt{R_1/R_L}$$
$$= \sqrt{3200/4}$$
$$= \sqrt{800}$$
$$= 28.3$$
3. The transformer must have a turns ratio of 28.3 : 1

You will find that supply catalogs usually list transformers in terms of impedance ratios. This transformer would be listed as "3200 Ω pri./4 Ω sec."

EXAMPLE 24-9:
An unknown transformer is tested. An input of 20 V is applied to one winding and an output of 0.5 V is measured across the terminals of the other winding. (a) What is the turns ratio? (b) What is the impedance transformation ratio?

Solution:
(a) $a = \dfrac{V_1}{V_2}$
$= \dfrac{20}{0.5} = 40$

The turns ratio is 40 : 1.
(b) The impedance transformation ratio is the square of a.
$$a^2 = 40^2 = 1600$$

We can make use of transformers in the application of the maximum power transfer theorem. Such an application is fine for power *transfer*. We *cannot* design an audio power amplifier for maximum possible power output because of the excessive distortion that would occur. Instead, we design for maximum undistorted output and cannot take advantage of the theorem. This is a topic you will study in advanced electronics courses.

24-6. AUTOTRANSFORMERS

Up to this point, we have considered transformers with windings that were conductively isolated from each other. There is a category of transformer with windings that are common to input and output circuits. These are known as *autotransformers*. Their circuit symbols are shown in Fig. 24-8.

Autotransformers have the advantage of requiring a much lower kVA rating for a given load when compared to a conventional transformer. The

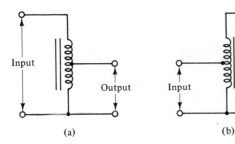

Figure 24-8 Autotransformer circuit symbols. (a) A step-down autotransformer. (b) A step-up autotransformer.

disadvantage is that there is no conductive isolation between secondary and primary. As a result, most applications are for voltage-level adjustments requiring low turns ratios.

In all applications, we make use of Kirchhoff's current law to determine currents in the windings. *In all cases, input kVA must equal load kVA.* The following examples will illustrate these concepts.

EXAMPLE 24-10:
An autotransformer is used to adjust a 280-V source to a 200-V load. Load resistance is 10 Ω. (a) Sketch the circuit and use the KCL to determine currents (Fig. 24-9). (b) Demonstrate that the kVA requirement of each winding is much less than input and output kVA.

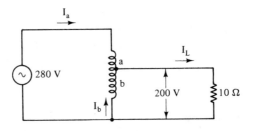

Figure 24-9

Solution:
(a) Current calculations:

1. $I_L = \dfrac{200 \text{ V}}{10 \text{ }\Omega}$

   $= 20 \text{ A}$

2. Solution for I_a:

$$\text{input VA} = \text{output VA}$$
$$280 \times I_a = 200 \times 20$$
$$I_a = \dfrac{4000}{280}$$
$$= 14.29 \text{ A}$$

3. Solution for I_b:
$$I_a + I_b = I_L$$
$$I_b = I_L - I_a$$
$$= 20 - 14.29$$
$$= 5.71 \text{ A}$$

(b) kVA relations:
 1. Input and output kVA:
$$4.0 \text{ kVA}$$
 2. Winding a:
$$80 \times 14.29 = 1143 \text{ VA}$$
Winding a kVA = 1.14.
 3. Winding b:
$$200 \times 5.71 = 1142 \text{ VA}$$
Winding b kVA = 1.14.

The kVA per winding is a fraction of the input and output kVA.

EXAMPLE 24-11:
An autotransformer is used to change a 220-V source to 250 V. The load is 5 Ω. (a) Sketch the circuit and use the KCL to determine currents. (b) Demonstrate that the kVA requirement of each winding is much less than input and output kVA.

Solution: Refer to Fig. 24-10.

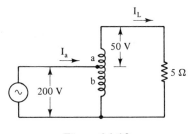

Figure 24-10

(a) Current calculations:
 1. $I_L = \dfrac{250}{5}$
 $= 50 \text{ A}$
 2. Solution for I_a:
$$\text{output VA} = 250 \times 50$$
$$= 12{,}500 \text{ VA}$$
$$= 12.5 \text{ kVA}$$

Sec. 24-6. Autotransformers

$$I_a \times 200 = 12{,}500 \text{ VA}$$
$$I_a = \frac{12{,}500}{200}$$
$$= 62.5 \text{ A}$$

3. Solution for I_b:
$$I_b = I_a - I_L$$
$$= 62.5 - 50$$
$$= 12.5 \text{ A}$$

(b) kVA relations:
  1. Input and output kVA:
$$12.5 \text{ kVA}$$

  2. Winding a:
$$50 \text{ V} \times 50 \text{ A} = 2.5 \text{ kVA}$$

  3. Winding b:
$$200 \text{ V} \times 12.5 \text{ A} = 2.5 \text{ kVA}$$

24-7. LOOSELY COUPLED TRANSFORMERS

This broad category of transformers is often called RF transformers. Recall that in all analysis of transformers, up to this point, turns ratio has been an important consideration. This is because we considered the transformers to be ideal, mainly due to the *very high* amount of coupling between the windings. In fact, we considered the coefficient of coupling to be equal to 1. We now must study transformers with such low coefficients of coupling that k ranges from 0.002 up to about 0.2.

Induced Voltage

1. We must calculate mutual inductance between windings. From our previous studies,
$$M = k\sqrt{L_1 L_2}$$

2. The induced voltage from primary to secondary is found by
$$v_s = -j\omega M i_p \qquad (24\text{-}8)$$

where i_p is the primary current at any instant. We do not consider turns ratio because of the low coefficient of coupling. Also, we treat v_s as a voltage in *series* with the secondary circuit, rather than in parallel.

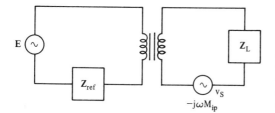

Figure 24-11 Impedance reflection in a loosely coupled transformer. Note that the reflected impedance is in series with the primary winding. Note also that the secondary *induced* voltage is a series voltage.

Reflected Impedance

The effect of secondary current upon the primary results in a *series* reflected impedance, as shown in Fig. 24-11.

$$Z_{ref} = \frac{(\omega M)^2}{Z_s} \qquad (24\text{-}9)$$

where Z_s is the *series* impedance of the secondary. It is important to note that inductance reflects as capacitance, while capacitance reflects as inductance.

We use RF transformers with resonant circuits. We have several general methods.

1. Tuned secondary, untuned primary: The secondary coil is made part of a *series* resonant circuit, as shown in Fig. 24-12. At resonance, the impedance is resistive. The reflected impedance is also purely resistive. The output voltage, taken across the capacitor, is Q times the induced voltage. We shall demonstrate this widely used coupling with an example.
2. Tuned primary, untuned secondary: The primary is a *parallel* resonant circuit load on a source, as in Fig. 24-13. We find the Q rise of circulating current for tank current. This results in a higher induced

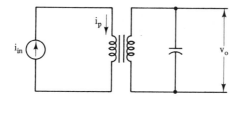

Figure 24-12 An RF transformer coupling circuit.

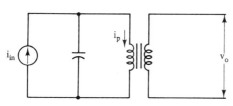

Figure 24-13 An RF transformer used in a tuned primary-untuned secondary coupling circuit.

Sec. 24-7. Loosely Coupled Transformers

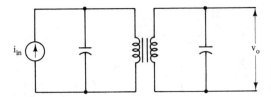

Figure 24-14 Tuned primary-tuned secondary RF transformer coupling.

voltage in the secondary at resonance. It can be shown that the voltage in the secondary is the same as the voltage across the capacitor in the tuned secondary circuit.

3. Tuned primary, tuned secondary: Both windings are part of resonant circuits, as in Fig. 24-14. This is a band-pass coupling method and results in a response that is flatter in the region of resonance. The rate of drop-off of output for frequencies outside the passband is much greater than with single tuned circuits. The circuit is widely used in the IF system of radio receivers.

EXAMPLE 24-12:
Given the circuit of Fig. 24-15, solve for the output voltage at the resonant frequency.

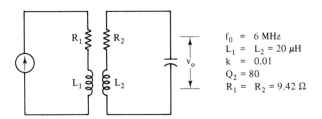

$f_0 = 6$ MHz
$L_1 = L_2 = 20\ \mu H$
$k = 0.01$
$Q_2 = 80$
$R_1 = R_2 = 9.42\ \Omega$

Figure 24-15

Solution:

1. Solve for mutual inductance.

$$M = k\sqrt{L_1 L_2}$$
$$= 0.01 \times \sqrt{400 \times 10^{-12}}$$
$$= 0.2\ \mu H$$

2. Solve for $\omega_0 M$.

$$\omega_0 M = 6.28 \times 6 \times 10^6 \times 0.2 \times 10^{-6}$$
$$= 7.536\ \Omega$$

3. Note that we are driving the primary from a current source, like a bipolar transistor. The current is independent of the reflected resistance. In any event, the reflected resistance is 6.03 Ω.

$$R_{\text{ref}} = \frac{7.536^2}{9.42} = 6.03\ \Omega$$

4. Solve for the induced voltage. This voltage is considered to be in series with the secondary.

$$v_s = -j\omega M i_p$$
$$= -j7.536 \times 0.5 \times 10^{-3}$$
$$= j3.768 \text{ mV}$$

5. Solve for the voltage across the capacitor. At f_0, the voltage is Q times v_s.

$$v_{out} = Q_2 v_s$$
$$= 80 \times 3.768 \text{ mV}$$
$$= -301.4 \text{ mV} \quad (\text{total phase shift} = 180°)$$

SUMMARY

1. Transformers are a broad category of devices both in physical size and frequency of operation.
2. Power and audio transformers are made with high-permeability cores, which result in a very high degree of coupling ($k \simeq 1$).
3. High-frequency (RF) transformers must use powdered iron or air as transformer cores, which results in high leakage loss and low coefficients of coupling ($k \ll 1$).
4. In power and audio transformers, the ratio of turns equals the ratio of voltages.
5. The ratio of currents in power and audio transformers equals the inverse of the turns ratio.
6. In audio and power transformers, secondary impedance is reflected back into the primary in *parallel* with the primary winding.
7. The reflected impedance in audio and power transformers is modified by the square of the turns ratio.
8. A step-down transformer reflects a high impedance across the primary.
9. The volt-ampere products of primary and secondary are assumed to be equal in power transformers.
10. Autotransformers do not have conductive isolation between windings.
11. The kVA requirements for an autotransformer are much lower than those for a conventional transformer for a given load kVA.
12. RF transformers are usually used with resonant circuits.
13. The reflected impedance in an RF transformer circuit appears in *series* with the primary winding.
14. The series reflected impedance in an RF transformer circuit varies directly with the square of the product of mutual inductance and frequency.

PROBLEMS

24-1. In the circuit of Fig. 24-16, $N = 13.2 : 1$ and $R_L = 15 \, \Omega$. Solve for V_L, I_L, and I_p.

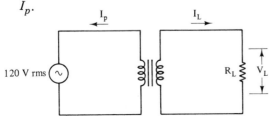

Figure 24-16

24-2. Given the data of Prob. 24.1, demonstrate that the volt-ampere products of the primary and secondary are equal.

24-3. Given the circuit of Fig. 24-16, $R_L = 800\ \Omega$ and $V_L = 480$ V. (a) Solve for the turns ratio. (b) Solve for primary current.

24-4. In the circuit of Fig. 24-16, $I_L = 2.5$ A. If $R_L = 9.6\ \Omega$, solve for V_L and I_p.

24-5. A transformer has a secondary voltage of 120 V when 480 V, 60 Hz is applied to the primary. If the secondary load current is 60 A, what is the primary current?

24-6. Given the circuit of Fig. 24-17, solve for primary current.

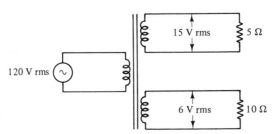

Figure 24-17

24-7. Given the data of Prob. 24-6, demonstrate that the primary volt-amperes is equal to the sum of the secondary volt-amperes.

24-8. A power transformer with a center-tapped secondary has a total turns ratio of 5 : 1 step-down. If 120 V, 60 Hz is applied to the primary, (a) what is the total secondary voltage? (b) what is each secondary voltage with respect to center tap?

24-9. A transformer with $a = 30.4$ is used for impedance transformation. If the secondary load resistance is 4 Ω, what is the reflected resistance?

24-10. We are required to couple a load for maximum power transfer. The source resistance is 200 Ω and the load resistance is 750 Ω. Specify the transformer turns ratio. Is this a step-up or step-down transformer?

24-11. An audio amplifier drives an 8-Ω load, as in Fig. 24-18. The required load as seen on the primary side is 1500 Ω. Specify the turns ratio of the output transformer.

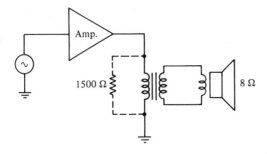

Figure 24-18

24-12. Calculate the mutual inductance:
    (a) $L_1 = 30\ \mu H$, $L_2 = 50\ \mu H$, $k = 0.3$
    (b) $L_1 = 100\ \mu H$, $L_2 = 95\ \mu H$, $k = 0.42$
    (c) $L_1 = L_2 = 15\ \mu H$, $k = 0.05$

(d) $L_1 = 75\ \mu H$, $L_2 = 80\ \mu H$, $k = 0.15$
(e) $L_1 = 150\ \mu H$, $L_2 = 120\ \mu H$, $k = 0.25$

24-13. Solve for the reflected impedance at resonance for the circuit of Fig. 24-19.

24-14. In the circuit of Fig. 24-19, solve for the induced voltage, v_s.

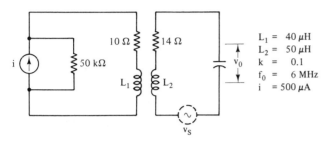

Figure 24-19

24-15. Given that the circuit of Fig. 24-19 is operating at resonance, solve for the output voltage.

25

Network Theorems and Filters

25-1. DELTA-TO-WYE CONVERSION

A form of circuit that cannot be solved by series–parallel circuit analysis is the *bridge* circuit (Fig. 25-1). Bridge circuits are used in many control and measurement applications. Also, variations of the basic circuit may appear in other networks. Common to the bridge circuit is the delta (Δ) arrangement of impedances. We also refer to this arrangement as a pi (π) network (Fig. 25-2).

We may use mesh analysis to solve delta circuits, but a more direct approach is to substitute an equivalent circuit for the delta. This circuit is called a wye (Y) or tee (T) circuit. If one circuit is equivalent to another, it must be so at all terminals of the circuits. Note that in Fig. 25-3 we identify impedances seen between pairs of terminals for the delta and wye circuits by the use of double subscripts. In each circuit, Z_{a-b} is the impedance between terminals a and b, Z_{b-d} is the impedance between terminals b and d, and Z_{c-d} is the impedance between terminals c and d.

$$\begin{array}{cc} \text{Wye} & \text{Delta} \\ Z_{a-b} = Z_a + Z_b = Z_1 \parallel (Z_3 + Z_2) \\ Z_{b-d} = Z_b + Z_c = Z_2 \parallel (Z_1 + Z_3) \\ Z_{c-d} = Z_a + Z_c = Z_3 \parallel (Z_1 + Z_2) \end{array}$$

When circuits are equivalent, the respective terminal impedances are equal.

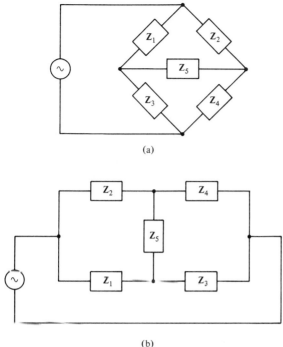

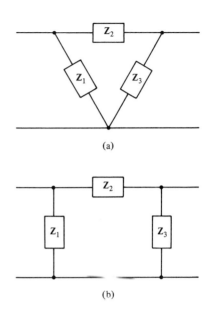

Figure 25-1 Bridge circuit diagrams. (a) Typical method of drawing the circuit. (b) Another way of representing the circuit.

Figure 25-2 Delta and pi networks. (a) Delta network. (b) Pi network. The network is exactly equivalent to a delta network.

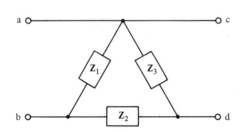

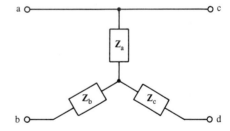

Figure 25-3 Terminal impedances of delta and wye networks.

$$Z_a + Z_b = \frac{Z_1(Z_2 + Z_3)}{Z_1 + Z_2 + Z_3}$$

$$Z_b + Z_c = \frac{Z_2(Z_1 + Z_3)}{Z_1 + Z_2 + Z_3}$$

$$Z_a + Z_c = \frac{Z_3(Z_1 + Z_2)}{Z_1 + Z_2 + Z_3}$$

Sec. 25-1. Delta-to-Wye Conversion

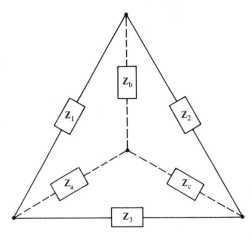

Figure 25-4 Delta to wye conversion. To convert delta to wye, simply sketch the wye circuit within the delta. The components of the wye are found as the product of adjacent delta impedances divided by the sum of the delta impedances.

Solving the equations for Z_a, Z_b, and Z_c in terms of Z_1, Z_2, and Z_3, we have the Y equivalents of the Δ impedances.

$$\Delta \text{ to Y:} \quad Z_a = \frac{Z_1 Z_3}{Z_1 + Z_2 + Z_3} \quad (25\text{-}1)$$

$$Z_b = \frac{Z_1 Z_2}{Z_1 + Z_2 + Z_3} \quad (25\text{-}2)$$

$$Z_c = \frac{Z_2 Z_3}{Z_1 + Z_2 + Z_3} \quad (25\text{-}3)$$

A convenient memory aid is given in Fig. 25-4.

EXAMPLE 25-1:
Given the circuit of Fig. 25-5, solve for the voltage drop across R_2.

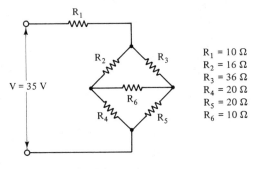

$R_1 = 10 \, \Omega$
$R_2 = 16 \, \Omega$
$R_3 = 36 \, \Omega$
$R_4 = 20 \, \Omega$
$R_5 = 20 \, \Omega$
$R_6 = 10 \, \Omega$

Figure 25-5

Solution:
1. Redraw the circuit, replacing R_4, R_5, and R_6 by a wye equivalent circuit (Fig. 25-6).

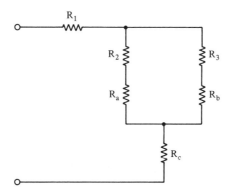

Figure 25-6

2. Solve for the resistances of the wye equivalent circuit.

$$R_a = \frac{R_4 R_6}{R_4 + R_5 + R_6}$$
$$= \frac{20 \times 10}{50}$$
$$= 4 \ \Omega$$

$$R_b = \frac{R_5 R_6}{R_4 + R_5 + R_6}$$
$$= \frac{20 \times 10}{50}$$
$$= 4 \ \Omega$$

$$R_c = \frac{R_4 R_5}{R_4 + R_5 + R_6}$$
$$= \frac{20 \times 20}{50}$$
$$= 8 \ \Omega$$

3. Redraw the circuit, and solve the resultant series-parallel circuit for the voltage drop across R_2 (Fig. 25-7).

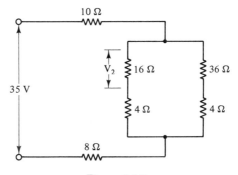

Figure 25-7

(a) $R_{eq} = (16 + 4) \parallel (36 + 4)$
$= \dfrac{20 \times 40}{40}$
$= 13.333 \, \Omega$

(b) Solve for the voltage drop across R_{eq}.

$$V_p = 35 \times \dfrac{13.333}{10 + 13.333 + 8}$$
$$= 14.89 \text{ V}$$

(c) Once the voltage drop across the parallel circuit is known, we find V_2.

$$V_2 = V_p \times \dfrac{16}{16 + 4}$$
$$= 14.89 \times \dfrac{16}{20}$$
$$= 11.9 \text{ V}$$

Proof:

1. Solve for the current in each branch of the parallel circuit.

$$I_2 = \dfrac{14.89}{20}$$
$$= 0.7445 \text{ A}$$

$$I_3 = \dfrac{14.89}{40}$$
$$= 0.3723 \text{ A}$$

2. The current in R_1 and R_c is the sum of the branch currents.

$$I_1 = I_2 + I_3$$
$$= 0.7445 + 0.3723$$
$$= 1.12 \text{ A}$$

3. We now return to the original circuit, redrawn in Fig. 25-8.

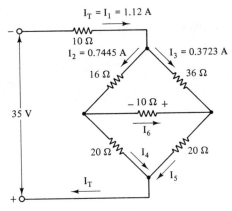

Figure 25-8

(a) The voltage drop across R_4 is found. The sum of the voltages across $R_1, R_2,$ and R_4 must equal applied voltage.

$$V_4 = 35 - (I_1 R_1 + I_2 R_2)$$
$$= 35 - (1.12 \times 10 + 0.7445 \times 16)$$
$$= 11.9 \text{ V}$$

(b) Solve for I_4.

$$I_4 = \frac{V_4}{R_4}$$
$$= \frac{11.9}{20}$$
$$= 0.595 \text{ A}$$

(c) Use the KCL to solve for I_6.

$$I_6 = I_2 - I_4$$
$$= 0.7445 - 0.595$$
$$= 0.15 \text{ A}$$

(d) We use the assumed current directions shown in the circuit diagram. The algebraic sum of all the voltage drops must equal zero.

$$V_2 + V_6 + V_3 = 0$$
$$-I_2 R_2 - I_6 R_6 + I_3 R_3 = 0$$
$$-0.7445 \times 16 - 0.15 \times 10 +$$
$$0.3723 \times 36 = 0$$
$$-11.91 - 1.5 + 13.4 = 0$$
$$0 = 0$$

4. We may also prove the results by demonstrating that the sum of the currents in R_4 and R_5 is equal to the total current.
   (a) $I_5 = I_6 + I_3$
   $= 0.15 + 0.3723$
   $= 0.5223 \text{ A}$
   (b) $I_T = I_1 = I_4 + I_5$
   $= 0.595 + 0.5223$
   $= 1.12 \text{ A}$

We may also convert a wye circuit to an equivalent delta circuit. We begin with the same set of impedance equations. However, we now solve for $Z_1, Z_2,$ and Z_3 in terms of $Z_a, Z_b,$ and Z_c.

Y to Δ:
$$Z_1 = \frac{Z_a Z_b + Z_a Z_c + Z_c Z_b}{Z_c} \tag{25-4}$$

$$Z_2 = \frac{Z_a Z_b + Z_a Z_c + Z_c Z_b}{Z_a} \tag{25-5}$$

Sec. 25-1. Delta-to-Wye Conversion

$$Z_3 = \frac{Z_a Z_b + Z_a Z_c + Z_c Z_b}{Z_b} \tag{25-6}$$

25-2. BRIDGE CIRCUITS

We stated that bridge circuits are used in some measurement and control applications. Consider the basic circuit of Fig. 25-1. The bridge is considered to be balanced if there is zero difference of potential across Z_5. This requires that the voltage drop across Z_1 be equal to the drop across Z_2 and that the drops across Z_3 and Z_4 be equal. As a result, the current in Z_5 is zero. Therefore,

$$I_1 = I_3, \quad V_1 = V_2$$
$$I_2 = I_4, \quad V_3 = V_4$$

and

$$I_1 Z_1 = I_2 Z_2$$
$$I_1 Z_3 = I_2 Z_4$$

By taking the ratio of the voltages, we have

$$\frac{I_1 Z_1}{I_1 Z_3} = \frac{I_2 Z_2}{I_2 Z_4}$$

and at balanced condition

$$\frac{Z_1}{Z_3} = \frac{Z_2}{Z_4} \tag{25-7}$$

$$Z_1 Z_4 = Z_2 Z_3 \tag{25-8}$$

In order for the requirements of Eq. (25-8) to be satisfied, the magnitudes and resulting phase angles must be equal.

$$|Z_1 Z_4| = |Z_2 Z_3|$$
$$\theta_1 + \theta_4 = \theta_2 + \theta_3$$

Wheatstone Bridge

Very accurate resistance measurements are made by using a resistance bridge arrangement, as in Fig. 25-9. A sensitive meter, a *galvanometer* with a zero center scale, is used in place of Z_5. The voltage source may be ac or dc. In most instruments a battery is used. The resistors R_1, R_3, and R_2 are known values, with R_x representing the unknown resistance that is to be measured.

Solving Eq. (25-8) for R_4, we have

$$R_4 = \frac{R_2}{R_1} R_3 = R_x$$

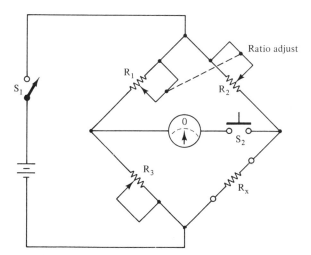

Figure 25-9 A schematic representation of a practical resistance bridge.

In commercial Wheatstone bridges the ratio of R_2/R_1 is adjusted so as to select some resistance range. This is done by a single selector switch. The resistance of R_3 is selected by a series of switches connected to calibrated dials. Usually, one dial indicates the digit to the left of the decimal point. The remaining dials indicate values to the right of the decimal point. In operation, the range is selected, and then the R_3 controls are adjusted for zero deflection of the galvanometer needle. The unknown resistance is the reading from the dials multiplied by the range switch. In many bridges, the range switch is called a multiplier.

EXAMPLE 25-2:
A Wheatstone bridge is used to measure an unknown resistance. The dials read 6.80314 and the multiplier is on the 100-kΩ range. What is the resistance?

Solution:

$$R_x = 6.80314 \times 100 \text{ k}\Omega$$
$$= 680.314 \text{ k}\Omega$$
$$= 680,314 \text{ }\Omega$$

Capacitance Bridge

Unknown capacitance, including resistance components, may be measured accurately by a bridge circuit. In Fig. 25-10, C_s is a capacitance standard; R_1 and R_2 are the ratio resistors and determine the capacitance range. R_3 is adjusted for the balance condition.

At balance,

$$R_1\left(R_x - j\frac{1}{\omega C_x}\right) = R_2\left(R_3 - j\frac{1}{\omega C_s}\right)$$

$$R_1 R_x - j\frac{R_1}{\omega C_x} = R_2 R_3 - j\frac{R_2}{\omega C_s}$$

Sec. 25-2. Bridge Circuits

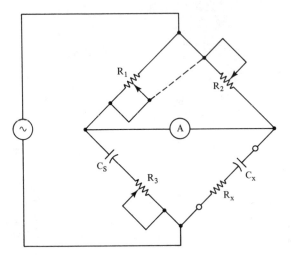

Figure 25-10 Capacitance bridge circuit.

Therefore, the resistance terms on each side of the equation must be equal, just as the reactance terms must equal each other.

$$R_1 R_x = R_2 R_3$$

$$-j\frac{R_1}{\omega C_x} = -j\frac{R_2}{\omega C_s}$$

Solving for R_x and C_x, we have

$$R_x = \frac{R_2}{R_1} R_3 \qquad (25\text{-}9)$$

$$C_x = \frac{R_1}{R_2} C_s \qquad (25\text{-}10)$$

Inductance Bridge

While inductance may be measured with an inductance substitution bridge by use of an inductance standard, it is more commonly done with a Maxwell bridge, as in Fig. 25-11. This permits us to use a capacitance standard, which results in a standard that is smaller and less affected by magnetic fields. The capacitance standard is used in parallel with a shunt resistance. The parallel combination is identified as Z_1.

At balance,

$$Z_4 = \frac{Z_2 Z_3}{Z_1}$$
$$= (Z_2 Z_3) Y_1$$

where $Z_4 = R_x + j\omega L_x$
$Z_3 = R_3$

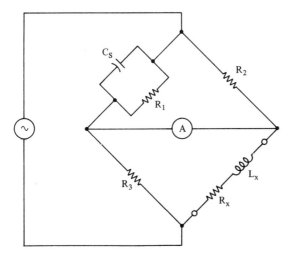

Figure 25-11 Inductance bridge circuit.

$$Z_2 = R_2$$
$$Y_1 = \frac{1}{R_1} + j\omega C_s$$

Therefore,

$$R_x + j\omega L_x = R_2 R_3 \left(\frac{1}{R_1} + j\omega C_s\right)$$

$$R_x + j\omega L_x = \frac{R_2 R_3}{R_1} + j\omega C_s R_2 R_3$$

As with the capacitance bridge, we solve for R_x and L_x.

$$R_x = \frac{R_2}{R_1} R_3 \qquad (25\text{-}11)$$

$$L_x = C_s R_2 R_3 \qquad (25\text{-}11)$$

25-3. THEVENIN'S THEOREM

In our earlier studies of Thevenin's and Norton's theorems, we worked with purely resistive circuits and dc sources. Both theorems have great value in ac circuit analysis, especially with amplifier circuits. We begin our studies in ac circuit analysis by restating Thevenin's theorem.

> The voltage across a two-terminal impedance may be found by connecting the impedance to an equivalent voltage source, V_{Th}. The voltage of the source is the voltage across the load terminals with the impedance removed. The internal impedance of the source, Z_{Th}, is the impedance found by looking back from the load terminals, with all energy sources replaced by their internal impedances.

EXAMPLE 25-3:

Given the circuit of Fig. 25-12, solve for the voltage drop across Z_3.

$$Z_1 = 180\underline{/-30°} = 156 - j90 \text{ } \Omega$$
$$Z_2 = 300\underline{/45°} = 212 + j212 \text{ } \Omega$$
$$Z_3 = 100\underline{/60°} = 50 + j86.7 \text{ } \Omega$$
$$Z_4 = 25\underline{/0°} = 25 + j0 \text{ } \Omega$$

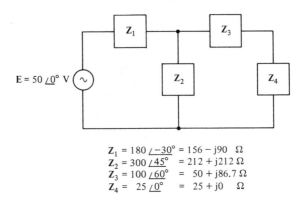

$Z_1 = 180\underline{/-30°} = 156 - j90$ Ω
$Z_2 = 300\underline{/45°} = 212 + j212$ Ω
$Z_3 = 100\underline{/60°} = 50 + j86.7$ Ω
$Z_4 = 25\underline{/0°} = 25 + j0$ Ω

Figure 25-12

Solution:

1. Remove Z_3 and sketch the circuit (Fig. 25-13). V_{Th} is the voltage across the open terminals.

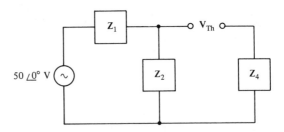

Figure 25-13

2. Solve for V_{Th}. Because there is no voltage drop across Z_4 (open circuit), V_{Th} equals the voltage drop across Z_2.

$$V_{Th} = E \frac{Z_2}{Z_1 + Z_2}$$

$$= 50 \times \frac{300\underline{/45°}}{(156 - j90) + (212 + j212)}$$

$$= 50 \times \frac{300\underline{/45°}}{368 + j122}$$

$$= 50 \times \frac{300\underline{/45°}}{388\underline{/18.34°}}$$
$$= 38.66\underline{/26.66°} \text{ V}$$

3. Solve for Z_{Th}.
$$Z_{Th} = Z_1 \| Z_2 + Z_4$$
$$Z_1 \| Z_2 = \frac{Z_1 Z_2}{Z_1 + Z_2}$$
$$= \frac{180\underline{/-30°} \times 300\underline{/45°}}{388\underline{/18.34°}}$$
$$= 139.2\underline{/-3.34°}$$
$$= 139 - j8.11 \cong 139\underline{/0°}$$
$$Z_{Th} = (139 + j0) + (25 + j0)$$
$$= 164\underline{/0°} \; \Omega$$

4. The Thevenin equivalent circuit is sketched in Fig. 25-14.

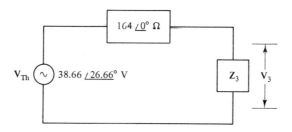

Figure 25-14

5. Solve for V_3.
$$V_3 = V_{Th} \frac{Z_3}{Z_{Th} + Z_3}$$
$$= 38.66\underline{/26.66°} \times \frac{100\underline{/60°}}{(164 + j0) + (50 + j86.7)}$$
$$= 38.7\underline{/26.7°} \times \frac{100\underline{/60°}}{214 + j86.7}$$
$$= \frac{38.7\underline{/26.7°} \times 100\underline{/60°}}{231\underline{/22°}}$$
$$= 16.8\underline{/64.7°} \text{ V}$$

EXAMPLE 25-4:
Solve for the Thevenin equivalent circuit of Fig. 25-15. Consider the load to be that circuit to the right of terminals a and b.

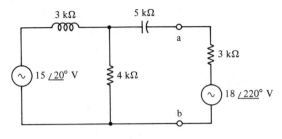

Figure 25-15

Solution:
1. Remove the branch to the right of terminals $a - b$.
2. Solve for V_{Th}.

$$V_{Th} = 15\underline{/20°} \times \frac{4\ k\Omega + j0}{4\ k\Omega + j3\ k\Omega}$$
$$= \frac{15\underline{/20^3} \times 4\ k\Omega\underline{/0°}}{5\ k\Omega\underline{/36.9°}}$$
$$= 12\underline{/-16.9°}\ V$$

3. Solve for Z_{Th}.

$$Z_{Th} = 5\ k\Omega\underline{/-90°} + 3\ k\Omega\underline{/90°}\ \|\ 4\ k\Omega\underline{/0°}$$
$$= 5\ k\Omega\underline{/-90°} + 2.4\ k\Omega\underline{/53.1°}$$
$$= (0 - j5000) + (1441 + j1919)$$
$$= 1441 - j3081$$
$$= 3401\underline{/-64.9°}\ \Omega$$

4. Draw the equivalent circuit (Fig. 25-16).

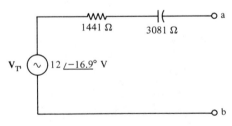

Figure 25-16

25-4. NORTON'S THEOREM

We restate the theorem for use with ac circuits.

Any two-terminal network can be replaced by a current source, I_N, and an equivalent internal impedance, Z_N. The source current I_N is the current between the terminals of the network when those terminals are *shorted*. The internal impedance

Z_N is the impedance between the terminals (short and load removed) with all sources replaced by their internal impedances. The Norton impedance is exactly the same as the Thevenin impedance.

EXAMPLE 25-5:
The equivalent circuit of a transistor amplifier is shown in Fig. 25-17. Use Norton's theorem to solve for current in the 2.2-kΩ resistance.

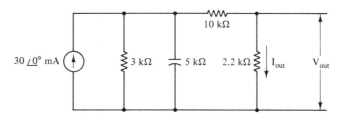

Figure 25-17

Solution:
1. Remove the 2.2-kΩ resistor and replace it by a short circuit.
2. Use the CDR to solve for I_N.

$$I_N = I \frac{G_{(10\ k\Omega)}}{Y_T}$$

$$G_{(10\ k\Omega)} = \frac{1}{10\ k\Omega} = 10^{-4}\ S$$

$$Y_T = \frac{1}{10\ k\Omega} + \frac{1}{3\ k\Omega} + j\frac{1}{5\ k\Omega}$$
$$= 4.33 \times 10^{-4} + j2 \times 10^{-4}\ S$$
$$= 4.77 \times 10^{-4}\underline{/24.8°}\ S$$

$$I_N = 30\underline{/0°}\ mA \times \frac{1 \times 10^{-4}\underline{/0°}}{4.77 \times 10^{-4}\underline{/24.8°}}$$
$$= 6.29\underline{/-24.8°}\ mA$$

3. Solve for Z_N.

$$Z_N = 10\ k\Omega + 3\ k\Omega\ \|\ -j5\ k\Omega$$
$$= 10\ k\Omega + \frac{3\ k\Omega\underline{/0°} \times 5\ k\Omega\underline{/-90°}}{3\ k\Omega - j5\ k\Omega}$$
$$= 10\ k\Omega + \frac{15 \times 10^6\underline{/-90°}}{5.83 \times 10^3\underline{/-59°}}$$
$$= 10\ k\Omega + 2.57\ k\Omega\underline{/-21°}$$
$$= 10\ k\Omega + (2.4\ k\Omega - j0.921\ k\Omega)$$
$$= 12.4\ k\Omega - j0.921\ k\Omega$$
$$= 12.4\underline{/0°}\ k\Omega$$

4. Sketch the circuit (Fig. 25-18).

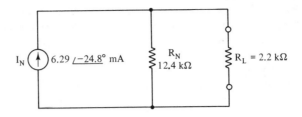

Figure 25-18

5. Solve for I_L.

$$I_L = I_N \frac{Z_N}{Z_N + Z_L}$$

$$= 6.29/\!-\!24.8° \text{ mA} \times \frac{12.4 \text{ k}\Omega}{14.4 \text{ k}\Omega}$$

$$= 5.42/24.8° \text{ mA}$$

25-5. MAXIMUM POWER TRANSFER

Earlier, in our studies of dc circuits, we learned that maximum power is transferred to a load when the load resistance is equal to the Thevenin equivalent resistance. We modify the theorem for use with ac circuits.

> Maximum power is delivered to a load when the load impedance is the conjugate of the Thevenin equivalent impedance.

This requires that the resistance of the load be equal to the resistance component of the Thevenin impedance. Also, the reactance of the load must have exactly the same magnitude, but be of opposite phase shift to the reactance of the Thevenin impedance.

Given

(a) $\mathbf{Z}_{Th} = R_{Th} + jX_{Th}$ then $\mathbf{Z}_L = R_L - jX$

(b) $\mathbf{Z}_{Th} = R_{Th} - jX_{Th}$ then $\mathbf{Z}_L = R_L + jX$

where $R_L = R_{Th}$ and $|X| = |X_{Th}|$. Under these conditions the power factor of the system is 1. The total impedance is twice R_L.

Power Relations

$$P_T = \frac{V_{Th}^2}{2R_L} \tag{25-13}$$

$$P_L = \frac{V_{Th}^2}{4R_L} \tag{25-14}$$

$$\eta = 0.5 = 50\% \tag{25-15}$$

EXAMPLE 25-6:
Given the circuit of Fig. 25-19, determine P_L under conditions of maximum power transfer.

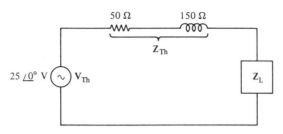

Figure 25-19

Solution:
1. Determine $\mathbf{Z}_L$. Given that $\mathbf{Z}_{Th} = 50 + j150$, then

$$Z_L = 50 - j150$$

2. Determine $\mathbf{Z}_T$.

$$\begin{aligned}\mathbf{Z}_T &= \mathbf{Z}_{Th} + \mathbf{Z}_L \\ &= (50 + j150) + (50 - j150) \\ &= 100 \ \Omega\end{aligned}$$

3. Solve for $\mathbf{I}_L$.

$$\begin{aligned}\mathbf{I}_L &= \frac{\mathbf{V}_{Th}}{\mathbf{Z}_T} \\ &= \frac{25}{100} \\ &= 0.25 \text{ A}\end{aligned}$$

4. Solve for P_L.

$$\begin{aligned}P_L &= I_L^2 R_L \\ &= 0.25^2 \times 50 \\ &= 3.125 \text{ W}\end{aligned}$$

Check:

$$P_L = \frac{V_{Th}^2}{4R_L} = \frac{25^2}{200} = 3.125 \text{ W}$$

EXAMPLE 25-7:
Given the circuit of Fig. 25-20, solve for the load components that will develop maximum output power.

Solution:
1. Solve for $\mathbf{Z}_{Th}$.

$$\begin{aligned}\mathbf{Z}_{Th} &= 2 \text{ k}\Omega \parallel 2 \text{ k}\Omega + j(6.28 \times 8 \times 10^3 \times 19.8 \times 10^{-3}) \\ &= 1000 + j995 \ \Omega\end{aligned}$$

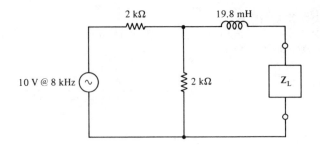

Figure 25-20

2. Z_L must be the conjugate of Z_{Th}.

$$Z_L = 1000 - j995 \ \Omega$$

Therefore, the load consists of 1 kΩ of resistance and a capacitance with a reactance of 995 Ω at 8 kHz.

3. Solve for the capacitance of the load.

$$C = \frac{1}{2\pi f X_C}$$
$$= \frac{1}{6.28 \times 8 \times 10^3 \times 0.995 \times 10^3}$$
$$= 2 \times 10^{-8}$$
$$= 0.02 \ \mu F$$

4. Draw the complete circuit (Fig. 25-21).

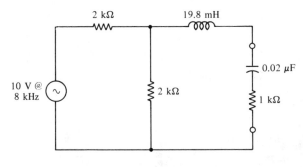

Figure 25-21

25-6. SOURCE CONVERSIONS

Circuit solutions are often simplified, especially with mixed sources, if we can change voltage sources to current sources or current sources to voltage sources.

Voltage Source to Current Source. Divide the source voltage by its internal impedance. This is the value of the current for the current source.

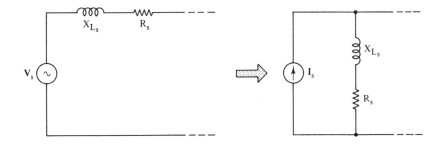

Figure 25-22 Converting a voltage source to a current source.

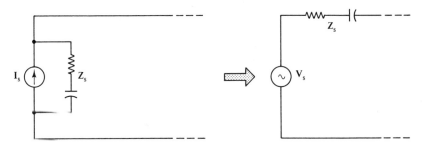

Figure 25-23 Converting a current source to a voltage source.

The internal impedance of the current source is the same as for the voltage source, but is in parallel with the current source, as in Fig. 25-22.

Current Source to Voltage Source. Source voltage is the product of source current and internal impedance. The internal impedance of the current source is used for the voltage source, but is in series, as shown in Fig. 25-23.

EXAMPLE 25-8:
Given the circuit of Fig. 25-24, replace the voltage source with a current source.

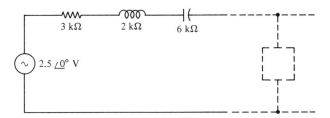

Figure 25-24

Solution:
1. Solve for Z_s.

$$Z_s = 3 \text{ k}\Omega + j2 \text{ k}\Omega - j6 \text{ k}\Omega$$
$$= 3 \text{ k}\Omega - j4 \text{ k}\Omega$$
$$= 5 \underline{/-53.1°} \text{ k}\Omega$$

Sec. 25-6. Source Conversions

2. Solve for I_s.

$$I_s = \frac{V_s}{Z_s}$$
$$= \frac{2.5\underline{/0°}\text{ V}}{5\underline{/-53.1°}\text{ k}\Omega}$$
$$= 0.5\underline{/53.1°}\text{ mA}$$

3. Sketch the circuit (Fig. 25-25).

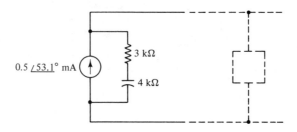

Figure 25-25

EXAMPLE 25-9:
Given the current source of Fig. 25-26, convert it to a voltage source.

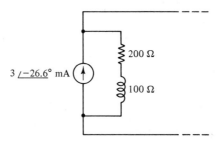

Figure 25-26

Solution:
1. Solve for Z_s.

$$Z_s = 200 + j100$$
$$= 223.6\underline{/26.6°}\text{ }\Omega$$

2. Solve for V_s.

$$V_s = I_s Z_s$$
$$= 3 \times 10^{-3}\underline{/-26.6°} \times 223.6\underline{/26.6°}$$
$$= 671\underline{/0°}\text{ mV}$$

3. Sketch the circuit (Fig. 25-27).

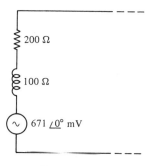

Figure 25-27

25-7. MESH CURRENT ANALYSIS

The use of Kirchhoff's laws in ac circuit analysis is similar to that for dc. However, we must work with phasors and must follow the rules associated with phasor arithmetic. All sources must be voltage sources. All impedances must be in polar form.

1. Assume some direction for mesh currents.
2. Assign polarities to voltage drops according to the assumed current directions.
3. Any current solution that is negative simply indicates that the assumed direction was not correct. The actual current is in the opposite direction.

EXAMPLE 25-10:
Use mesh analysis to solve for the current in each impedance of the circuit of Fig. 25-28.

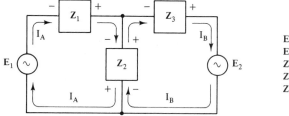

$E_1 = 25 \angle 0°$ V
$E_2 = 40 \angle 0°$ V
$Z_1 = 30 \angle 40°$ Ω $= 23 + j19.3$ Ω
$Z_2 = 100 \angle 20°$ Ω $= 94 + j34.2$ Ω
$Z_3 = 20 \angle 60°$ Ω $= 10 + j17.3$ Ω

Figure 25-28

Solution:
1. Solve for the polar form of the following sums:

$$Z_1 + Z_2 = 30 \angle 40° + 100 \angle 20°$$
$$= 128.7 \angle 24.6° \text{ Ω}$$
$$Z_2 + Z_3 = 100 \angle 20° + 20 \angle 60°$$
$$= 116 \angle 26.3° \text{ Ω}$$

2. Assign mesh currents. In this example we use clockwise mesh currents, as shown on the diagram.
3. Write the loop equations. Begin each mesh at the lower left-hand corner.

$$-E_1 - I_A Z_1 - I_A Z_2 + I_B Z_2 = 0$$
$$-I_B Z_2 - I_B Z_3 + I_A Z_2 + E_2 = 0$$

Rearranging, we have

$$I_A(Z_1 + Z_2) - I_B Z_2 = -E_1$$
$$-I_A Z_2 + I_B(Z_2 + Z_3) = E_2$$
$$(128.7\underline{/24.6°})I_A - (100\underline{/20°})I_B = -25$$
$$-(100\underline{/20°})I_A + (116\underline{/26.3°})I_B = 40$$

4. Use determinants to find I_A and I_B.

$$I_A = \frac{\begin{vmatrix} -25 & -100\underline{/20°} \\ 40 & 116\underline{/26.3°} \end{vmatrix}}{\begin{vmatrix} 128.7\underline{/24.6°} & -100\underline{/20°} \\ -100\underline{/20°} & 116\underline{/26.3°} \end{vmatrix}}$$

$$= \frac{-2900\underline{/26.3°} + 4000\underline{/20°}}{14{,}930\underline{/50.9°} - 10{,}000\underline{/40°}}$$

$$I_A = \frac{1160\underline{/0°}}{5448\underline{/71.2°}}$$

$$= 0.213\underline{/-71.2°} \text{ A}$$

$$I_B = \frac{\begin{vmatrix} 128.7\underline{/24.6°} & -25 \\ -100\underline{/20°} & 40 \end{vmatrix}}{5448\underline{/71.2°}}$$

$$= \frac{5148\underline{/24.6°} - 2500\underline{/20°}}{5448\underline{/71.2°}}$$

$$= \frac{2663\underline{/28.9°}}{5448\underline{/71.2°}}$$

$$= 0.489\underline{/-42.3°} \text{ A}$$

5. Sketch the circuit, showing currents in all impedances as in Fig. 25-29.

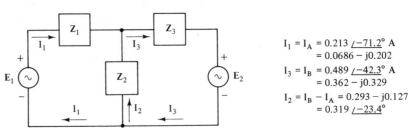

$I_1 = I_A = 0.213 \underline{/-71.2°}$ A
$ = 0.0686 - j0.202$
$I_3 = I_B = 0.489 \underline{/-42.3°}$ A
$ = 0.362 - j0.329$
$I_2 = I_B - I_A = 0.293 - j0.127$
$ = 0.319 \underline{/-23.4°}$

Figure 25-29

$$I_1 = I_A = 0.213\underline{/-71.2°}\text{ A}$$
$$= 0.0686 - j0.202$$
$$I_3 = I_B = 0.489\underline{/-42.3°}\text{ A}$$
$$= 0.362 - j0.329$$
$$I_2 = I_B - I_A = 0.293 - j0.127$$
$$= 0.319\underline{/-23.4°}$$

25-8. SUPERPOSITION

The analysis of any network containing many sources is simplified by use of the principle of superposition.

1. The response of the network is analyzed *one* source at a time. All other sources are replaced by their internal impedances.
2. The phasor sum of the individual responses is the actual response of the network with all sources connected.

We repeat the network used with mesh current analysis in order to compare results and methods.

EXAMPLE 25-11:
Given the circuit of Fig. 25-28, use superposition and solve for the current in each impedance.

Solution:
1. Remove E_2 and redraw the circuit indicating current directions (Fig. 25-30).

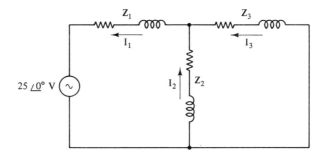

Figure 25-30

2. Solve the circuit for I_1, I_2, and I_3.

$$Z_1 = 23 + j19.3 = 30\underline{/40°}\ \Omega$$
$$Z_2 = 94 + j34.2 = 100\underline{/20°}\ \Omega$$
$$Z_3 = 10 + j17.3 = 20\underline{/60°}\ \Omega$$

(a) $Z_T = Z_1 + Z_2 \| Z_3$.

$$Z_2 \| Z_3 = \frac{100\underline{/20°} \times 20\underline{/60°}}{(94 + j34.2) + (10 + j17.3)}$$

$$= \frac{2000\underline{/80°}}{116\underline{/26.3°}}$$

$$= 17.24\underline{/53.7°} \; \Omega$$

$$= 10.2 + j13.9 \; \Omega$$

$$Z_T = Z_1 + Z_2 \| Z_3$$
$$= (23 + j19.3) + (10.2 + j13.9)$$
$$= 33.2 + j33.2$$
$$= 47\underline{/45°} \; \Omega$$

(b) Solve for total current.

$$I_T = I_1 = \frac{25\underline{/0°}}{47\underline{/45°}}$$

$$= 0.532\underline{/-45°} \; A$$

(c) Solve for I_2 and I_3.

$$I_2 = I_T \frac{Z_3}{Z_2 + Z_3}$$

$$= 0.532\underline{/-45°} \times \frac{20\underline{/60°}}{116\underline{/26.3°}}$$

$$= 0.0917\underline{/-11.3°} \; A$$

$$I_3 = I_T \frac{Z_2}{Z_2 + Z_3}$$

$$= 0.532\underline{/-45°} \times \frac{100\underline{/20°}}{116\underline{/26.3°}}$$

$$= 0.457\underline{/-51.3°} \; A$$

3. Remove E_1, using E_2 as the only source, and redraw the circuit. Indicate current directions (Fig. 25-31).

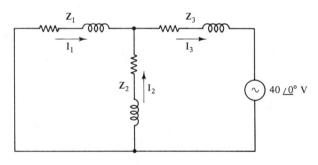

Figure 25-31

4. Solve the circuit for I_3, I_1, and I_2.
   (a) $Z_T = Z_3 + Z_1 \| Z_2$.

   $$Z_1 \| Z_2 = \frac{Z_1 Z_2}{Z_1 + Z_2}$$
   $$= \frac{30\underline{/40°} \times 100\underline{/20°}}{(23 + j19.3) + (94 + j34.2)}$$
   $$= \frac{3000\underline{/60°}}{128.7\underline{/24.6°}}$$
   $$= 23.3\underline{/35.4°}\ \Omega$$
   $$= 19 + j13.5\ \Omega$$

   $$Z_T = (10 + j17.3) + (19 + j13.5)$$
   $$= 29 + j30.8\ \Omega$$
   $$= 42.3\underline{/46.7°}\ \Omega$$

   (b) Solve for total current.

   $$I_T = I_3 = \frac{40\underline{/0°}}{42.3\underline{/46.7°}}$$
   $$= 0.946\underline{/-46.7°}\ A$$

   (c) Solve for I_1 and I_2.

   $$I_1 = I_T \frac{Z_2}{Z_1 + Z_2}$$
   $$= 0.946\underline{/-46.7°} \times \frac{100\underline{/20°}}{128.7\underline{/24.6°}}$$
   $$= 0.735\underline{/-51.3°}\ A$$

   $$I_2 = I_T \frac{Z_1}{Z_1 + Z_2}$$
   $$= 0.946\underline{/-46.7°} \times \frac{30\underline{/40°}}{128.7\underline{/24.6°}}$$
   $$= 0.219\underline{/-31.3°}\ A$$

5. The actual currents are found as phasor sums and differences.
   (a) I_1 is the phasor difference between the currents found in steps 2(b) and 4(c).

   $$I_1 = 0.735\underline{/-51.3°} - 0.532\underline{/-45°}$$
   $$= \begin{array}{r} 0.46 - j0.574 \\ -(0.376 - j0.376) \end{array}$$
   $$= 0.084 - j0.198)$$
   $$= 0.215\underline{/-67°}\ A$$

   (b) I_2 is the phasor sum of the currents found in steps 2(c) and 4(c).

Sec. 25-8. Superposition

$$I_2 = 0.0917\underline{/-11.3°} + 0.219\underline{/-31.3°}$$
$$= \begin{array}{c} 0.09 - j0.018 \\ +(0.187 - j0.114) \end{array}$$
$$= 0.277 - j0.132$$
$$= 0.307\underline{/-25.5°} \text{ A}$$

(c) I_3 is the phasor difference between the currents found in steps 4(b) and 2(c).

$$I_3 = 0.946\underline{/-46.7°} - 0.457\underline{/-51.3°}$$
$$= \begin{array}{c} 0.649 - j0.688 \\ -(0.286 - j0.357) \end{array}$$
$$= 0.363 - j0.331$$
$$= 0.491\underline{/-42.4°}$$

There are slight differences between the results of Example 25-10 and this example. We have been rounding off as calculations have been made and this causes insignificant differences between results. Certainly, the results are accurate for all practical applications.

25-9. NODE VOLTAGE ANALYSIS

We make use of Kirchhoff's law of currents in the node voltage circuit analysis method, often referred to as *nodal* analysis. A node is any point in a circuit where two or more circuit components are joined. The *reference node* is that node in a circuit common to the major portion of the circuit. This is usually ground.

We write the Kirchhoff current equations in terms of voltages and impedances and solve for unknown voltages. Compare this with our work on mesh current analysis, where we used currents and impedances to solve for currents.

Once again, we work with the same circuit so that the reader may compare the methods of analysis.

EXAMPLE 25-12:

Given the circuit of Example 25-10 (Fig. 25-28), use nodal analysis to find the currents. The circuit has been redrawn in Fig. 25-32 to identify the nodes we shall work with.

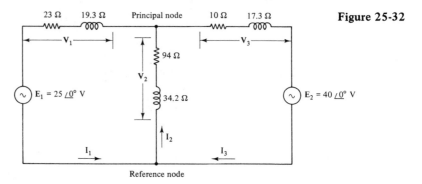

Figure 25-32

Solution:
1. Define the KCL relationships.
$$I_1 + I_3 - I_2 = 0$$
$$I_1 + I_3 = I_2$$

2. Write the equation for each current.
$$I_1 = \frac{V_1}{Z_1}$$
$$I_2 = \frac{V_2}{Z_2}$$
$$I_3 = \frac{V_3}{Z_3}$$

3. Define V_1 and V_3 in terms of the sources and V_2.
$$V_1 = E_1 - V_2$$
$$V_3 = E_2 - V_2$$

Therefore,
$$I_1 = \frac{E_1 - V_2}{Z_2}$$
$$= Y_1(E_1 - V_2)$$
$$I_3 = \frac{E_2 - V_2}{Z_3}$$
$$= Y_3(E_2 - V_2)$$
$$I_2 = Y_2 V_2$$

4. Rewrite the KCL equation.
$$I_1 + I_3 = I_2$$
$$Y_1(E_1 - V_2) + Y_3(E_2 - V_2) = Y_2 V_2$$

5. Solve the KCL equation for V_2.
$$V_2 = \frac{Y_1 E_1 + Y_3 E_2}{Y_1 + Y_2 + Y_3}$$

where
$$Y_1 = \frac{1}{Z_1} = 0.0333\underline{/-40°}\text{ S}$$
$$= 0.0255 - j0.0214 \text{ S}$$
$$Y_2 = \frac{1}{Z_2} = 0.01\underline{/-20°}\text{ S}$$
$$= 0.0094 - j0.0034 \text{ S}$$
$$Y_3 = \frac{1}{Z_3} = 0.05\underline{/-60°}\text{ S}$$
$$= 0.0250 - j0.0433 \text{ S}$$

$$Y_1 + Y_2 + Y_3 = 0.0599 - j0.0681 \text{ S}$$
$$= 0.0907\underline{/-48.7°} \text{ S}$$

$$Y_1 E_1 = 0.833\underline{/-40°} \text{ A}$$
$$= 0.638 - j0.535 \text{ A}$$

$$Y_3 E_2 = 2.0\underline{/-60°} \text{ A}$$
$$= 1.0 - j1.732 \text{ A}$$

$$Y_1 E_1 + Y_3 E_2 = 1.638 - j2.267$$
$$= 2.8\underline{/-54.2°} \text{ A}$$

We are now prepared to solve for V_2. Note that we shall divide current by admittance, which results in voltage.

$$V_2 = \frac{2.8\underline{/-54.2°}}{0.0907\underline{/-48.7°}}$$
$$= 30.84\underline{/-5.5°} \text{ V}$$
$$= 30.7 - j3 \text{ V}$$

6. Solve for I_2.

$$I_2 = \frac{V_2}{Z_2}$$
$$= \frac{30.84\underline{/-5.5°}}{100\underline{/20°}}$$
$$= 0.308\underline{/-25.5°} \text{ A}$$

7. Solve for I_1.

$$I_1 = \frac{E_1 - V_2}{Z_1}$$
$$= \frac{(25 + j0) - (30.7 - j3)}{30\underline{/40°}}$$

Note that $V_2 > E_1$; therefore, the assumed direction for I_1 must be reversed. We rewrite the equation subtracting E_1 from V_2.

$$I_1 = \frac{(30.7 - j3) - (25 + j0)}{30\underline{/40°}}$$
$$= \frac{5.7 - j3}{30\underline{/40°}}$$
$$= \frac{6.44\underline{/-27.8°}}{30\underline{/40°}}$$
$$= 0.215\underline{/-67.8°} \text{ A}$$

8. Solve for I_3.

$$I_3 = \frac{E_2 - V_2}{Z_3}$$
$$= \frac{(40 + j0) - (30.7 - j3)}{20\underline{/60°}}$$

$$= \frac{9.3 + j3}{20\underline{/60°}}$$
$$= \frac{9.77\underline{/17.9°}}{20\underline{/60°}}$$
$$= 0.489\underline{/-42.1°} \text{ A}$$

25-10. MILLMAN'S THEOREM

Whenever sources drive the same load, we may analyze the circuit by the use of Millman's theorem (Fig. 25-33). It is a way for us to replace parallel current sources with a single source.

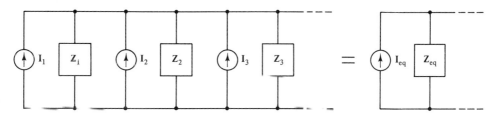

Figure 25-33 Millman's theorem can be used when several current sources supply a common load.

Given n current sources in parallel, they may be replaced by a single source, with a current that is the phasor sum of the individual source currents.

$$\mathbf{I}_{eq} = \mathbf{I}_1 + \mathbf{I}_2 + \cdots + \mathbf{I}_n \qquad (25\text{-}16)$$

The internal impedance of the single source ($\mathbf{I}_{eq}$) is the equivalent impedance of all source impedances in parallel.

$$\mathbf{Z}_{eq} = \frac{1}{1/\mathbf{Z}_1 + 1/\mathbf{Z}_2 + \cdots + 1/\mathbf{Z}_n} \qquad (25\text{-}17)$$

We use the same circuit as in the previous examples in demonstrating Millman's theorem.

EXAMPLE 25-13:
Given the circuit of Fig. 25-28, solve for the current in $\mathbf{Z}_2$.

Solution:
1. Change the voltage sources to current sources.

$$\mathbf{I}_1 = \frac{\mathbf{E}_1}{\mathbf{Z}_1}$$
$$= \frac{25\underline{/0°}}{30\underline{/40°}}$$
$$= 0.833\underline{/-40°} \text{ A}$$
$$= 0.638 - j0.535 \text{ A}$$

$$I_3 = \frac{E_2}{Z_3}$$
$$= \frac{40\underline{/0°}}{20\underline{/60°}}$$
$$= 2.0\underline{/-60°} \text{ A}$$
$$= 1.0 - 1.732 \text{ A}$$

2. Solve for the Millman equivalent source.

$$\mathbf{I_{eq}} = \mathbf{I_1} + \mathbf{I_3}$$
$$= (0.638 - j0.535) + (1.0 - j1.732)$$
$$= 2.8\underline{/-54.15°} \text{ A}$$

$$\mathbf{Z_{eq}} = \mathbf{Z_1} \parallel \mathbf{Z_3}$$
$$= \frac{Z_1 Z_3}{Z_1 + Z_3}$$

$$Z_1 + Z_3 = 33 + j36.6 \; \Omega$$
$$= 49.32\underline{/48°} \; \Omega$$

$$Z_{eq} = \frac{30\underline{/40°} \times 20\underline{/60°}}{49.32\underline{/48°}}$$
$$= 12.17\underline{/52°} \; \Omega$$
$$= 7.49 + j9.59 \; \Omega$$

3. Use the CDR to find $\mathbf{I_2}$.

$$\mathbf{I_2} = \mathbf{I_{eq}} \frac{Z_{eq}}{Z_{eq} + Z_2}$$

$$Z_{eq} + Z_2 = 101.5 + j43.8 \; \Omega$$
$$= 110.5\underline{/23.4°} \; \Omega$$

$$\mathbf{I_2} = 2.8\underline{/-54.15°} \times \frac{12.17\underline{/52°}}{110.5\underline{/23.4°}}$$
$$= 0.308\underline{/-25.6°} \text{ A}$$

25-11. DECIBELS

Power ratios in electricity and electronics can be very large. For example, the power delivered by an antenna to a stereo FM receiver may be as little as 0.0133 picowatts (1.33 × 10^{-14} W). The output power to the speakers may be 60 W. The *ratio* of the powers is 4.5 × 10^{15}. We have a way to work with power ratios without the need for such large numbers. We work with decibels (dB), a method first devised for use with telephone transmission lines and amplifiers.

$$\text{dB} = 10 \log \frac{P_1}{P_2} \tag{25-18}$$

where $P_1 < P_2$; dB are negative.
$P_1 > P_2$; dB are positive.
$P_1 = P_2$; dB = 0

Calculator sequence:

$$P_1 \boxed{\div} P_2 \boxed{=} \log \boxed{\times} 10 \boxed{=}$$

Because power varies as the square of voltage or current, we may substitute voltage or current ratios, provided that:

1. The measurements are made in the same point of the circuit, or
2. If measured at input and output, that the input and output impedances are *exactly* equal.

$$dB = 10 \log \left(\frac{V_1}{V_2}\right)^2$$

Therefore,

$$dB = 20 \log \frac{V_1}{V_2} \qquad (25\text{-}19)$$

Similarly,

$$dB = 20 \log \frac{I_1}{I_2} \qquad (25\text{-}20)$$

EXAMPLE 25-14:

An amplifier has an ouput power of 3 W when the input power is 2 mW. What is the power gain in decibels?

Solution:

$$dB = 10 \log \frac{P_1}{P_2}$$

In a gain calculation, let P_1 be the output power.

$$\begin{aligned} dB &= 10 \log \frac{3}{2 \times 10^{-3}} \\ &= 10 \log 1.5 \times 10^3 \\ &= 10 \times 3.176 \\ &= 31.76 \text{ dB} \end{aligned}$$

EXAMPLE 25-15:

A circuit has an ouput voltage of 2.5 V at 5 kHz. The output at 17 kHz is 1.6 V. If we use the output at 5 kHz as the reference, what is the loss in decibels at 17 kHz?

Solution:

$$\begin{aligned} dB &= 20 \log \frac{V_1}{V_2} \\ &= 20 \log \frac{1.6}{2.5} \\ &= 20 \times -0.194 \\ &= -3.88 \text{ dB} \end{aligned}$$

When decibel gain or loss is known, we can solve the equation for other unknowns. For example, given that

$$dB = 10 \log \frac{P_1}{P_2} = 20 \log \frac{V_1}{V_2}$$

Therefore,

$$\log \frac{P_1}{P_2} = \frac{dB}{10}$$

$$\log \frac{V_1}{V_2} = \frac{dB}{20}$$

At this point, we have a logarithm. We need only take the *antilog* of the logarithm to find the numerical value of the ratio.

$$\frac{P_1}{P_2} = \text{antilog} \frac{dB}{10} \qquad (25\text{-}21)$$

$$\frac{V_1}{V_2} = \text{antilog} \frac{dB}{20} \qquad (25\text{-}22)$$

EXAMPLE 25-16:
Given the decibel column, solve for the voltage and power ratios.

Solution:

$$\text{power ratio} = \text{antilog} \frac{dB}{10}$$

$$\text{voltage ratio} = \text{antilog} \frac{dB}{20}$$

Power ratio:

$$\boxed{dB} \boxed{\div} \boxed{10} \boxed{=} \boxed{\text{2ndF}} \boxed{\log}$$

Voltage ratio:

$$\boxed{dB} \boxed{\div} \boxed{20} \boxed{=} \boxed{\text{2ndF}} \boxed{\log}$$

For example, at −4 dB

$$4 \boxed{+/-} \boxed{\div} \boxed{20} \boxed{=} \boxed{\text{2ndF}} \boxed{\log} \; 0.630957\ldots$$

dB	Power Ratio	Voltage Ratio
6	3.981	1.995
3	1.995	1.413
1	1.259	1.122
0	1.000	1.000
−1	0.794	0.891
−3	0.501	0.708
−6	0.251	0.501

The + or −3 dB points are referred to as half-power points. A +3-dB change means power has doubled, while a −3-dB change means power is halved. We make extensive use of this knowledge in many applications.

Note, also, the voltage ratio is the square root of the power ratio. This is to be expected, since power varies as the square of the voltage.

EXAMPLE 25-17:
An amplifier has a power gain of 65 dB. If the output power is 35 W, what is the input power?

Solution:
1. Solve for the power ratio.

$$\frac{P_1}{P_2} = \text{antilog}\frac{65}{10}$$
$$= \text{antilog } 6.5$$
$$= 3.162 \times 10^6$$

2. Let P_1 = output power and P_2 = input power. Solve for P_2.

$$P_2 = \frac{35}{3.162 \times 10^6}$$
$$= 11.07 \, \mu W$$

EXAMPLE 25-18:
A filter circuit causes a loss of 12 dB at a certain frequency. If the input signal voltage is 700 mV, what is the output voltage? The input and output impedances are equal.

Solution:
The word *loss* tells us that decibels is a negative quantity. Let the output voltage be the numerator of the fraction.

$$\frac{V_{out}}{V_{in}} = \text{antilog}\frac{dB}{20}$$
$$= \text{antilog}\frac{-12}{20}$$
$$= 0.251$$

Solve for V_{out}.

$$\frac{V_{out}}{V_{in}} = 0.251$$
$$V_{out} = 0.251 \times V_{in}$$
$$= 0.251 \times 700 \, mV$$
$$= 176 \, mV$$

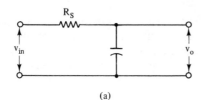

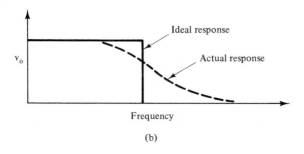

Figure 25-34 A single section RC low pass filter. (a) RC low pass filter. (b) Ideal and actual response curves.

25-12. LOW-PASS FILTERS

We may make use of elementary *RC* circuits as low-pass filter circuits, as in Fig. 25-34. Output is taken across the capacitor. At low frequencies, capacitive reactance is very much greater than the series resistance, and all the input appears as output. At high frequencies, the reactance is very much less than the resistance. Output is very much less than the input. At a frequency defined as the *corner* frequency, the output is at −3 dB of the low-frequency output. This a half-power point. It is also a voltage ratio such that the output voltage is approximately 0.707 of the input voltage. Figure 25-35 shows equivalent circuits for low- and high-frequency response and for the conditions that occur at the corner frequency.

It can be demonstrated that at the corner frequency (f_2), resistance and reactance are equal.

$$\text{At } f_2: \quad |X| = R$$

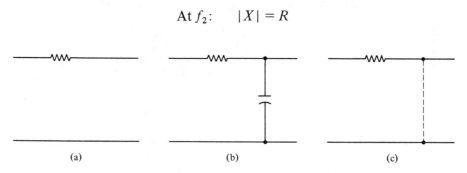

Figure 25-35 Equivalent circuits for the low pass RC filter. (a) Equivalent circuit accurate for frequencies that are *less* than 0.1 f_2. (b) Equivalent circuit for frequencies within a range above and below the corner frequency, f_2. (c) Equivalent circuit accurate for frequencies that are much greater than f_2.

In the low-pass filter,

$$\frac{1}{2\pi f_2 C} = R_s \tag{25-23}$$

Therefore,

$$f_2 = \frac{1}{2\pi R_s C} \tag{25-24}$$

$$C = \frac{1}{2\pi f_2 R_s} \tag{25-25}$$

We define the ratio of the output voltage at *any* frequency to the output at the low frequencies as k_2.

$$k_2 = \frac{V_{out}}{V_{out\ \ell}}$$

where $V_{out} = V_{in} \dfrac{1/j\omega C}{R_s - 1/j\omega C}$

$V_{out\ \ell} = V_{in}$

Then

$$k_2 = \frac{-j(1/2\pi fC)}{R_s - j(1/2\pi fC)}$$

Dividing numerator and denominator by $-j(1/2\pi fC)$ and substituting $1/2\pi f_2 C$ for R_s, we obtain

$$k_2 = \frac{1}{1 + j(f/f_2)} \tag{25-26}$$

This equation is known as the *universal high-frequency response* formula. The very same formula is used in the frequency response analysis of *LR* low-pass filters.

An inductive filter is shown in Fig. 25-36, along with a universal high-frequency response graph. At low frequencies, the output equals the input because inductive reactance is very low compared to the output resistor. As frequency increases, so does the inductive reactance. At the corner frequency (f_2), reactance equals resistance.

At f_2: $|X| = R$

$$2\pi f_2 L = R \tag{25-27}$$

$$L = \frac{R}{2\pi f_2} \tag{25-28}$$

$$f_2 = \frac{R}{2\pi L} \tag{25-29}$$

It is not common to use inductive filters. They are more expensive, take up greater volume than *RC* filters, and can cause interference with other parts of the system because of their magnetic fields.

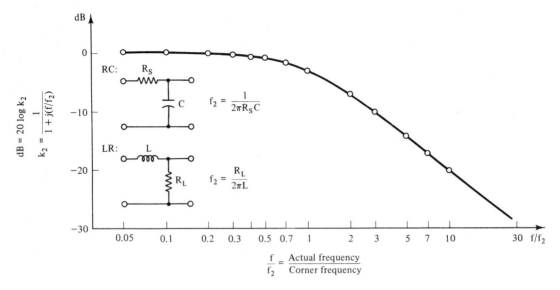

Figure 25-36 Universal Frequency Response Curve (low pass).

We now use one example to cover a great deal about low-pass filters.

EXAMPLE 25-19:
Given a low-pass filter, with $R = 15$ kΩ and $C = 425$ pF, solve for (a) the corner frequency, f_2; (b) the highest frequency for which the output is approximately equal to the input; (c) the k_2 factor at $f = 10f_2$.

Solution:

(a) $f_2 = \dfrac{1}{2\pi RC}$

$= \dfrac{1}{6.28 \times 15 \times 10^3 \times 425 \times 10^{-12}}$

$= 25$ kHz

(b) In order for the output to equal the input, k_2 must equal 1.

$$1 = \frac{1}{1 + j(f/f_2)}$$

$$1 = 1 + j(f/f_2)$$

For this to be true, the j term cannot be greater than 0.1.

$$0.1 = f/f_2$$

$f = 0.1 f_2$
$= 0.1 \times 25$ kHz
$= 2.5$ kHz

(c) Solution for k_2 when $f/f_2 = 10$.

$$k_2 = \frac{1}{1 + j10}$$

$\approx 0.1 \underline{/-90°}$

Decibel ratios: at f_2, $k_2 = 0.707$; dB $= -3$
at $0.1f_2$, $k_2 = 1$; dB $= 0$
at $10f_2$, $k_2 = 0.1$; dB $= -20$

25-13. HIGH-PASS FILTERS

An ideal high-pass filter has zero output for all frequencies below the corner frequency (f_1). For frequencies above f_1, the output equals the input. A practical RC filter and its response are shown in Fig. 25-37.

1. For frequencies much less than the corner frequency, the output is very low. This is because the reactance of the capacitor is very much greater than R_L.
2. At frequencies much greater than the corner frequency, the capacitor appears as a short circuit, and the output equals the input.
3. At the corner frequency, the output is 0.707 of the input. This is a -3 dB (half-power) point.

At f_1, the resistance and reactance are equal.

$$R_L = \frac{1}{2\pi f_1 C} \quad (25\text{-}30)$$

$$f_1 = \frac{1}{2\pi R_L C} \quad (25\text{-}31)$$

$$C = \frac{1}{2\pi f_1 R_L} \quad (25\text{-}32)$$

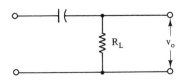

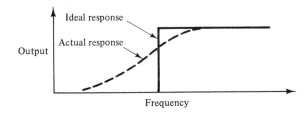

Figure 25-37 A single section RC high pass filter. The ideal and actual frequency responses are sketched.

We define the ratio of the output at any frequency to the output where $V_{out} = V_{in}$ as k_1.

$$k_1 = \frac{V_{out}}{V_{out\,h}}$$

where

$$V_{out\,h} = V_{in}$$

$$V_{out} = V_{in} \frac{R_L}{R_L - j(1/2\pi fC)}$$

Therefore,

$$k_1 = \frac{R_L}{R_L - j(1/2\pi FC)}$$

If we divide numerator and denominator by R_L and substitute $1/2\pi f_1 C$ for R_L, we obtain the *universal low-frequency response* formula.

$$k_1 = \frac{1}{1 - j(f_1/f)} \qquad (25\text{-}33)$$

The universal low-frequency response formula is equally accurate for inductive high-pass filters. In this case, f_1 is found by the following formula:

$$f_1 = \frac{R_s}{2\pi L} \qquad (25\text{-}34)$$

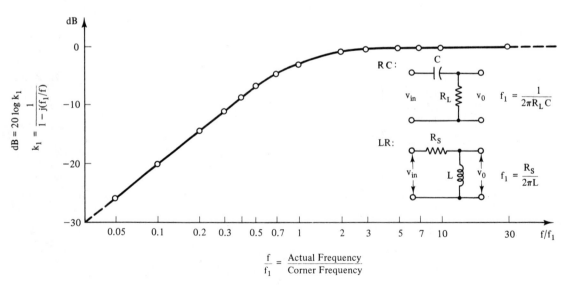

Figure 25-38 Universal Frequency Response Curve (high pass).

As with the low-pass filter, we can establish the frequency range where output equals input when we know f_1. Recall that in the low-pass filter the *highest* frequency for which output was equal to input was $0.1f_2$. In the high-pass filter the *lowest* frequency for which the output is equal to the input is $10f_1$. The universal low-frequency response curve of Fig. 25-38 illustrates these concepts.

SUMMARY

1. Delta circuit forms can be difficult to solve. A wye equivalent circuit greatly simplifies circuit analysis.
2. Delta-to-wye conversions are particularly useful in circuit solutions of parts of bridge circuits.
3. A bridge circuit is balanced when the ratio of impedances is the same on both sides of the bridge.
4. Bridge circuits have many applications, including R, L, and C measurement.
5. We use Thevenin's and Norton's theorems with ac circuits in the same way as in dc circuits. One must be aware of the rules for working with vector quantities.
6. For maximum power in a load, the load impedance must be the *conjugate* of the Thevenin equivalent impedance.
7. An ac voltage source is converted to a current source if $\mathbf{I}_s = \mathbf{V}_s/\mathbf{Z}_s$, with $\mathbf{Z}_s$ placed in parallel with $\mathbf{I}_s$.
8. An ac current source is converted to a voltage source if $\mathbf{V}_s = \mathbf{I}_s\mathbf{Z}_s$, with $\mathbf{Z}_s$ placed in series with $\mathbf{V}_s$.
9. Use of any of the circuit analysis methods, whether Kirchhoff's laws, Ohm's law, superposition, or the like, is the same as in dc, *except that we must work with polar and rectangular values.*
10. Millman's theorem allows us to resolve several parallel sources (current or voltage) into one equivalent source.
11. Decibels (dB) are a convenient way of working with very large ratios.
12. Decibels are based upon the logarithm of a power ratio.
13. The corner frequency in untuned filters is the frequency for which the output is -3 dB of the maximum output.
14. For low-pass filters the output is maximum and constant from very low frequencies to a frequency that is one-tenth of the corner frequency. Beyond the corner frequency (f_2), the output falls off at a rate of 20 dB per decade.
15. For high-pass filters the output reaches maximum at a frequency that is 10 times the corner frequency (f_1). For frequencies above $10f_1$, the output remains maximum and constant. At $0.1f_1$, the output is at -20 dB.

PROBLEMS

25-1. If 30 V is applied to the circuit of Fig. 25-39, solve for the currents in R_5 and R_7.
25-2. Reduce the pi network of Fig. 25-40 to an equivalent wye.

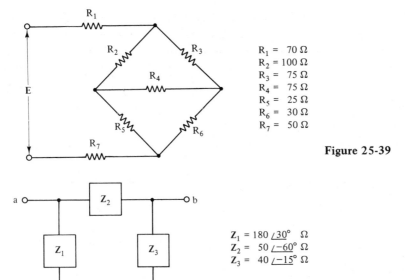

Figure 25-39

$R_1 = 70\ \Omega$
$R_2 = 100\ \Omega$
$R_3 = 75\ \Omega$
$R_4 = 75\ \Omega$
$R_5 = 25\ \Omega$
$R_6 = 30\ \Omega$
$R_7 = 50\ \Omega$

$Z_1 = 180\ \underline{/30°}\ \Omega$
$Z_2 = 50\ \underline{/-60°}\ \Omega$
$Z_3 = 40\ \underline{/-15°}\ \Omega$

Figure 25-40

25-3. Convert the tee network of Fig. 25-41 to an equivalent delta.

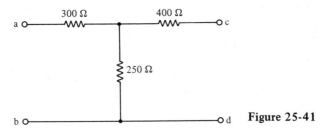

Figure 25-41

25-4. In the circuit of Fig. 25-42, $Z_1 = 100\underline{/0°}\ \Omega$, $Z_2 = 300\underline{/0°}\ \Omega$, $Z_3 = 200\underline{/0°}\ \Omega$, and Z_4 is unknown. The bridge is in balance so that $V_{a-b} = 0$. Solve for Z_4.

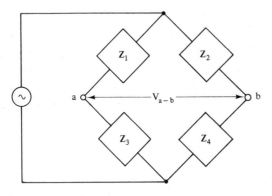

Figure 25-42

25-5. In the circuit of Fig. 25-42, $Z_1 = 80\underline{/30°}\ \Omega$, $Z_2 = 50\underline{/-40°}\ \Omega$, and $Z_3 = 200\underline{/50°}\ \Omega$. Solve for Z_4 so that the bridge is balanced.

25-6. In the circuit of Fig. 25-43, determine the Thevenin equivalent circuit for terminals $x-y$.

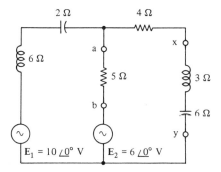

Figure 25-43

25-7. In the circuit of Fig. 25-43, solve for the Thevenin equivalent circuit for terminals $a-b$.

25-8. Solve for the Norton equivalent circuit of Fig. 25-44, with $a-b$ as the load terminals.

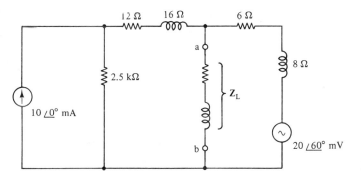

Figure 25-44

25-9. What value of Z_L would result in maximum power transfer in the circuit of Fig. 25-44?

25-10. Determine the load for maximum power transfer in the circuit of Fig. 25-45.

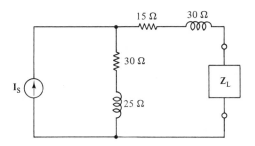

Figure 25-45

25-11. Given the data of Prob. 25-6, what value of load impedance should be connected at the $x-y$ terminals in place of L and C for maximum power transfer?

25-12. A voltage source has an internal impedance of $2.1 + j1.5$ kΩ. The effective value of the source is 15 V. What is the equivalent current source?

25-13. Calculate the current source needed to replace the voltage source in Fig. 25-46.

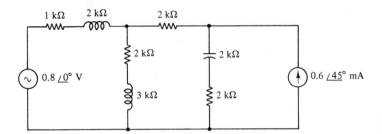

Figure 25-46

25-14. Given the circuit of Fig. 25-46, replace the current source with an equivalent voltage source.

25-15. The equivalent circuit of a transistor amplifier is sketched in Fig. 25-47. Redraw the circuit so that only current sources are used.

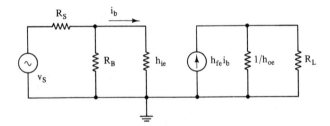

Figure 25-47

25-16. Use mesh current analysis to determine each current in the circuit of Fig. 25-48.

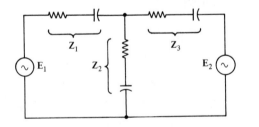

$E_1 = 20 \angle 0°$ V
$E_2 = 30 \angle 0°$ V
$Z_1 = 50 - j40$ Ω
$Z_2 = 75 - j60$ Ω
$Z_3 = 20 - j30$ Ω

Figure 25-48

25-17. Solve for the currents of the circuit of Fig. 25-48 by use of the superposition method.

25-18. Use nodal analysis to determine each current in the circuit of Fig. 25-48.

25-19. Solve for the current in Z_2 of Fig. 25-48 by use of Millman's theorem.

25-20. An amplifier system requires an input signal of 2 mW in order to produce an output of 50 W. What is the decibel gain of the amplifier?

25-21. An amplifier system is sketched in Fig. 25-49. The loss or gain of each component is indicated. What is the overall gain?

Figure 25-49

25-22. If the input to the system of Fig. 25-49 is 0.3 mW, what is the output power?

25-23. A technician measures an output voltage of 20 V at 1 kHz. At 15 kHz the output drops to 5 V. Using 20 V as 0 dB, what is the output at 15 kHz in decibels?

25-24. A filter produces a loss of 6 dB for each doubling of frequency. If the loss is 18 dB at 20 kHz, what is the loss at 80 kHz?

25-25. A manufacturer claims a signal-to-noise ratio of 82 dB for a tape deck. What power ratio does this represent?

25-26. Given the filter circuit of Fig. 25-50, solve for the corner frequency f_2 and the highest frequency for which $v_{in} = v_{out}$.

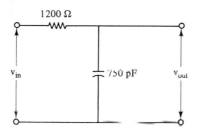

Figure 25-50

25-27. Change the resistor in the filter of Fig. 25-50 so that the corner frequency is 200 kHz.

25-28. Given the circuit of Fig. 25-51, what value of C will result in a corner frequency of 30 Hz?

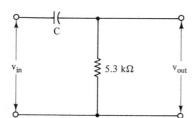

Figure 25-51

25-29. Given the data of Prob. 25-28, with a corner frequency of 30 Hz, (a) at what frequency does the output voltage first become equal to input voltage? (b) How many decibels down from the high-frequency reference level is the output at 3 Hz?

25-30. Given a corner frequency of 40 kHz in a low-pass filter, at what frequency is the output down 20 dB?

Answers to Even-Numbered Problems

CHAPTER 1

1-2.	(a) 6.974	(b) 20.608	
1-4.	0.0005	9421.18	
	19.22	12.54	
	0.0032	638.81	
	0.166	44.82	
	0.049	270.9	
1-6.	0.28%	76.8%	
	83.5%	235%	
	1.5%	105%	
	25%	68%	
1-8.	180	27	
	2.25	30	
	112.5	0.42	
	0.675	0.564	
1-10.	30%		
1-12.	3/4 = 9/12		
	5/6 = 15/18		
	4/6 = 20/30		
1-14.	5.63×10^{-4}	3.86×10^{-3}	
	1.76×10^{2}	2.51×10^{-2}	
	8.76×10^{3}	1.61×10^{3}	
	9.05×10^{6}	6.31×10^{-6}	

1-16. 1.5×10^{-1}
3×10^{-2}
5×10^{5}
3×10^{6}
1-18. 150 MHz
55 μH
31 mA
40 μA
550 kW
10.7 MW
98 mV
1-20. 25 A
1-22. 5.075 A
1-24. 16.8 mA

CHAPTER 2

2-2. Positive
2-4. Atomic number 6; 2 1-shell and 4 m-shell electrons.
2-6. 0.018
2-8. 0.762 inches
2-10. 2956 V

CHAPTER 3

3-2. 1.4 kV
3-4. 2 MHz
3-6. 66.7 ns
3-8. (a) 550 kHz   (b) 1 MHz
3-10. 200 mA

CHAPTER 4

4-2. 0.12 mA, 0.00012 A
4-4. (a) alternating, (b) alternating with dc content, (c) pulsating dc, (d) pure dc
4-6. 186.5 W
4-8. no
4-10. 155 W
4-12. 74.6%
4-14. 1.026 kW

CHAPTER 5

5-2. 680 Ω   1 kΩ   15 kΩ
     20 Ω   2.2 MΩ   470 Ω
5-4. 0.4147 Ω
5-6. 4.56 in.
5-8. 10 gauge
5-10. 2.7 kΩ ± 10%   red, violet, red, silver
      56 kΩ ± 20%   green, blue, orange, silver
      0.47 Ω ± 5%   yellow, violet, silver, gold
      10 MΩ ± 10%   brown, black, blue, silver
      150 Ω ± 5%   brown, green, brown, gold
5-12.

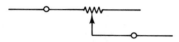

5-14. R_1: brown, black, brown, gold
      R_2: orange, orange, red, silver
      R_3: brown, green, orange, gold

CHAPTER 6

6-2. 1.5 A
6-4. less load resistance
6-6. increase load resistance
6-8. 10 mA
6-10. 0.5 A
6-12. 50.5 mA
6-14. 10 W
6-16. 250 W, 500 W, 62.5 W
6-18.

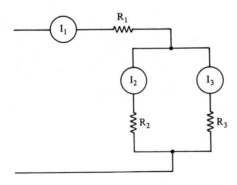

6-20 & 6-22.

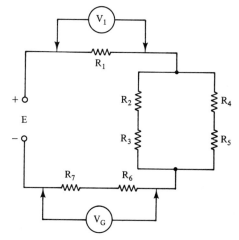

- 6-24. R_1 is open.
- 6-26. 3.08 V
- 6-28. 48 V
- 6-30. 10 V = 5 V + 3 V + 2 V

CHAPTER 7

- 7-2. I = 250 mA, 500 mA, 1 A, 1.5 A, 2 A.
- 7-4. I = 200 mA
- 7-6. (a) 20 mA (b) 6 μA (c) 16 A (d) 1.5 A
- 7-8. (a) 1 kΩ (b) 469 Ω (c) 66.7 Ω (d) 133.3 kΩ
- 7-10. 1680 MΩ
- 7-12. (a) 0.25 V (b) 1.64 V (c) 157.5 V (d) 164.5 V
- 7-14. 24 V
- 7-16. 0.521 μA
- 7-18. 2.5 mA
- 7-20. 8.8 kV
- 7-22.

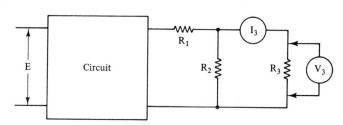

7-24. R_x is greater than 15 $\Omega(R_2)$.

Answers to Even-Numbered Problems

CHAPTER 8

- 8-2. 220 ma; current is the same in all parts of a series circuit.
- 8-4. $R_2 = 1 \text{ k}\Omega$
- 8-6. R_1 in series with R_2, in series with R_3.
- 8-8. $I = 270$ mA
- 8-12. $R_T = 880 \ \Omega; R_T = nR$
- 8-14. 2.99 V
- 8-16. 4.5 V
- 8-18. 3.3 kΩ
- 8-20. $I = 1.06$ mA; counterclockwise
- 8-22. 3.1 V
- 8-24. $V_1 = 9.17$ V, $V_2 = 23.5$ V, $V_3 = 13.16$ V, $V_4 = 5.17$ V
- 8-26. $V_A = 6$ V, $V_B = 17$ V, $V_C = 40.5$ V, $V_D = 48$ V
- 8-28. Yes (R_3 open).
- 8-30. R_1 resistance reduced from 10 kΩ to 2 kΩ.

CHAPTER 9

- 9-4. 5 V
- 9-6. 1.4 A
- 9-8. $I_1 = 35$ mA
- 9-10. (a) $R_T = 13.79 \ \Omega$
  (b) $I_1 = 4$ A, $I_2 = 2$ A, $I_3 = 1.25$ A
  (c) $R_3 = 80 \ \Omega$
- 9-12. Node A: $4 + 0.05 - 1 - 2 - 1.5 = 0$
  Node B: $1.5 + 1.5 - 2.5 - 0.5 = 0$
- 9-14. (a) 5 Ω (b) 6.25 Ω (c) 23.3 kΩ
  (d) 5.85 Ω (e) 47 Ω
- 9-16. (a) 20 Ω (b) 120 Ω (c) 4 kΩ (d) 8 kΩ (e) 20 Ω
- 9-18. (a) 0.03 S (b) 0.086 S
- 9-20. (a) $I_1 = 39.62$ mA (b) $I_1 = 17.9 \ \mu$A
  $I_2 = 27.38$ mA $I_2 = 10.1 \ \mu$A
- 9-22. (a) $I = 3.6$ mA
  (b) $R = 1$ kΩ
  (c) $I = 0.536$ mA
  (d) $I = 4.5$ mA, $R = 10$ kΩ

CHAPTER 10

- 10-4. (a) $I_T = I_2 + I_3 = I_1$
  (b) $I_T = I_1 + I_2 = I_3 + I_4 + I_5 = I_6$
  (c) $I_T = I_b + I_d = I_b + I_c = I_a$
  (d) $I_T = I_5 + I_6 = I_4 + I_3 = I_4 + I_2 = I_1$

10-6. $I_1 = 200$ mA
10-8. 40 V
10-10. 40 V
10-12. $R_{eq(A-A')} = 2.2$ kΩ
10-16. $I_2 = 3.6$ A
10-18. $R_2 = 4.7$ kΩ
10-20. $R_1 = 75$ kΩ
$R_2 = 25$ kΩ
10-22. $V_C = 72$ V
$V_A = 24$ V
$V_B = 8$ V
10-24. $V_{10} = 60$ V
10-26. $I_9 = 6.25$ mA

CHAPTER 11

11-2. $24 - 18 - 6 = 0$
$24 - 9 - 15 = 0$
11-4. $V_2 = E(\dfrac{R_2}{R_1 + R_2 + R_3 \parallel R_4})$
11-6. $V_\parallel = 3.41$ V
11-8. (a) $I_1 = 8.7$ mA
$I_2 = 18.5$ mA
$I_3 = 4.83$ mA
11-10. $10 - 7I_1 - 3I_3 = 0$
$15 - 10 - 6I_2 - 3I_3 = 0$
11-12. (a) $I_1 = 8$A
$I_2 = -4$A
(b) $I_a = 0.424$ A
$I_b = 0.605$ A
(c) $I_1 = 45.05$ mA
$I_2 = 61.07$ mA
11-14. $I_1 = 97.1$ mA
$I_2 = 154.2$ mA
$I_3 = 57.1$ mA
11-16. $I_1 = 0.437$ A
$I_2 = 0.072$ A
$I_3 = 0.509$ A
11-18. $I_L = 0.628$ mA
$V_L = 0.753$ V
11-20. (a) $I_L = 167$ mA
$V_L = 16.7$ V
(b) $I_L = 2.73$ mA
$V_L = 16.4$ V

Answers to Even-Numbered Problems

CHAPTER 12

12-2. (a) 4 W (d) 8.8 mW
 (b) 11.25 kW (e) 995 μW
 (c) 5.625 kW
12-4. (a) 2.25 W (d) 60 W
 (b) 3.32 μW (e) 4.79 nW (nanowatts)
 (e) 500 kW
12-6. (a) 1 W
 (b) 4.5 W
 (c) 53.3 mW
12-8. (a) 46.5% (b) 35.7% (c) 21.1%
12-10. 23.26 mW
12-12. 314 Ω
12-14. 303 Ω
12-16. (a) 44.81 V (b) 43.9 V

CHAPTER 13

13-2. 1250 μF
13-4. 12.5 μC
13-6. 12 kV
13-8. 0.0997 μF ≅ 0.1 μF
13-10. titanium dioxide
13-12. 2.86 μF
13-14. 25 μF
13-16. 0.115 μF
13-18. 35 V
13-20. 0.023 J

CHAPTER 14

14-2. 0.4 Wb/m^2
14-4. 1703
14-6. 746.4 rels
14-8. 13.5 At
14-10. 68.7 μWb; B = 1.37 Wb/m^2

CHAPTER 15

15-2. 40 V
15-4. 150 V
15-6. 0.375
15-8. (a) 1000 Ω/V (b) 10,000 Ω/V
 (c) 20,000 Ω/V (d) 50,000 Ω/V

15-10. 0 – 500 µA: 275 Ω
0 – 5 mA: 22.45 Ω
0 – 50 mA: 2.236 Ω

CHAPTER 16

16-2. 1500 µH 165,000 µH
30 µH 850 µH
8 µH 10,000 µH
16-4. 60 V (aiding)
16-6. 1 H
16-8. 3.46 H
16-10. 1.38 H
16-12. 26.4 mH
16-14. 0.09 J

CHAPTER 17

17-2. 10 s
17-4. 29.6 mV, 5.251 V, 17.017 V, 21.968 V, 24 V
17-6. $v_C = 16$ mV, $v_R = 23.84$ V; 1.55 V, 22.45 V; 15.17 V, 8.83 V; 22.33 V, 1.76 V; 23.97 V, 30 mV
17-8. 17.85 V, 10.92 V, 3.4 V, 4 mV
17-10. 17.85 V, 10.92 V, 3.4 V, 4 mV
17-12. (a) 0.201 s (b) 1.1 s
17-14. (a) 371 ns (b) 1.078 µs
17-16. yes, $\tau < T$
17-18. no, $T > \tau$
17-20. 100 ns 5 ns
5 µs 2 ms
10 µs 3.33 ms

CHAPTER 18

18-2. (a) 6 µs (b) 1.5 µs
18-4. 10.5 kHz, 17.5 kHz, 24.5 kHz
18-6. (a) 17.68 mA (b) −17.68 mA
18-8. 220 V
18-10. $i = 40 \sin(\theta - 50°)$ mA
18-12. (a) 73°
(b)

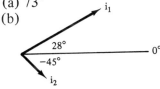

Answers to Even-Numbered Problems

18-14. 2.12 A
18-16. (a) 34.64 V (b) 48.98 V
18-18. 8 W
18-20. 20 V
18-22. (a) 60 Hz (b) 118 V
18-24. $v = 50 \sin(6280t + 35°)$ V

CHAPTER 19

19-2. 100 mA
19-4. 3.95 mA
19-6. (a) 626 mW (b) 25.03 V
19-8. $i_3 = 55.8 \sin 377t$ mA
19-10. 25.94 V
19-12. 108 mW
19-14. 17 $V_{p\text{-}p}$ = 6.01 V_{rms} 62 $V_{p\text{-}p}$ = 21.9 V_{rms}
       28 V_{max} = 19.8 V_{rms} 339 $V_{p\text{-}p}$ = 120 V_{rms}
       99 $V_{p\text{-}p}$ = 35 V_{rms} 40 V_{max} = 28.3 V_{rms}
19-16. 0.0188 Ω, 94.22 Ω, 0.565 Ω, 471 Ω, 1.13 Ω, 1884 Ω, 18.84 Ω, 4710 Ω
19-18. $i = -1.46$ mA
19-20. 1.59 MΩ, 79.6 kΩ, 3.18 kΩ, 1.06 kΩ, 199 Ω, 133 Ω, 15.9 Ω, 0.796 Ω
19-28. 24/60°
       0.2/60°
       15/−40°
       4/120°
       360/50°
19-30. 22.76/44.3°
       102.66/−76.9°
       1239.8 − j404
19-32. 10/60° V
19-34. 430.12/−35.54° mA
19-36. 6.63 mH

CHAPTER 20

20-2. 894.43/26.57° Ω
20-4. 1260/85.45° Ω
20-6. 1265/−71.57° Ω
20-8. 23.7/71.57° mA
20-10. 7.55/74.36° V
20-12. 9.94/0° V
20-14. V_R = 67.08/−93.43° mV
       V_C = 167.71/−183.43° mV
       V_L = 301.87/−3.43° mV

20-16. $6/{-90°}$ kΩ
20-18. 11.17 kΩ
20-20. 10.47 Ω
20-22. 104.7 Ω
20-24. 105.2 Ω

CHAPTER 21

21-2. $1766/23.7°$ $I_2 = 21.8/{-50°}$ mA
21-4. $E = 60/0°$ V $I_2 = 21./{-50}$
21-6. $278/0°$ Ω, $278/0°$ Ω, $319/27.3°$ Ω, $2179/82°$ Ω, $13.5/90°$ kΩ
21-8. $I_1 = 469/{-55°}$ μA $I_2 = 185/55°$ μA $I_T = 441/{-31.7°}$ μA
21-10. $15/0°$ V (approx.)
21-12. 0.942 mS, 12.56 mS, 157 mS, 2.355 mS
21-14. (a) Parallel: $R = 1.5$ kΩ (b) Series: $R = 1.024$ kΩ
              $X_L = 2.2$ kΩ $X_L = 698$ Ω
21-16. $12 - j425$ Ω
21-18. 7.1
21-20. $3.87/{-5.6°}$ V
21-22. 531 μF
21-24. 948 μF

CHAPTER 22

22-2. 576 kVA
22-4. 0.799 (lagging)
22-6. (a) 0.53 (lagging) (b) 3.212 kvars
      (c) 2.017 kW   (d) 3.793 kVA
22-8. 784 μF
22-10. (a) 80 kvars (b) 227 A (c) 136.32 A

CHAPTER 23

23-2. 58.7 μH
23-4. 101.4 pF
23-6. 105
23-8. (a) $11.8/0°$ Ω (b) $13.7/30.7°$ Ω (c) $12.6/{-20.9°}$ Ω
23-10. $162.6/{-90°}$ mV
23-12. 153.5
23-14. 53.5
23-16. 45 Ω
23-18. $Z_0 = 46.38/0°$ kΩ; BW = 341 kHz
23-20. $16.6/0°$ μA
23-22. $1.41/0°$ V
23-24. $200/0°$ mV

Answers to Even-Numbered Problems

CHAPTER 24

24-2. $V_P I_P = 5.51$ VA; $V_S I_S = 5.51$ VA
24-4. $V_L = 24$ V; $I_P = 0.5$ A
24-6. 0.405 A
24-8. (a) 24 V (b) $V_{S1} = 12$ V; $V_{S2} = -12$V
24-10. 1:1.94 (step-up)
24-12. (a) 11.62 μH (b) 40.94 μH (c) 0.75 μH
   (d) 11.62 μH (e) 33.54 μH
24-14. $80.88\underline{/-90°}$ mV

CHAPTER 25

25-2. $\mathbf{Z}_a = 40.4\underline{/-39.4°}$ Ω $\mathbf{Z}_b = 8.99\underline{/-84.4°}$ Ω
   $\mathbf{Z}_c = 32.4\underline{/5.61°}$ Ω
25-4. $600\underline{/0°}$ Ω
25-6. $6.43\underline{/22.3°} = 5.95 + j2.44$ Ω
25-8. $10\underline{/53.1°}$ Ω
25-10. $45 - j55 = 71.06\underline{/-50.71°}$ Ω
25-12. $5.81\underline{/-35.5°}$ mA
25-14. $E = 1.7\underline{/0°}$ V, $Z_{int} = 2000 - j2000$ Ω
25-16. $I_1 = 29.1\underline{/86°}$ mA
   $I_2 = 222\underline{/42.6°}$ mA
   $I_3 = 245\underline{/47.2°}$ mA
25-18. See problem 16.
25-20. 44 dB
25-22. 75.3 W
25-24. 30 dB
25-26. 17.7 kHz
25-28. 1 μF
25-30. 400 kHz

Index

A

ac waves, 334, 335
Admittance, 406, 407
Alternator, 280
Ammeter, 85
  shunt, 286, 287
Angular velocity, 346
Armature, 279
Atomic number, 31
Atomic structure, 31-33

B

Bandwidth, 444-47, 454, 455
Bridge circuits, 490-93
  capacitance, 491, 492
  inductance, 492, 493
  Wheatstone, 490, 491

C

Capacitance, 242-55
  in ac circuits, 363-65
  stray, 244
  unit of, 244
Capacitive coupling, 306, 307
Capacitive reactance, 363-64
Capacitor(s),
  ceramic, 249
  electrolytic, 249-51
  energy storage, 255
  mica, 248
  paper, 249
  in parallel, 254, 255
  in series, 251-53
  variable, 251
Circuits
  ac, parallel RC, 407-11
  ac, parallel RL, 412-14
  ac, parallel RLC, 415
  ac, series, 379-91
  dc, parallel, 126-44
  dc, series, 108-21
  dc, series-parallel, 151-73
  series equivalent, 157-60
Circuit analysis,
  parallel ac, 398-406, 415-21
  parallel dc, 139-43

Circuit analysis, (cont.)
    series ac, 389–92
    series dc, 116–19
    series-parallel, 160–65
Circulating current, 448, 456
Closed circuit, 79, 80
Commutator, 279
Conductance, 61, 62
Conduction, 35, 36
Conductors, 36
Cosine wave, 354
Coulomb, 19
Current, 18, 35, 50–54
    capacity of wire, 69, 70
    direction, 52, 53
    and magnetic flux, 263–65
    measurement, 85, 86
    and power, 84, 85
    source conversion, 236, 237
    source (defined), 233, 234
    types of, 53, 54
    units of, 50–52
Current division rule (CDR), 184
Cycle, 42

D

D'Arsonval movement, 285
Decibels (dB), 512–15
Decimals,
    arithmetic, 3–5
    defined, 2, 3
Delta-to-wye conversion, 484–89
Determinants, 189–92
Dielectric materials, 246, 248
Differentiator,
    RC, 317, 318
    LR, 319, 320
Distance units, 18

E

Efficiency, 56–58, 223–29
Electricity,
    dynamic, 34
    static, 33, 34
Electron, 19, 31
Electrostatic field, 242, 243
Elements, 30
Energy, 34
Engineering notation, 16–18
Equivalent resistance,
    equal branch resistors, 134, 135
    parallel circuit, 128–39
    two branch circuit, 132, 133

F

Faraday's law, 274
Filters,
    band-pass, 458–60
    band-reject, 461, 462
    high-pass, 519–21
    low-pass, 516–19
Flux linkages, 274, 275
Frequency, 44, 45, 331, 332

G

Generated voltages, 275–79
Generator, 279, 280
Ground, 167, 168

H

Harmonics, 335
Horsepower, 55

I

Impedance,
    series RC circuit, 383, 384
    series RL circuit, 379–83
Induced voltage, 291, 292
Inductance, 290–94
    in ac circuits, 360–62
    energy storage, 302
Inductive reactance, 361, 362

Inductors, 296, 297
    in parallel, 301
    in series, 297–301
Insulators, 36, 37
Integrator,
    RC, 315, 316

J

Joule, 54, 55

K

Kilo, 17
Kilowatt-hour (kWh), 58, 59
Kirchhoff's laws,
    branch current analysis, 185
    current law, 180
    voltage law, 180–82
    mesh current analysis, 185
    node voltage analysis, 185, 186

L

Lenz's law, 275
Load current, 80
LR circuits,
    current rise, 317
    discharge, 319
    pulse response, 319, 320

M

Magnetic circuit, 268–70
Magnetic field, 260, 261
Magnetic flux, 260
    density, 263
Magnetic materials, 261, 262
Mass units, 19
Maximum power transfer,
    ac circuits, 498–500
    dc circuits, 230, 231
Mesh current analysis, 503, 504
Meters, 285–88

Mho, 61
Micro, 17
Millman's theorem, 511, 512
Molecule, 30
Motors, 281–83
Multimeter, 85, 86
Mutual inductance, 294–96

N

Nano, 17
Node voltage analysis, 508–11
Norton's theorem, 212–16, 496–98

O

Ohm's law, 93–101
    computational aids, 102, 103
    current formula, 93, 94
    resistance formula, 95, 96
    voltage formula, 96, 97
Open circuit, 81, 82

P

Peak-to-peak value, 359
Percentage, 6–9
Period, 44, 331, 332
Periodic functions, 346, 347
Periodicity, 305
Permeability, 266
Permittivity, 245, 246
Phase angle, 339–42
Pico, 17
Potentiometer (see Resistors,
    variable)
Power, 18, 34, 35, 54–58, 221, 222,
    359, 360, 427
    apparent, 428, 429
    and current, 84, 85
    true, 429
Power factor, 429–32
    correction, 432, 433
Powers of 10, 11–16
Proportion, 10, 11

Q

Q, 292, 293
Q, effective, 393, 394
Q of a parallel circuit, 422

R

Ratio, 10
RC circuit,
    discharge, 311–13
    pulse response, 313–17
    voltage rise, 307–10
Reactive volt-ampere (var), 430
Rectangular wave, 42
Regulation (percent),
    current, 235
    voltage, 234
Relay, 273, 274
Reluctance, 267
Resistance, 61–76
    color code, 72, 73
    temperature effects, 67–69
    of wire, 65–67
Resistors, 70–76
    power ratings, 76, 76
    types, 70, 71
    variable, 74, 75
Resonance, 436–57
    parallel (ideal), 447, 448
    parallel (practical), 449–57
    series, 436–47
Resonant circuit applications, 457–62
Retentivity, 262
Rheostat (*see* Resistors, variable)
Rotor, 280

S

Scientific notation, 15, 16
Short circuit, 83, 84
Siemen, 61
Sine wave, 42, 336–38
    average value, 344, 345
    effective value, 342–44
    equation, 337
    phasor representation, 338–40
    rate of change of, 354–56
Solenoid, 272, 273
Source conversions,
    current to voltage, 501, 502
    voltage to current, 500, 501
Square wave, 43
Stator, 280
Superposition theorem, 193–98, 505–8
Susceptance, 406, 407

T

Thevenin's theorem, 199–212, 493–96
Time constant,
    L/R, 317, 318
    RC, 310, 311
Transformers, 283, 284, 466–68
    autotransformer, 475–78
    current ratios, 470–72
    impedance ratios, 472–75
    impedance reflection, 479, 480
    loosely coupled, 478–80
    voltage ratios, 468–70
Transient, 305
Triangular wave, 42
Troubleshooting,
    parallel circuit, 143, 144
    series circuit, 119–121

V

Var, 430
Voltage, 20, 39–47
    alternating, 42
    dividers, 168–73
    measurement, 86, 87
    pulsating dc, 41, 42
    pure dc, 41
    series-aiding sources, 114, 115
    source conversion, 235, 236

source (defined), 231, 232
sources of, 46, 47
units, 40
Voltage division rule (VDR), 183
Volt-ampere, 428
Voltmeter, 86, 87
multiplier, 288

W

Waveshape, 82
Work, 18, 34
Wye-to-delta conversion, 489, 490